Data-Centric Systems and Applications

Vipul Kashyap · Christoph Bussler ·
Matthew Moran

The Semantic Web

Semantics for Data and Services on the Web

With 61 Figures and 18 Tables

Vipul Kashyap
Partners HealthCare System
Clinical Informatics R&D
93 Worcester St, Suite 201
Wellesley, MA 02481
USA
vkashyap1@partners.org

Christoph Bussler
Merced Systems, Inc.
333 Twin Dolphin Drive
Suite 500
Redwood Shores, CA 94065
USA
chbussler@aol.com

Matthew Moran
Nortel
Mervue Industry Park
Galway
Ireland
maitiu_moran@hotmail.com

ISBN 978-3-540-76451-9 e-ISBN 978-3-540-76452-6

DOI 10.1007/978-3-540-76452-6

Library of Congress Control Number: 2008931474

ACM Computing Classification (1998): H.4, I.2, J.3

Cover design: KünkelLopka GmbH, Heidelberg

Printed on acid-free paper

9 8 7 6 5 4 3 2 1

springer.com

Preface

A decade ago Tim Berners-Lee proposed an extraordinary vision: despite the phenomenal success of the Web, it would not, and could not, reach its full potential unless it became a place where automated processes could participate as well as people. This meant the publication of documents and data to the web in such a way that they could be interpreted, integrated, aggregated and queried to reveal new connections and answer questions, rather than just browsed and searched. Many scoffed at this idea, interpreting the early emphasis on language design and reasoning as AI in new clothes. This missed the point. The Grand Challenge of the Semantic Web is one that needs not only the information structure of ontologies, metadata, and data, but also the computational infrastructure of Web Services, P2P and Grid distributed computing and workflows. Consequently, it is a truly whole-system and multi-disciplinary effort.

This is also an initiative that has to be put into practice. That means a pragmatic approach to standards, tools, mechanisms and methodologies, and real, challenging examples. It would seem self-evident that the Semantic Web should be able to make a major contribution to clinical information discovery. Scientific communities are ideal incubators: knowledge-driven, fragmented, diverse, a range of structured and unstructured resources with many disconnected suppliers and consumers of knowledge. Moreover, the clinicians and biosciences have embraced the notions of annotation and classification using ontologies for centuries, and have demanding requirements for trust, security, fidelity and expressivity.

This book is the first to describe comprehensively the two main characteristics of the Semantic Web – its information and its processes – and to apply it not to toy, artificial examples but to a challenging application that matters, namely translational medicine. As such, it will become a key text for all of those serious about discovering the many facets of the Semantic Web, those who need to understand the current state of the art as it really is in practice, and those who need to be knowledgeable about its future.

Professor Carole Goble

University of Manchester

Acknowledgments

Vipul would like to dedicate this book to his son Ashish and daughter Aastha. A significant portion of this book was written during paternity leave, when Ashish was born! Vipul also expresses his gratitude to his wife Nitu for tolerating the demands on their family time. The support of Vipul's parents, Yogendra Singh and Vibha, during these times was very valuable and helped reduce stress for Vipul and his family. Vipul would like to acknowledge Dr. Tonya Hongsermeier for her spirited evangelism of the Semantic Web at Partners Healthcare and the W3C Healthcare and Life Sciences Interest Group (HCLSIG). She helped create opportunities to work on Semantic-Web-related projects and pointed me to the translational medicine use case, discussed in Chapter 2 of the book. Vipul also valued his interactions with Dr. Howard Goldberg, which helped him understand pragmatic applications of Semantic Web technologies in the healthcare area. Dr. Blackford Middletonís support for work related to Semantic Web technologies at Partners is also gratefully acknowledged. Discussions and debates with Olivier Bodenreider at the National Library of Medicine were very instructive and introduced Vipul to the area of biomedical informatics. The interactions with various colleagues in HCLSIG, such as Eric Neumann, Kei Cheung, Bill Bug, Susie Stephens, Don Doherty, June Kinoshita, Alan Ruttenberg and others, were also very thought provoking and intellectually stimulating. With Stefan Decker, Vipul had the pleasure of co-presenting a tutorial on the Semantic Web at VLDB 2003, on which this book is based. Vipul has also enjoyed tremendously the follow-on collaboration with Christoph Bussler and Matt Moran on writing this book. Finally, Vipul would like to acknowledge Dr. Amit Sheth, who has been a great mentor and was responsible for introducing me to the research area of semantics and the Semantic Web.

Chris wants to acknowledge a few people he met during his Semantic Web Services journey in Ireland. First there is Dieter Fensel, who in a nutshell was instrumental in opening up the field of semantics to him during many years of collaboration. Then there is the WSMX team that started making Semantic Web Services a reality in various research projects and through a still ongoing open source implementation that started to make its way into standards and industry: David Aiken, Emilia Cimpian, Enrica Dente, Doug Foxvog, Armin Haller, Thomas Haselwanter, Juan Miguel Gomez, Mick Kerrigan, Edward Kilgariff, Adrian Mocan, Matt Moran, Manuel Ohlendor, Eyal Oren, Ina O'Murchu, Brahmananda Sapkota, Laurentiu Vasiliu, Jana Viskova, Tomas Vitvar, Maciej Zaremba, and Michal Zaremba. In context of this book, Vipul and Matt were great co-authors and

Chris enjoyed working together with them a lot during this project. Last but not least he would like to give a huge thanks to his family, Barbara and Vernon, for joining him and supporting him during their time in Ireland; the memories about that time shall remain with them forever.

Matt wants to thank the folks who have encouraged him in his transition into research over the last four years. The WSMX team that Chris has already listed has been fantastic to work with. Matt enjoyed being in the team and thinks its energy and positive approach has been a strong factor in keeping up the momentum in Semantic Web Services research. He would like to particularly thank Tomas Vitvar and Maciej Zaremba with whom he has established a great relationship over the last two years. He is very grateful to Chris and Dieter Fensel, who gave him the chance to step into Semantic Web Services research with DERI. He has learned a lot from them both and hopes to keep learning from them in the future. Writing this book with Vipul and Chris has been a great experience for Matt. The back-and-forth conversations have been generous and enlightening. Finally, he would like to thank his sisters Maeve, Claire, Joanne and Mags, and his parents, for their biased praise and keeping him supplied with Irish stew and wild salmon through the course of his working on this book.

The three authors would like to thank the team at Springer Verlag for its support, especially Ralf Gerstner, who was always engaged, supportive, patient as well as available for discussion and advice.

June 2008

Vipul Kashyap, Boston, MA, USA
Christoph Bussler, San Jose, CA, USA
Matthew Moran, Galway, Ireland

Contents

Part I
Preliminaries

1 Introduction

Semantics in the Webster dictionary is defined as *meaning or relationship of meanings of a sign or a set of signs* [4]. From an Information Systems perspective, semantics of information can be defined as *the meaning and use of information* [1]. The Semantic Web is defined as an extension of the current Web in which information is given a well-defined meaning, better enabling people and computers to work together [2]. The Semantic Web is a vision: the idea of having data on the Web defined and linked in a way that it can be used by machines not just for display purposes, but also for automation, integration and reuse of data across various applications. The goal of the Semantic Web initiative is as broad as that of the Web: to create a universal medium for the exchange of data. It is envisaged to smoothly interconnect personal information management, enterprise application integration, and the global sharing of commercial, scientific and cultural data.

These descriptions, though espousing the same vision and goal, however give rise to multiple interpretations, reflecting the perspectives of various fields of computer science and informatics, which can be enumerated as follows:

- Researchers in the database and information systems communities have developed conceptual, logical and physical data and process models to capture semantics of information and processes. These communities have focussed on efficient and scalable storage, indexing and querying of data on one hand, and efficient choreography and orchestration of workflows and services on the other.
- Researchers in the knowledge representation community have developed expressive knowledge representation schemes and theories to capture semantics of information and processes in a declarative manner. This community has focussed on implementation of reasoners and inference mechanisms for checking validity and satisfiability of knowledge specifications.
- Researchers in the information retrieval community have developed thesauri and taxonomies to capture semantics of information. These thesauri and taxonomies have been used to guide search and browsing of documents in document collections. Statistical approaches to capture latent semantics by computing co-occurrence frequencies of terms in a corpus have also been developed.
- Researchers in the machine learning and natural language processing committees have focussed on semantic annotations of data and documents with respect to a well defined set of categories and concepts. Recently there have been efforts related to learning ontologies or taxonomies of concepts.

- Semantics of information has been used to support efficient distributed computing related to location of relevant resources. Researchers in the peer-to-peer communities have proposed approaches that use semantic annotations and localized mappings to locate resources and perform data integration.
- Researchers in the agent systems communities have used ontologies to represent both the semantics of the messages exchanged between agents and the protocols followed by a community of agents for performing a set of tasks.
- Recently researchers in the Web Services communities have proposed process models and ontologies to capture the semantics of services and to a limited extent the semantics of computations to enable reuse and interoperability of applications.

Information technology and the Web has become ubiquitous in our day-to-day lives. Web-based approaches have become the default mode for implementing business processes and operations. Different communities and market verticals such as telecommunication, manufacturing, healthcare and the life sciences are using information and Web-based systems in significant ways to streamline their processes and gain competitive advantage in the market. Approaches that put machine understandable data on the Web are becoming more prevalent.

We discuss next the motivations behind and the vision and goals of the Semantic Web. A framework for characterization of the technologies behind the Semantic Web is presented. A use case for the Semantic Web from the domain of healthcare and the life sciences is presented. This will be the underlying scenario used to motivate the various Semantic Web technologies that will be discussed in this book. Finally, the organization of this book into various units and chapters will be presented.

1.1 Motivation: Why Semantic Web?

With Internet connectivity everywhere and an overabundance of available information the infrastructure for communication is in place. Still, information and services are distributed, often hard to find and hard to integrate. This results in a higher cost to find relevant information and get value from it. Several aspects are fueling the Semantic Web effort:

- Increased cost pressure and competition require businesses to reduce costs by interconnecting workflows and business processes and simplifying the effort of data and service sharing.
- Portal implementations within organizations and e-Government in almost every developed country are aiming to unify the access to government information and services.
- Organizations are attempting to increase automation and interoperability by publishing machine-interoperable data on the Web. Scalability and interopera-

bility across multiple information systems within and across communities has become an urgent priority.

- Scientific progress requires a stronger collaboration and intra- and inter-community information sharing. Various efforts in these areas are aiming to enable data and services sharing: examples are, among many more, the Gene Ontology and Bio Ontology Working Groups for genomics data, the CME project of the Southern California Earthquake Center for seismological data and services among seismologists, and GEON, a geosciences project aiming to interlink and share multi-disciplinary data-sets.
- E-business efforts have focussed on creating business-specific vocabularies for information and process interoperation across enterprises within and across industry boundaries. Some examples of these vocabularies are BPMI, XML-HR and CIM/DMTF.
- There have been efforts in the healthcare and life sciences areas to create specialized vocabularies for different domains. The Unified Medical Language Thesaurus (UMLS®) is a collection of various biomedical vocabularies used for capturing diagnoses (SNOMED), billing information (ICD9) and search queries (MeSH). Life sciences researchers have developed specialized vocabularies such as Gene Ontology for capturing information about genomic structure, function and processes; BioPax for capturing information about Biological Pathways; and MAGE-ML for representing micro-array data in a standardized format.
- The Web is increasingly used as a collaboration forum, using blogs, wikis and other tools for sharing information and tasks such as collaborative metadata annotation and ontology building.

However, this has led to the design and implementation of a multitude of data and metadata schemes along with specialized workflows and processes within and across communities, giving rise to the problem of information overload and the "Tower of Babel" phenomenon. The Web can reach its full potential only if it becomes a place where data can be shared and processed by automated tools as well as by people. For the Web to scale, tomorrow's programs must be able to share and process data even when these programs have been designed totally independently. This is one of the main goals and motivation behind the Semantic Web vision.

1.2 A Framework for Semantic Web

There is a widespread misperception that the Semantic Web is primarily a rehash of existing AI and database work focussed on encoding KR formalisms in markup languages such as RDF(S), DAML+OIL or OWL. We seek to dispel this notion by presenting the broad dimensions of this emerging Semantic Web and the multi-disciplinary technological underpinnings. In fact, we argue that it is absolutely critical to be able to seek, leverage and synergize contributions from a wide variety of technologies and sub-fields of computer science. A framework for presenting the Semantic Web viewed from different perspectives is:

- **Information Aspects of the Semantic Web**:
  Semantic Web Content. This refers to the myriad forms of data that will be presented on the Semantic Web along with the metadata descriptions embedded in the data. This is best exemplified by the equation:
  Semantic Web Content = Data + Metadata.
  Data. This includes structured (e.g., relational) data, semi-structured (e.g., RDF, XML) data and unstructured (e.g., raw text multimedia) data consisting of metadata descriptions embedded in the data.
  Metadata and Annotations. This refers to the various types of domain- or application- specific metadata descriptions that will be used to annotate data on the Semantic Web. Annotation is fundamental to the creation of the Semantic Web.
  Ontologies and Schemas. This refers to the underlying vocabulary and semantics of the metadata annotations. Collections of domain-specific concepts may be used to create domain- and application-specific views on the underlying content. Schemas are a special case of metadata that are structured and may contain semantic information. In the cases where the metadata is explicit (e.g., database schemas, XML schemas), these metadata may be mapped to other related metadata or ontological concepts.

Table 1.1. Framework for the Semantic Web

	Information Aspects			**Process Aspects**
	Content	**Metadata and Annotations**	**Ontologies and Schemas**	
DB and CM Systems[a]	X	X		
KR Systems[b]		X	X	
Machine Learning		X	X	
Statistical Clustering		X	X	
Information Retrieval	X	X		
NLP[c]	X	X	X	
Distributed Computing				X
SOA	X	X	X	X
Agents	X	X	X	X

Table 1.1. Framework for the Semantic Web

	Information Aspects			Process Aspects
	Content	**Metadata and Annotations**	**Ontologies and Schemas**	
P2P[d]		X	X	X
Grid Computing				X

a. DB and CM systems refer to database and content management systems respectively
b. KR Systems refer to Knowledge Representation Systems
c. NLP refers to Natural Language Processing Technologies
d. P2P refers to Peer-to-Peer Computing Infrastructures

- **Computational Aspects of the Semantic Web**:
  Computing Infrastructures. This refers to various computing infrastructures that support the communication between computational entities such as agents, P2P, Web Services and the different styles of communication.
  Direct Communication. Direct communication enables computational entities to send synchronous and asynchronous messages to each other through direct communication channels.
  Mediated Communication. Mediated communication supports a mediator that acts as intermediary for two or more computational entities to communicate.
  Service Description. This refers to the definition of the interfaces of computational entities that will communicate with each other. The description needs to ensure interoperability of applications at the semantic level through the definition of service behavior as well as dynamic service composition and invocation. Service-level agreements governing the scalability and performance are part of service descriptions.

1.3 Use Case: Translational Medicine Clinical Vignette

The field of translational medicine may be defined as in [3]: (a) validation of theories emerging from preclinical experimentation on disease-affected human subjects and; (b) refinement of biological principles that underpin human disease heterogeneity and polymorphisms by using information obtained from preliminary human experimentation. This is a new emerging field which straddles the health ecosystem, consisting of diverse market sectors such as healthcare delivery, drug discovery and life sciences. In this section, we present a high-level description of a clinical vignette which has ramifications across the health ecosystem. This clinical vignette will be used throughout the book to motivate solutions based on Semantic Web technologies and discuss details related to the same. A detailed description of the clinical vignette, identifying high-level functional requirements will be presented in the next chapter.

The use case begins when a patient enters a doctor's office complaining of fainting-like symptoms and pain in the chest. The doctor performs a clinical examination of the patient that reveals abnormal heart sounds. On further discussion with the patient, the doctor learns that the patient has a family history of sudden death, with the patient's father dying at the age of 40. His two younger brothers are apparently normal. The doctor then decides to order an ultrasound based on his clinical examination of the patient. The echocardiogram reveals hypertrophic cardiomyopathy. This could lead to the sequencing of various genomes such as MYH7 and MYBPC3 which could result in the doctor's ordering various types of therapies, drugs and monitoring protocols.

This clinical use case will be used to motivate the need for various aspects of semantics-related technologies:

- Semantics-rich data and information models that can capture phenotypic information related to the physical exam, structured image reports and family history on one the hand; and on the other hand genotypic information such as the results of molecular diagnostic tests such as mutations, expression levels, and so on.
- Semantic mappings that would enable integration of data and information across clinical and genomic data sources. Some of these mappings may be complex and require execution of rule-based specifications.
- The ability to specify semantic rules for implementing clinical decision support that could suggest appropriate molecular diagnostic tests based on the phenotypic characteristics of the patient and propose therapies based on the patient's genotype.
- The ability to semi-automate knowledge acquisition of genotypic and phenotypic associations and other knowledge that could inform decision support and provide a substrate for information integration. Statistical and Machine Learning techniques are specially relevant in this context.
- The ability to manage change in knowledge due to the rapid rate of knowledge discovery in the healthcare and life sciences. The role of semantic inferences based on expressive ontologies and information models will be crucial in this context.
- The ability to represent processes such as therapeutic protocols and biological pathways and possible ways of combining them. Semantic Web process models are likely to be useful in this context.
- The ability to orchestrate clinical workflows across multiple clinical and genomic contexts.

1.4 Scope and Organization

In this book, we will discuss the state of the art related to the use and deployment of Semantic Web specifications. These discussions will be presented in the context of a framework presented in Section 1.2. We focus on a very pragmatic view of the Semantic Web, viz., that of a way of standardizing data, information, knowledge and process specifications to achieve enhanced information and process interoperability, decision support and knowledge change management. We will focus on how these semantic technologies and specifications can work in the context of a use case discussed in the previous section. We will not discuss issues related to the "strong AI" such as higher order logics, modeling of consciousness and cognitive abilities of human beings.

The organization of the book is as follows:

- Chapter 2 presents a detailed discussion of the clinical vignette discussed earlier. High-level use case description and functional requirements and architectural assumptions are presented. These will be used to motivate the various Semantic Web technologies discussed in this book.
- Chapter 3 discusses various examples of different types of "Semantic Web" content spanning across structured, semi-structured and unstructured data.
- Chapter 4 discusses various metadata frameworks based on W3C recommendations such as XML, RDF and OWL. Issues related to the data models of markup specifications such as XML, RDF and OWL; and associated query languages such as XML Schema and SPARQL are discussed with examples drawn from a solution approach to the clinical use case .
- Chapter 5 discusses the broad question of *What is an Ontology?* and various artifacts created by different communities such as thesauri, schemas and classification schemes will be discussed. Specifications for representation of schemas and ontologies on the Web such as XML Schema, RDF Schema and OWL are discussed with examples drawn from a solution approach to the clinical use case.
- Chapter 6 discusses issues related to ontology authoring, bootstrapping and management. Tools for ontology authoring, merging, versioning and integration are discussed follow by a discussion of ontology versioning and change management issues that arise in the context of the clinical use case.
- Applications enabled by the use of metadata descriptions and ontologies on the Semantic Web, are presented in Chapter 7. Tools for metadata annotation and are discussed followed by a discussion on approaches for ontology-based information integration.
- A discussion of communication models such agents, P2P, client/server and their relationship to Web Services and the Semantic Web is presented in Chapter 8.
- A discussion of the current standards and the state of the art in Web Services is presented in Chapter 9. The impact of semantic inadequacies on web services based on current web standards will also be discussed.

- Dynamic Services Composition, a key enabler of functionality reuse on the Semantic Web, is discussed in Chapter 10
- Chapter 11 discusses the idea of Semantic Web Services and how the inadequacies of current web standards can be addressed.
- Chapter 12 enumerates standards in the area of Semantic Web and Semantic Web Services.
- Chapter 13 presents a solution based on Semantic Web technologies for the use case requirements presented in Chapter 2.

Finally, the list of references and the list of index entries follow.

2 Use Case and Functional Requirements

The success of new innovations and technologies are very often disruptive in nature. At the same time, they enable novel next-generation infrastructures and solutions. These solutions often give rise to creation of new markets and/or introduce great efficiencies. For example, the standardization and deployment of IP networks resulted in introducing novel applications that were not possible in older telecom networks. The Web itself has revolutionized the way people look for information and corporations do business. Web-based solutions have dramatically driven down operational costs both within and across enterprises. The Semantic Web is being proposed as the next-generation infrastructure, which builds on the current Web and attempts to give information on the Web a well-defined meaning [2]. This may well be viewed as the next wave of innovation being witnessed in the information technology sector.

On the other hand, the healthcare and life sciences sectors is playing host to a battery of innovations triggered by the sequencing of the human genome. A significant area of innovative activity is the area of translational medicine which aims to improve the communication between basic and clinical science so that more therapeutic insights may be derived from new scientific ideas and vice versa. Translational research [3] goes from bench to bedside, where theories emerging from preclinical experimentation are tested on disease-affected human subjects, and from bedside to bench, where information obtained from preliminary human experimentation can be used to refine our understanding of the biological principles underpinning the heterogeneity of human disease and polymorphisms. The products of translational research, such as molecular diagnostic tests, are likely to be the first enablers of personalized medicine (see an interesting characterization of activity in the healthcare and life sciences areas in [5]). We will refer to this activity as Translational Medicine in the context of this book.

We are witnessing a confluence of two waves of innovation, Semantic Web activity on the one hand, and translational medicine activity on the other. Informatics and Semantic Web technologies will play a big role in realizing the vision of translational medicine. The organization of this chapter is as follows. We begin (Section 2.1) with a detailed discussion of the clinical vignette described in the introduction. This illustrates the use of molecular diagnostic tests in a clinical setting. We believe that initially translational medicine will manifest itself in clinical practice in this manner. This is followed in Section 2.2 with an analysis of various stakeholders in the fields of healthcare and life sciences, along with their respective needs and requirements. In Section 2.3, we discuss conceptual architectures for translational medicine followed by identification of key functional requirements that need to be supported. Finally, in Section 2.4, we discuss research issues motivated by the functional requirements with pointers to chapters in the book where these issues will be discussed in more detail.

2.1 Detailed Clinical Use Case

We anticipate that one of the earliest manifestations of translational research will be the introduction and hopefully accelerated adoption of therapies and tests gleaned from genomics and clinical research into everyday clinical practice. The weak link in this chain is obviously the clinical practitioner. The worlds of genomic research and clinical practice have been separate until now, though there are efforts underway that seek to utilize results of genomic discovery in the context of clinical practice. We now present a detailed discussion of the clinical use vignette presented in the introduction.

Consider a patient with shortness of breath and fatigue in a doctor's clinic. Subsequent examination of the patient reveals the following information:

- A clinical examination of the patient reveals abnormal heart sounds which could be documented in a structured physical exam report.
- Further discussion of the family history of the patient reveals that his father had a sudden death at the age of 40, but his two brothers are normal. This information needs to be represented in a structured family history record.
- Based on the finding of abnormal heart sounds, the doctor may decide (or an information system may recommend him) to order an ultrasound for the patient. The results of this ultrasound can be represented as structured annotations on an image file.

The finding of the ultrasound may reveal cardiomyopathy, based on which the doctor may decide (or the information system may recommend him) to order molecular diagnostic tests to screen the following genes for genetic variations:

1. beta-cardiac Myosin Heavy Chain (MYH7)
2. cardiac Myosin-Binding Protein C (MYBPC3)
3. cardiac Troponin T (TNNT2)
4. cardiac Troponin I (TNNI3)
5. alpha-Tropomyosin (TPM1)
6. cardiac alpha-Actin (ACTC)
7. cardiac Regulatory Myosin Light Chain (MYL2)
8. cardiac Essential Myosin Light Chain (MYL3)

If the patient tests positive for pathogenic variants in any of the above genes, the doctor may want to recommend that first and second degree relatives of the patient consider testing. The doctor in charge can then select treatment based on all data. He can stratify the treatment by clinical presentation, imaging and noninvasive physiological measures in the genomic era, e.g., noninvasive serum proteomics. The introduction of genetic tests introduces further stratification of the patient population for treatment. For instance, a patient is considered at high risk for sudden death if the following hold (based on recommendations by the American College of Cardiologists [6]):

1. Previous history of cardiac arrest
2. Mass hypertrophy (indicated by a septal measurement of 3.0 or higher)
3. Significant family history
4. Serious arrhythmias (documented)
5. Recurrent syncope
6. Adverse blood pressure response on stress test

Whenever a patient is determined to be at a high risk for sudden death, he is put under therapeutic protocols based on drugs such as Amiadorone or Implantable Cardioverter Defibrillator (ICD). It may be noted that the therapy is determined purely on the basis of phenotypic conditions which in the case of some patients may not have held to be true. In this case while molecular diagnostic tests may indicate a risk for cardiomyopathy, phenotypic monitoring protocol may be indicated.

2.2 Stakeholders and Information Needs

We now present an analysis of the clinical use case, by specifying an information flow (illustrated in Figure 2.1). Various stakeholders and their information needs and requirements in the context of the information flow are also presented. The information needs in the contexts of clinical diagnosis and therapeutic intervention are presented, which include (a) aggregation of data for identifying patients for clinical trials and tissue banks, (b) data-driven knowledge acquisition for creation of knowledge bases for decision support, and (c) mappings between genotypic and phenotypic traits.

An enumeration of the information requirements is presented in Table 2.1. The key stakeholders involved in this flow are patients and clinicians (physicians and nurses who treat these patients) on the one hand and clinical and life science researchers on the other. Life science researchers and clinical trials designers are interested in information (a database for genotypic and phenotypic associations) that gives them new ideas for drugs (target validation and discovery, lead generation) whereas a clinical trials designer would be interested in patient cohorts and results of therapies for effective clinical trials design. The clinical practitioner would need this information to help him decide which tests to order and what therapies to prescribe (decision support). Some of this information would also be stored in the electronic medical record. Clinical researchers would like to use therapeutic information in conjunction with test results to develop clinical guidelines for appropriate use of these therapies on the one hand, and knowledge bases for decision support (typically represented as rules) on the other. An interesting class of stakeholders is healthcare institutions, which would like to use this information to enable translational medicine by integrating clinical and genomic data, and to monitor the quality of clinical care provided. A knowledge engineer is an interesting stakeholder who is interested in the information to construct knowledge bases for decision support and to design clinical guidelines.

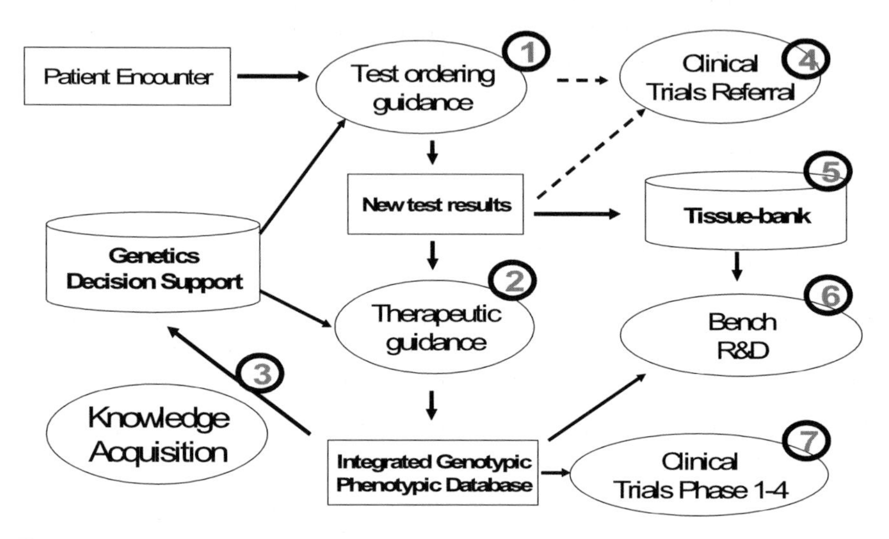

Fig. 2.1. Translational medicine information flows

Table 2.1. Information requirements

Step Number	Information Requirement	Application	Stakeholders
1	Description of Genetic Tests, Patient Information, Decision Support KB	Decision Support, Electronic Medical Record	Clinician, Patient
2	Test Results, Decision Support KB	Decision Support, Database of Genotypic Phenotypic associations	Clinician, Patient, Healthcare Institution
3	Database with Genotypic Phenotypic Associations	Knowledge Acquisition, Decision Support, Clinical Guidelines design	Knowledge Engineer, Clinical Researcher, Clinician
4	Test Orders, Test Results	Clinical Trials Management Software	Clinical Trials Designer
5	Tissue and Specimen Information, Test Results	Laboratory Information Management System	Clinician, Life Science Researcher

Table 2.1. Information requirements

Step Number	Information Requirement	Application	Stakeholders
6	Tissue and Specimen Information, Test Results, Database with Genotypic Phenotypic associations	Lead Generation, Target Discovery and Validation, Clinical Guidelines Design	Life Science Researcher, Clinical Researcher
7	Database with Genotypic and Phenotypic Associations	Clinical Trials Design	Clinical Trials Designer

2.3 Conceptual Architecture

In the previous section, we presented an analysis of information requirements for translational medicine. Each requirement identified in terms of information items, has multiple stakeholders, and is associated with different contexts, such as:

- Different domains such as genomics, proteomics or clinical information
- Research (as required for drug discovery) as opposed to operations (for clinical practice)
- Different applications such as the electronic medical record (EMR), and laboratory information systems (LIMS)
- Different services such as decision support, data integration and knowledge-provenance-related services

In this section, we build upon this analysis and present a conceptual architecture required to support a cycle of learning from innovation and its translation into the clinical care environment. The components of the conceptual architecture illustrated in Figure 2.2 are as follows.

Portals: This is the user interface layer and exposes various personalizable portal views to various applications supported by the architecture. Different stakeholders such clinical researchers, lab personnel, clinical trials designers, clinical care providers, hospital administrators and knowledge engineers can access information through this layer of the architecture.

Applications: Various stakeholders will access a wide variety of applications through their respective portals. The two main applications, viz. the Electronic Health Record (EHR) system and Laboratory Information Management Systems (LIMS) are illustrated in the architecture. We anticipate the emergence of novel applications that integrate applications and data across the healthcare and life science domains , which are collectively identified as "translational medicine" applications.

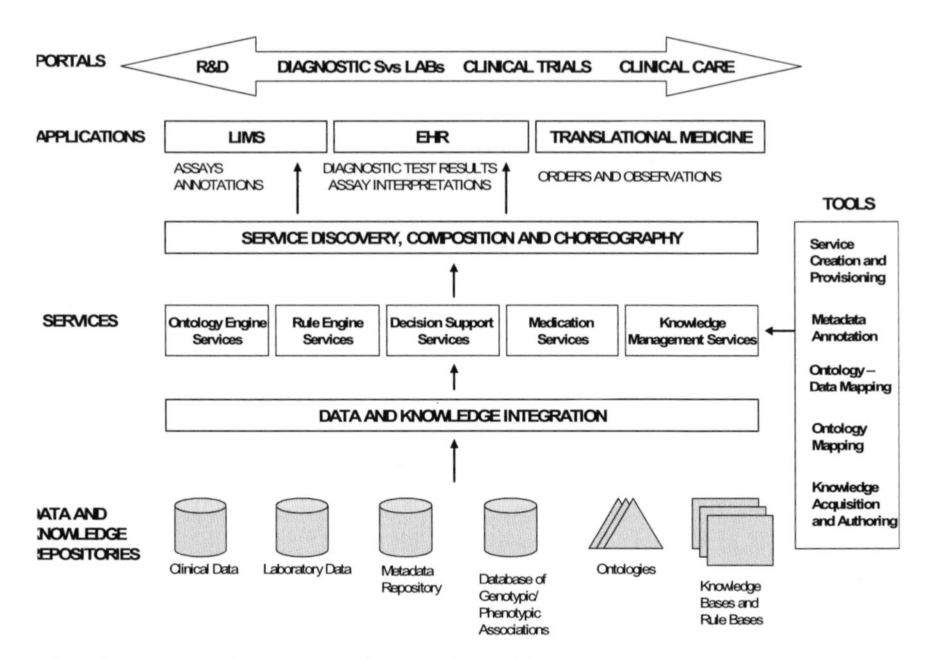

Fig. 2.2. Conceptual Architecture for Translational Medicine

Service Discovery, Composition and Choreography: Newly emerging applications are likely to be created via composition of pre existing services and applications. This component of the architecture is responsible for managing service composition and choreography aspects. Tools that support annotations of services and help define and create new services are used to create service descriptions that can be consumed by this component to enable service discovery, composition and choreography.

Services: The services that need to be implemented for enabling Translational Medicine applications can be characterized as (a) business or clinical services and (b) infrastructural or technological services. Examples of clinical services are clinical decision support services and medication services, which address some aspect of functionality needed to support clinical systems requirements. Some examples of technological services are ontology and rules engine services, which provide specific informatics services such as classification or inferencing and may be invoked by clinical services to implement their functionality. Another example is a knowledge management service which implements functionality for creation and maintenance of various ontologies and knowledge bases. Tools that support knowledge authoring are used to create knowledge that can be consumed by these services.

Data and Knowledge Integration: This component of the architecture enables integration of genotypic and phenotypic patient data and reference information data. This integrated data could be used for enabling clinical care transactions, knowledge acquisition of clinical guidelines and decision support rules, and for hypothesis discovery for identifying promising drug targets. Examples of knowledge integration would be merging of ontologies and knowledge bases to be used for clinical decision support. Tools that support creation of mappings and annotations are used to create mappings across knowledge and data sources that are consumed by the data and knowledge integration component.

Data and Knowledge Repositories: These refer to the various data, metadata and knowledge repositories that exist in healthcare and life sciences organizations. Some examples are databases containing clinical information and results of laboratory tests for patients. Metadata related to various knowledge objects (e.g., creation data, author, category of knowledge) are stored in a metadata repository. Knowledge such as genotypic-phenotypic associations can be stored in a database, whereas ontologies and rule bases can be stored in specialized repositories managed by ontology engines and rule bases. Tools that support metadata annotation can be used to create metadata annotations that can be stored in the metadata repository.

2.4 Functional Requirements

The conceptual architecture discussed in the earlier section helps us identify crucial functional requirements that need to be supported for enabling translational medicine, discussed next.

Service Discovery, Composition and Choreography: The ability to rapidly provision new services is crucial for enabling new emerging applications in the area of translational medicine. This involves the ability to define and develop new services from pre existing services on the one hand and the ability to provision and deploy them in an execution environment on the other. This may involve composition of infrastructural and business services. For instance, one may want to develop a new service that composes an ontology engine and a rules engine service for creating new semantics-based decision support services. From a clinical perspective, one may want to compose clinical protocol monitoring services and notification services (which monitor the state of a patient and alert physicians if necessary) with appropriate clinical decision support services and medication dosing services to offer sophisticated decision support.

Data and Knowledge Integration: This represents the ability to integrate data across different types of clinical and biological data repositories. In the context of the clinical use case discussed in Section 2.1, there is a need for integration and correlation of clinical and phenotypic data about a patient obtained from the EMR with molecular diagnostic test results obtained from the LIMS. Furthermore, the integrated information product will need to be used in different contexts in different ways as identified in the information requirements enumerated in Table 2.1. Effective data integration would require effective knowledge integration where clinically oriented ontologies such as SNOMED [7] and ICD-10 [8] may need to be integrated with biologically oriented ontologies such as BioPAX [9] and Gene Ontology [10].

Decision Support: Based on the clinical use case discussed in Section 2.1, there is a need for providing guidance to a clinician for ordering the right molecular diagnostic tests in the context of phenotypic observations about a patient and for ordering appropriate therapies in response to molecular diagnostic test results. The decision support functionality spans both the clinical and biological domains and depends on effective integration of knowledge and data across data repositories containing clinical and biological data and knowledge.

Knowledge Maintenance and Provenance: All the functional requirements identified above (service composition, data integration and decision support) critically depend on domain-specific knowledge that could be represented as ontologies, rule bases, semantic mappings (between data and ontological concepts), and bridge ontology mappings (between concepts in different ontologies). The healthcare and life sciences domains are experiencing a rapid rate of new knowledge discovery and change. A knowledge change "event" has the potential of introducing inconsistencies and changes in the current knowledge bases that inform semantic data integration and decision support functions. There is a critical need to keep knowledge bases current with the latest knowledge discovery and changes in the healthcare and life sciences domains.

2.5 Research Issues

We now discuss research issues motivated by functional requirements presented in the previous section.

Semantic Web Services: Annotation of services using semantic and standardized descriptions is an active area of research and is used to enable service discovery and composition. Models and languages for specifying Semantic Web Services with the goal of automating service composition and choreography have been proposed. A discussion of Semantic Web Services and their role in enabling service discovery, composition and choreography is presented in Chapter 11.

Ontology-Driven Decision Support: Ontology and rule integration is emerging as a new area of research activity. This is specially relevant in the context of scalable and extensible decision support implementations, where ontologies can be used to enhance the scalability and extensibility of rule-based approaches. A discussion of these research issues is presented in Chapter 5.

Ontology-Driven Knowledge Maintenance and Provenance: Knowledge maintenance and provenance in the context of ontology evolution, versioning and management has been an extensive area of research. Issues related to ontology bootstrapping and creation are also closely related to ontology evaluation and versioning. A discussion of these research issues is presented in Chapter 5.

Semantic Information Integration: Approaches for integration of information across diverse data sources has been an active area of research across various fields, particularly database systems and knowledge representation. A large number of approaches have used various forms of semantics to achieve the goal of information integration. Approaches based on ontology integration have also been proposed in the context of information integration. A discussion of these research issues is presented in Chapter 7.

Semantic Metadata Annotation: The ability to annotate Web resources and services with semantic descriptions is crucial to supporting the various functional requirements discussed in the previous section. An interesting form of annotation is the ability to create mappings between schema describing the metadata and concepts and relationships in an ontology, or across concepts and relationships in multiple ontologies. There has been extensive work presented in database, machine learning and Semantic Web literature on tools and approaches for metadata annotation and schema and ontology matching. A discussion of these research issues in presented in Chapter 7.

2.6 Summary

In this chapter, we presented a detailed clinical use case that serves as an example and provides the motivation for various Semantic Web technologies discussed in this book. The use case describes a typical clinical scenario when a patient visits a physician with some symptoms. The patient is evaluated by the physician who may then decide to order some tests and prescribe some therapies based on the results of those tests. Various stakeholders over and above the physician and the patient, such as the healthcare institution, the knowledge engineer, the clinical/life science researcher and the clinical trials designer are identified. The applications involved in enabling the use case such as electronic medical records, clinical and genomic decision support, knowledge management, clinical trials management systems and laboratory information management systems are also identified. A conceptual architecture and abstract functional requirements based on the use case are presented. Research issues that arise in the context of the use case are finallt identified.

Part II
Information Aspects of the Semantic Web

3 Semantic Web Content

In this chapter, we begin with an understanding of the nature of data and content available on the web today and discuss preliminary approaches for representing metadata. We will then methodically enumerate and discuss various types of Semantic Web content spanning structured, unstructured and semi-structured data, with examples from the healthcare and life science domains. The basic premise of the Semantic Web is to build up on the current web to give *each piece of data a well-defined meaning*. We will illustrate with examples the notion of "self-describing" data and the role of metadata descriptions and ontologies in achieving this goal.

3.1 Nature of Web Content

There are two groups of web content. One, which we would call the "surface" Web is what everybody knows as the "Web," a group that consists of static, publicly available web pages, and which is a relatively small portion of the entire web. The surface web is defined as those web pages whose links are visible in search results obtained from various search engines. Another group is called the "deep" Web, and it consists of specialized Web-accessible databases and dynamic Web sites, which are not widely known by "average" surfers, even though the information available on the "deep" Web is 400 to 550 times larger than the information on the "surface."

The "surface" Web consists of approximately 2.5 billion documents with a growth of 7.3 million page per day. Most of these documents are static HTML pages with embedded images, audio and video content. With improving search engine technology, more of the "deep Web" pages can now be indexed. Pages in non-HTML formats (Portable Document Format (PDF), Word, Excel, Corel Suite, etc.) are "translated" into HTML now in most search engines and can "seen" in search results. Script-based pages, whose links contain a "?" or other script coding, no longer cause most search engines to exclude them. Pages generated dynamically by other types of database software (e.g., Active Server Pages, ColdFusion) can be indexed if there is a stable URL somewhere that search engine spiders can find. There are now many types of dynamically generated pages like these that are found in most general web search engines. Some of these documents are created from XML-based content by applying appropriate style sheets and XSLT transforms and also contain basic metadata such as keywords and authors represented as "meta tags" in HTML.

The "deep Web" primarily consists of content stored in structured (relational or XML) databases which are retrieved by Web servers and converted to HTML pages dynamically on the fly. As discussed above, most of this information can now be indexed and searched via stable URLs. Other examples of deep Web content are real-time streaming data such as stock quotes and streaming media such as audio and video content. Sites that require forms to be filled out and require user login and authentication also belong to the deep Web.

Various types of unstructured, semi-structured and structured content are now available on the web. A limited collection of metadata is also being represented in current Web pages. We now discuss various types of metadata and how they can be used to provide meaning to content on the Semantic Web.

3.2 Nature of Semantic Web Content

The nature of Semantic Web content, with its emphasis on providing every piece of Web content a well-defined meaning, is likely to be very different from current Web content. Some characteristics or requirements of Semantic Web content are listed below:

- Semantic Web content should support dual requirements of understandability to the human reader and machine processability at the same time. This requires the ability to precisely and unambiguously specify the meaning. This has led to the creation of various XML-based specifications such as RDF and various human readable notations such as N3 for the same.
- The key to machine processability of content on the Semantic Web is that it should be self-describing. This is achievable partly by producing a common language to specify data and metadata on the Web.
- The current state of the art is limited when it comes to understanding, which is still an area of research in the artificial intelligence community. Thus, standardization is the key element for enhancing machine understanding of data and content on the Web. This is achievable by grounding terms used in creating metadata descriptions in well-defined ontologies on the Web, which could potentially lead to enhanced information interoperability.
- Finally, there is a widespread tendency in Web-based applications to hard code the semantics within application code. Whereas current Web content has implemented the separation of content from presentation to a large extent, the Semantic Web aims to externalize the inherent semantics from syntax, structure and other considerations. This has led to a layered characterization of metadata that have been used to capture these various aspects of information.

Thus, metadata forms a key aspect of enabling the Semantic Web vision, and as discussed above, Semantic Web data and content items will require metadata descriptions associated with the data in some form or other. This observation may be formalized with the following equation:

Semantic Web Content = Data + Metadata

Metadata descriptions on the other hand should be grounded in terms, concepts and relationships present in well-defined ontologies that can be referenced to on the Web via URIs. This is illustrated in the Semantic Web mind map illustrated in Figure 3.1.

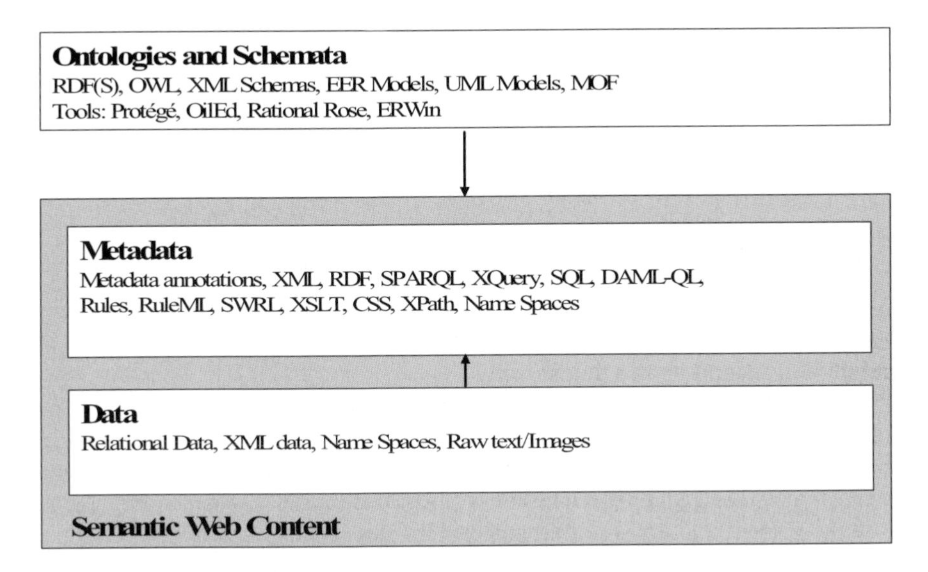

Fig. 3.1. Semantic Web mind map

We will now discuss various types of data and metadata found in the healthcare and life science domains. Furthermore, we characterize metadata based on the captured semantics.

3.3 Metadata

Metadata in its most general sense is defined as data or information about data. For structured databases, the most common example of metadata is the schema of the database. However with the wide variety of datatypes on the Web, we shall refer to an expanded notion of metadata of which the schema of structured databases is a (small) part. Metadata may be used to store derived properties of media useful in information access or retrieval. They may describe or be a summary of the information content of the data described in an intensional manner. They may also be used to represent properties of or relationships between individual objects of heterogeneous types and media. The function of metadata descriptions is twofold:

- To enable the abstraction of representational details such as the format and organization of data, and capture the information content of the underlying data independent of representational details. These expressions may be used to represent useful relationships between various pieces of data within a repository or web site.
- To enable representation of domain knowledge describing the information domain to which the underlying data belongs. This knowledge may then be used to make inferences about the underlying data to determine the relevance and identify relationships across data stored in different repositories and web sites.

We now discuss issues related to metadata from two different perspectives identified in [13], viz., the usage of metadata in various applications and the information content captured by the metadata.

3.3.1 Metadata Usage in Various Applications

We now discuss a set of application scenarios that require functionality for manipulation and retrieval of digital content that are relevant to the Web and the role of metadata in supporting this functionality.

Navigation, Browsing and Retrieval from Image Collections: An increasing number of applications, such as those in healthcare, maintain large collections of images. There is a need for semantic-content-based navigation, browsing, and retrieval of images. An important issue is to associate a user's semantic impression with the images, e.g., image of a brain tumor. This requires knowledge of spatial content of the image, and the way it changes or evolves over time, which can be represented as metadata annotations.

Video: In many applications relevant to news agencies, there exist collections of video footage which need to be searched based on semantic content, e.g., videos containing field goals in a soccer game, or the video of an echocardiogram. This gives rise to the same set of issues as described above, such as the change in the spatial positions of various objects in the video images (spatial evolution). However, there is a temporal aspect to videos that was not captured above. Sophisticated time-stamp-based schemes can be represented as a part of the metadata annotations.

Audio and Speech: Radio stations collect many, if not all, of their important and informative programs, such as radio news, in archives. Parts of such programs are often reused in other radio broadcasts. However, to efficiently retrieve parts of radio programs, it is necessary to have the right metadata generated from, and associated with, the audio recordings. An important issue here is capturing, in text, the essence of the audio, in which vocabulary plays a central role. Domain-specific vocabularies can drive the metadata extraction process making it more efficient.

Structured Document Management: As the publishing paradigm is shifting from popular desktop publishing to database-driven web-based publishing, processing of structured documents becomes more and more important. Particular document information models, such as SGML [14] and XML, introduce structure- and content-based metadata. Efficient retrieval is achieved by exploiting document structure, as the metadata can be used for indexing, which is essential for quick response times. Thus, queries asking for documents with a title containing "Computer Science" can be easily optimized.

Geographic and Environmental Information Systems: These systems have a wide variety of users who have very specific information needs. Information integration is a key requirement, which is supported by provision of descriptive information to end users and information systems. This involves issues of capturing descriptions as metadata and reconciling the different vocabularies used by the different information systems in interpreting the descriptions.

Digital Libraries: Digital libraries offer a wide range of services and collections of digital documents, and constitute a challenging application area for the development and implementation of metadata frameworks. These frameworks are geared toward description of collections of digital materials such as text documents, spatially referenced data-sets, audio, and video. Some frameworks follow the traditional library paradigm with metadata like subject headings [15] and thesauri [16].

Mixed-media Access: This is an approach which allows queries to be specified independently of the underlying media types. Data corresponding to the query may be retrieved from different media such as text and images, and "fused" appropriately before being presented to the user. Symbolic metadata descriptions may be used to describe information from different media types in a uniform manner.

Knowledge-Based Decision Support Systems: Rule-based systems are being increasingly used for providing decision support functionality in various domains such as finance and clinical decision support [17] [18]. Rules may be viewed as specialized metadata descriptions that provide further description or elaboration of a business object model or schema.

Knowledge Management Systems: Metadata descriptions play a critical role in managing large amounts of knowledge and data. Important functionality enabled by metadata descriptions are effective and precise searching, browsing and categorization of information artifacts, and change management via encoding and representation of provenance-related knowledge.

3.3.2 Metadata: A Tool for Describing and Modeling Information

We now characterize various types of metadata based on the amount of information content they capture, and present a classification of various types of metadata used by researchers and practitioners in the healthcare and life science fields. A brief classification based on the one presented in [12] is discussed and illustrated in Figure 3.2.

Content-Independent Metadata: This type of metadata captures information that does not depend on or capture the content of the document or information artifact it is associated with. Examples of this type of metadata could be identifying information such as gene or other life science identifiers, patient medical record number, and sample accession numbers used by laboratories to track patient (and animal) samples in laboratories. Other types of metadata could be date of dictation and transcription of patient clinical notes, date of collection and processing of patient samples for laboratory tests. Still other types of metadata could be system-level or Web-related identifiers such as locations, URIs and mime types. There is no information content captured by these metadata but they are still useful for identifying units of information that might be of interest, for retrieval of data from actual physical locations and for checking for currency or obsolescence. This type of metadata is used to encapsulate information units of interest and organize them according to an information or object model.

Content-Based Metadata: This type of metadata depends on the content of the document or information artifact it is associated with. Examples of content-based metadata are the size of a document, the coordinates of a spot on a gene chip and the number of rows and columns in a radiological image. Other popular examples of content-based metadata are inverted tree indices and document vectors based on the text of a document and the shape, color and texture of image-based data. These metadata typically capture representational and structural information, and enable interoperability through support for browsing and navigation of the underlying data. Content based metadata can be further subdivided as follows:

Structural Metadata: These metadata primarily capture structural information of the document or information artifact. An example of such metadata is one that characterizes various sections in a clinical note for a patient, such as *History of Present Illness, Review of Symptoms*, and *Medications*. Different subdomains of medicine may have different structural metadata associated with them. Also structural metadata might be independent of the application or subject domain of the application, for example, C/C++ parse trees and XML DTDs.

Domain-Specific Metadata: Metadata of this type is described in a manner specific to the application or subject domain of the information. Issues of vocabulary become very important in this case, as the metadata terms have to be chosen in a domain-specific manner. This type of metadata, that helps abstract out representational details and capture information meaningful to a particular application or subject domain, is domain specific metadata. Examples of such metadata are patient state descriptors such as contraindications, allergies and medical subject headings (MeSH) in the healthcare domain and gene mutations, variants, SNPs, proteins and transcription factors in the life science domain. Domain-specific metadata can be further characterized as follows:

Intra-domain-specific Metadata: These types of metadata capture relationships and associations between data within the context of the same information domain. For example, the relationship between a patient and the allergies and contraindications suffered by him belong to a common clinical information domain.

Inter-domain-specific Metadata: These types of metadata capture relationships and associations between data across multiple information domains. For example the relationship between the clinical findings and the phenotypes of a person belong to the healthcare information domain and his or her genes and mutations or genotype belong to the life science information domain.

Table 3.1. Metadata for digital data

Metadata	Media/Metadatatype
Q-Features	Image, Video/Domain Specific
R-Features	Image, Video/Structural
Impression Vector	Image/Content Based
NDVI, Spatial Registration	Image/Domain Specific
Speech feature index	Audio/Content Based
Topic change indices	Audio/Content Based
Inverted Indices	Text/Content Based
Document Vectors	Text/Content Based
Content Classification Metadata	Multimedia/Domain Specific
Document Composition Metadata	Multimedia/Structural
Metadata Templates	Media Independent/Domain Specific
Land Cover, Relief	Media Independent/Domain Specific
Parent-Child Relationships	Text/Structural
Contexts	Structured Database/Domain Specific
Concepts from Cyc	Structured Database/Domain Specific
User's Data Attributes	Text. Structured Database/Domain Specific
Medical Subject Headings	Text/Domain Specific
Domain Ontologies	Media Independent/Domain Specific

We present in Table 3.1 above, a brief survey of different types of metadata used by various researchers. Q-Features and R-Features were used for modeling image and video data [24]. Impression vectors were generated from text descriptions of images [26]. NDVI and spatial registration metadata were used to model geospatial maps, primarily of different types of vegetation [30]. Interesting examples of mixed-media access are the speech feature index [23] and topic change indices [20]. Metadata capturing information about documents are document vectors [22], inverted indices [25], document classification and composition metadata [19] and parent-child relationships (based on document structure) [31]. Metadata templates [29] have been used for information resource discovery. Semantic metadata such as contexts [32] [33], land-cover, relief [35], Cyc concepts [21], and concepts from domain ontologies [34] have been constructed from well-defined and standardized vocabularies and ontologies. Medical Subject Headings (MeSH) [15] are used to annotate biomedical research articles in MEDLINE [28]. These are constructed from biomedical vocabularies available in the UMLS [16]. An attempt at modeling user attributes is presented in [36].

The above discussion suggests that domain-specific metadata capture information which is more meaningful with respect to a specific application or a domain. The information captured by other types of metadata primarily reflect the format and organization of underlying data. Thus, domain-specific metadata is the most appropriate for capturing meaning on the Semantic Web.

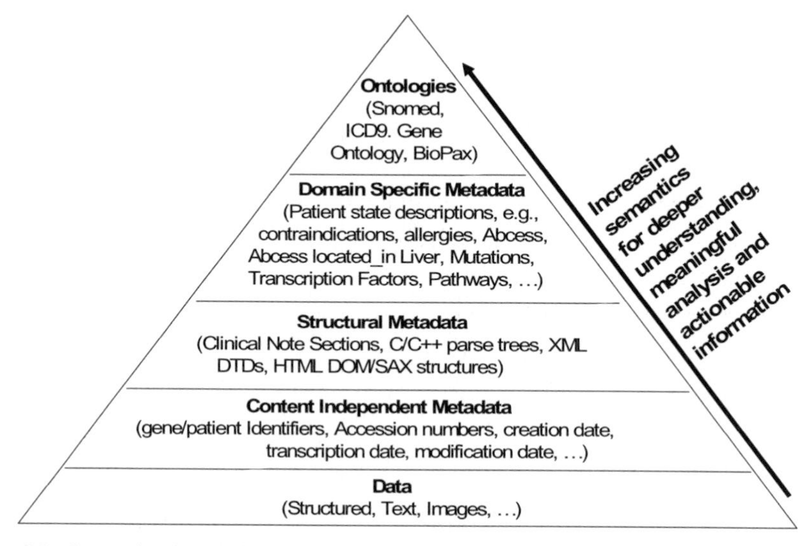

Fig. 3.2. Semantics-based characterization of metadata

Domain-specific metadata can be constructed from terms in a domain-specific ontology, or terms in concept libraries describing information in an application or subject domain. Thus, we view ontologies as metadata, which themselves can be viewed as a vocabulary of terms for construction of more domain-specific metadata descriptions. We discuss various types of ontologies in the next section.

3.4 Ontologies: Vocabularies and Reference Terms for Metadata

We discussed in the previous section that the role of metadata-based descriptions are for describing and modeling information on the Web. The degree of semantics depends on the nature of these descriptions, i.e., whether they are domain-specific. A crucial aspect of creating metadata descriptions is the vocabulary used to create them. The key to utilizing the knowledge of an application domain is identifying the basic vocabulary consisting of terms or concepts of interest to a typical user in the application domain and the interrelationships among the concepts in the ontology.

An ontology may be defined as the specification of a representational vocabulary for a shared domain of discourse which may include definitions of classes, relations, functions and other objects [37]. However, there exist a variety of standards that are being currently used for enabling high quality search, categorization and interoperation of information across multiple information systems. These standards do not have the formal sophistication and logical underpinnings of an ontology, but are currently in extensive use and provide an extensive collection of concepts and relationships that can serve as a substrate for rich metadata descriptions, and thus can be viewed as "ontology-like" artifacts. We present a characterization of these artifacts from an informatics perspective with examples from the healthcare and life sciences and other domains.

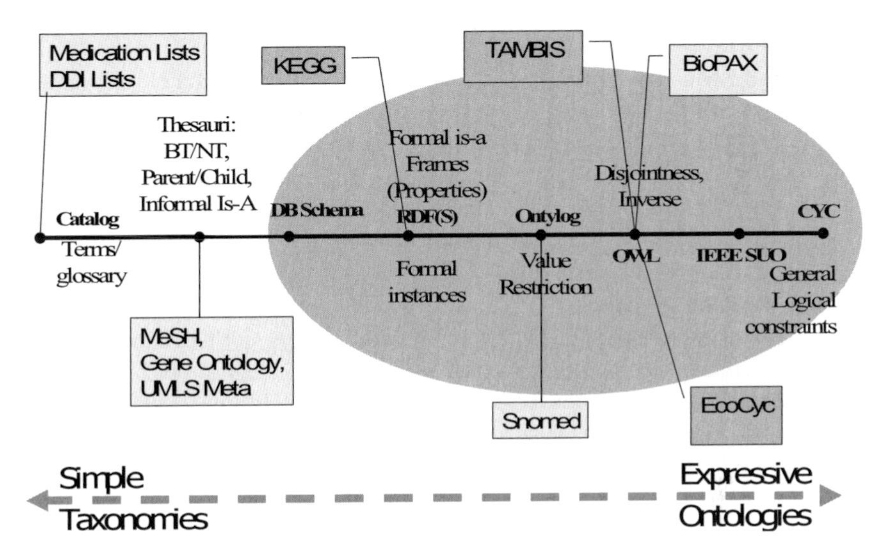

Fig. 3.3. Ontology characterization

An informal categorization of ontology-like artifacts illustrated in Figure 3.3 has the following categories:

- **Term Lists:** These are collections of terms or tokens that play the role of standardized dictionaries in an organization. Examples of term lists are standardized dictionaries of medications, patient problems and drug interactions used in healthcare organizations.
- **Thesauri/Taxonomies:** These are collections of concepts in which the concepts are organized according to interrelationships with each other. The relationships represented in taxonomies and thesauri are typically limited to *synonyms* and hierarchical relationships such as *broader than/narrower than*, *parent/child* and informal *is-a*. The semantics of these relationships are not clearly defined and

the hierarchical relationships typically incorporate *hyponymy/hypernyms*, *partonomy* and *instance-of* relationships. Some examples of thesauri and taxonomy being used in biomedicine are the Medical Subjects Heading [15], the Gene Ontology [10], the International Classification of Diseases [8], and the UMLS Metathesaurus [16]. Concepts from thesauri and taxonomies are typically used in metadata annotations of publication documents such as those available in PubMed MEDLINE [28].

- **Database Schemas:** These are collections of tables and columns that are used to implement underlying information models represented in entity-relationship [39] and UML models [40]. The semantics of database schemas are well defined in terms of the relational model [38], where tables typically represent concepts and their properties are represented by table columns. Relationships between concepts are represented by foreign keys. An interesting example of an information model used in healthcare for interoperability across information systems is the HL7 RIM [41].
- **RDF and XML Schemas:** These are collections of concepts that have complex interrelationships and nesting within each other. Concepts and relationships in RDF Schemas [42] are instantiated in RDF based on a graph-based data model with reification, whereas XML Schema [43] documents are instantiated in XML using a tree-based data model. KEGG [44] is an example of a genomic database which uses an XML/RDF-based markup language to represent and store genomic data.
- **OWL Ontologies:** Ontologies represented using the Web Ontology Language (OWL) [45] capture specialized constraints and axioms. OWL-DL, a variant of OWL, is equivalent in expressiveness to Description Logics. OWL reasoners can use these constraints and axioms to infer subsumption and equivalence relationships between two concepts and mutual contradictions between concepts if they exist. The interesting property of OWL-DL is that it is a highly expressive and tractable subset of first-order logic for a large number of important use cases. SNOMED [7] is a widely used standard in the healthcare domain and is represented using Ontylog [48], a less expressive variant of OWL-DL. Other examples of ontologies in use are in the TAMBIS [46] system for biomedical data integration and the BioPax [47] ontology for representing biological pathways.
- **Logical First Order Theories:** More sophisticated ontologies such as the IEEE Standard Upper Ontology (IEEE SUO) [49] use a first-order logic language such as the Knowledge Interchange Format (KIF) [50] to represent knowledge. These languages, though more expressive than OWL-DL, are for the most part intractable and are not implemented in practical information systems.
- **Higher-Order Logics**: There are ontologies such as the Cyc Ontology [51] which use higher-order logics to represent language. The goal of such ontologies is to capture common sense knowledge and reason with it to a limited extent. Such ontologies are not considered practical, given the state of the art today.

3.5 Summary

In this chapter, a discussion on the nature of content present found on the Web was presented. The nature of Semantic Web content in contrast to Web content today was discussed and the critical role of metadata was identified. A three layered framework based on data, metadata and ontology was presented. The usage contexts and information content captured in different types of metadata was discussed. Ontologies were identified as a specialized type of metadata with a rich and domain-specific content. The different types of ontologies in the vocabulary schema logical-theory continuum were also discussed and presented.

4 Metadata Frameworks

We present a discussion of various frameworks and schemes proposed for representation, storage and manipulation of data and metadata. Standards and specifications proposed by the World Wide Web Consortium (W3C) such the eXtensible Markup Language (XML) [43], Resource Description Framework (RDF) [42] and the Web Ontology Language (OWL) [45] are presented. The frameworks are described and contrasted along various dimensions such as the data model and expressiveness of the specification and query languages for manipulation of metadata specifications and repositories. We will also discuss relationships of these metadata frameworks with rule-based and ontology standards, where applicable.

4.1 Examples of Metadata Frameworks

We define a metadata framework as consisting of a set of specifications that address various needs for creating, manipulating and querying metadata descriptions. Typically a metadata framework would consist of:

- **Data Model:** A data model may be viewed as a collection of datatypes that can be used to construct an abstract view of a Web document (or collection of documents), along with functions that allow access to the information contained in values belonging to these types [52].
- **Semantics:** In some cases, languages for specifying metadata descriptions have well-defined semantics. These semantics are typically specified by using model-theoretic semantics, a branch of mathematics that is used to provide meaning for many logics and representation formalisms, and has recently been applied to several Semantic-Web-related formalisms, namely RDF [53] and OWL [54].
- **Serialization Format:** A serialization format typically provides a meta-language and syntactic constructs for encoding metadata descriptions. It may be noted that multiple data models can be serialized using the same format. For instance, RDF and OWL have different underlying data models but are serialized using the XML syntax.
- **Query Language**: A query language provides a language for expressing the information needs of a user and is typically based on constructs provided by the data model in conjunction with a boolean expressions and specialized operators such as for ordering, sorting and computing transitive closure.

We now present the core metadata frameworks that have been developed and deployed in the context of Web and Semantic-Web-based applications.

4.1.1 XML-Based Metadata Framework

This is one of the earliest metadata frameworks proposed for capturing Web content. The initial goal of XML-based specifications was to enable separation of content and presentation of Web content, but this later evolved into representing both content and metadata. The XML-based metadata framework proposed by the W3C consists of the following components:

- Specifications for XML itself, including namespaces and InfoSets which are used to represent content and metadata. This also includes the XLink and XPointer specifications which seek to represent hypertext links in XML documents.
- Specifications for the Extensible Stylesheet Language (XSL) including both XSL Transformations (XSLT) for conversion of an XML document into alternate XML descriptions or presentation elements for display on a Web page; and XSL Formatting Objects (XSL/FO) for specification of formatting semantics.
- Specifications for XQuery and XPath which are used to extract XML content from real and virtual XML documents on the Web.
- Specification for XML Schemas that provide mechanisms to define and describe the structure, content, and to some extent semantics of XML documents.

4.1.2 RDF-Based Metadata Framework

The RDF specification was designed from the ground up as a language for representing metadata about resources on the Web. The RDF-based metadata framework proposed by the W3C consists of the following components:

- Specifications for RDF itself based on an XML syntax called RDF/XML in terms of XML Namespaces, XML Information Sets and XML Base specifications. This includes a well-defined graph-based data model (discussed later in this chapter). Alternative syntaxes such as N3 which represent RDF expressions as collections of triples have also been proposed.
- Specifications for RDF Schema (RDF(S)), which is used to define RDF vocabularies or models of which RDF statements are instances or extensions.
- The SPARQL specification, which contains a protocol and query language for accessing RDF data stored in RDF data stores.

4.1.3 OWL-Based Metadata Framework

The OWL specification (in particular OWL Full) contains constructs that enable expression of different types of constraints and axioms at both the schema and the data levels. OWL specifications come in these dialects: OWL-Lite, OWL-DL and OWL Full. Any RDF graph is typically OWL Full unless it has been restricted (via its RDF schema) to the other dialects.

- At the data level, OWL specifications help us represent class membership of an instance and the values of its various properties, along with descriptions of anonymous individuals without actually specifying who they are. Relationships between instances such same-as and different-from are also supported.
- At the schema level, OWL specifications enable representation of relationships between classes such as subclass-of, disjointedness or equivalence. The dialects of OWL are more geared toward specifications at the schema level as opposed to the data level.
- Query languages based on OWL, such as OWL-QL have been proposed and are still in their infancy. Approaches that combine SPARQL and OWL have also been proposed to leverage the use of OWL-based inferences in the context of query processsing.

4.1.4 WSMO-Based Metadata Framework

WSMO is a meta-ontology in terms of the Object Management Group (OMG) Meta-Object Facility (MOF) [326] specification for an abstract language and framework to represent its meta-models. MOF provides the constructs of classes and their generalization through subclasses as well as attributes with declarations of type and multiplicity. MOF defines four layers:

- Information layer: the data to be described
- Model layer: metadata describing the data in the information layer
- Meta-model layer: defines the structure and semantics of the metadata
- Meta-meta-model layer: defines the structure and semantics of the meta-meta-data

The four layers of MOF and how they relate to WSMO are described in detail in [325]. Briefly, the language defining WSMO corresponds to the meta-meta-model layer. The WSMO model itself represents the meta-model layer. WSMO has four top level elements: Ontologies, Web Services, Goals and Mediators. These constitute the model layer. Finally the actual data described by the ontologies and exchanged between Web Services, for example, make up the information layer.

4.2 Two Perspectives: Data Models and Model-Theoretic Semantics

We now present a discussion on the data models and semantics underlying the various metadata frameworks discussed in the previous section. As discussed in [52], there are several fundamental differences between data models and model-theoretic semantics:

Information retention: XML data models tend to retain all the text information from the input document, such as comments, white spaces and text representations of typed values. In model-theoretic semantics, on the other hand, there is a decision made on just which kind of information to retain, and which kind of information to ignore. For example, it is typical in model-theoretic semantics to ignore the order in which information is presented.

Direction of flow: In the data model approach, there is a process of generating a data model from an input document and thus the result is constructed from the input. In model-theoretic semantics, on the other hand, the interpretations are simply mathematical objects that are not constructed at all. Instead there is a relationship between syntax constructs and interpretations that determines which interpretations are compatible with a document. Generally there are many interpretations that are so compatible, not just one.

Schema vs. Data: XML data models usually make a fundamental distinction between schema and data. In model-theoretic semantics, both schema and data are part of a model on which one can perform reasoning.

In this section, we will focus on data model issues related to our discussion on XML/RDF and model-theoretic issues related to our discussion on RDF/OWL. In particular, we attempt to reconcile these two perspectives by:

- Presenting a discussion of the underlying data models of both XML and RDF. Similarities and differences between the XML and RDF data models are also discussed.
- Presenting an introductory discussion of model-theoretic semantics, including notions of interpretations and entailments. A small discussion of how model theory can be used to describe the semantics of RDF is presented and pointers to more detailed specifications of RDF, RDF(S) and OWL are provided.

4.2.1 Data Models

We begin with a discussion on XML data models presented in [55]. The key distinction between data in XML and data in traditional models is that XML is not rigidly structured. In the relational and object-oriented models, every data instance has a schema, which is separate from and independent of the data. In XML, the schema exists with the data as tag names. This is consistent with our observation in Chapter 2, that Semantic Web content would include both data and metadata and also be self-describing.

For example, in the relational model, a schema might define the relation person with attribute names name and address, e.g., `person(name, address)`. An instance of this schema would contain tuples such as `("Smith", "Philadelphia")`. The relation and attribute names are separate from the data and are usually stored in a database catalog. In XML, the schema information is stored with the data. Structured values are called elements. Attributes, or element names, are called tags, and elements may also have attributes whose values are always atomic. For instance, the following XML is well-formed.

```
<person>
  <name>Smith</name>
  <address>Philadelphia</address>
</person>
```

Thus, XML data is self-describing and can naturally model irregularities that cannot be modeled by relational or object-oriented data. For example, data items may have missing elements or multiple occurrences of the same element; elements may have atomic values in some data items and structured values in others; and collections of elements can have heterogeneous structures.

Self-describing data has been considered recently in the database research community. Researchers have found this data to be fundamentally different from relational or object-oriented data, and called it semi-structured data. Semi-structured data is motivated by the problems of integrating heterogeneous data sources and modeling sources such as biological databases, Web data, and structured text documents, such as SGML and XML. Research on semi-structured data has addressed data models, query language design, query processing and optimization, schema languages, and schema extraction. Consider the XML data below.

```
<patients>
  <patient ID="patient1">
      <!-- This is a MGH patient -->
      <lastname>Doe</lastname>
      <firstname>John</firstname>
      <physician><lastname>Kashyap</lastname></physician>
      <hospital><name>Massachusetts General<name></hospital>
  </patient>
  <patient ID="patient2">
      <!-- This is a BWH patient -->
      <lastname>Doe</lastname>
      <firstname>Jane</firstname>
      <physician><lastname>Bussler</lastname></physician>
      <physician><lastname>Moran</lastname></physician>
      <hospital><name>Brigham and Womens</name></hospital>
  </patient>
</patients>
```

One approach is to view XML data from an "unordered" perspective. The first record has four elements (one last name and first name, one physician, and one hospital) and the second record has two physicians. The XML document, however, contains additional information that is not directly relevant to the data itself, such as the comment at the beginning of the patient elements, the fact that lastname and firstname precedes physicians, and the fact that the patient with the id "patient1" precedes the patient with the id "patient2." This information is not always relevant to the semantics of the information, which are independent of comments and the order in which the data values are displayed. We assume that a distinction can be made between information that is intrinsic to the data and information, such as document layout specification, that is not. The unordered model ignores comments and relative order between elements, but preserves all other essential data. An unordered XML Graph consists of:

- A graph, G, in which each node is represented by a unique string called an object identifier (OID),
- G's edges are labeled with element tags,
- G's nodes are labeled with sets of attribute-value pairs,
- G's leaves are labeled with one string value, and
- G has a distinguished node called the root.

The example patient data is represented by the XML graph in Figure 4.1. Attributes are associated with nodes and elements are represented by edge labels.T he terms node and object interchangeably object identifiers are omitted for clarity.

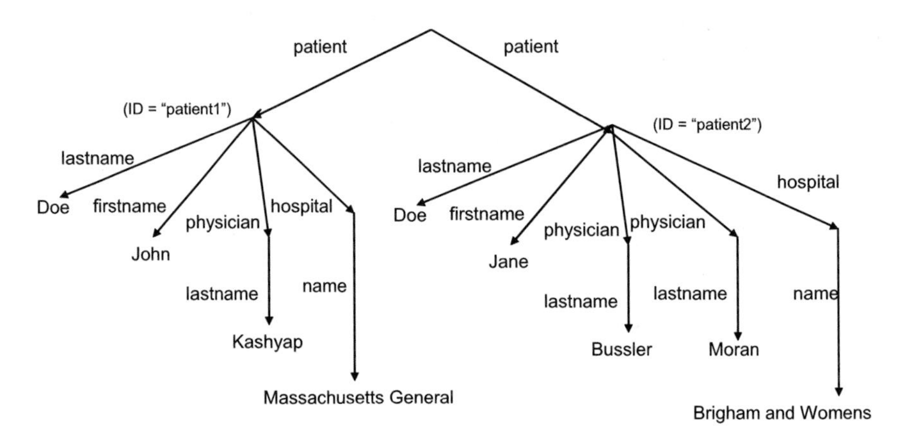

Fig. 4.1. Unordered XML Graph for example XML data

The data model allows several edges between the same two nodes, but with the following restriction. A node cannot have two outgoing edges with the same labels and the same values. Here value means the string value in the case of a leaf node, or the OID in the case of a non-leaf node. Restated, this condition says that (1) between any two nodes there can be at most one edge with a given label, and (2) a node cannot have two leaf children with the same label and the same string value. It is important to note that XML graphs are not only derived from XML documents, but are also generated by queries. An ordered XML graph is an XML graph in which there is a total order on all nodes in the graph. For graphs constructed from XML documents a natural order for nodes is their document order. Given a total order on nodes, we can enforce a local order on the outgoing edges of each node. The Example data would be represented by the ordered graph in Figure 4.2. Nodes are labeled with their index (parenthesized integers) in the total node order and edge labels are labeled with their local order (bracketed integers).

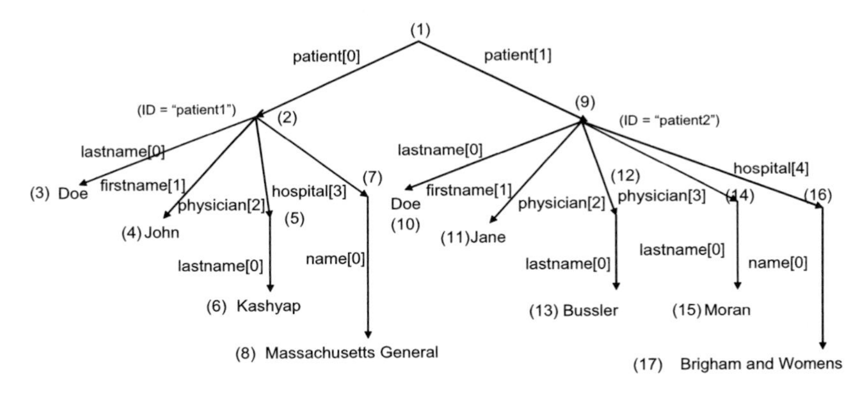

Fig. 4.2. Ordered XML Graph for example data

To support element sharing, XML reserves an attribute of type `ID` (often called ID) to specify a unique key for an element. An attribute of type `IDREF` allows an element to refer to another element with the designated key, and an attribute of type `IDREFS` may refer to multiple elements. In the data model, these attributes are treated differently from all others. For example, assume attributes `ID` and `physician` have types `ID` and `IDREFS` respectively:

```
<!ATTLIST patient ID ID #REQUIRED>
<!ATTLIST patient treated-by IDREFS #IMPLIED>
```

For example, in the XML fragment below, the two `<patient>` elements are the same as the previous example, and the `<physician>` element which is refered to by the `treated-by` attribute.

```
<patient ID="patient1" treated-by="physician1">
...
</patient>
<patient ID="patient2" treated-by="physician2 physician3">
. . .
</patient>
<physician ID="physician1">
<lastname>Kashyap</lastname>
</physician>
```

```
<physician ID="physician2">
<lastname>Bussler</lastname>
</physician>
<physician ID="physician3">
<lastname>Moran</lastname>
</physician>
```

An IDREF attribute is represented by an edge from the referring element to the referenced element; the edge is labeled by the attribute name. ID attributes are also treated specially, because they become the node's OID. The elements above are represented by the XML graph in Figure 4.3.

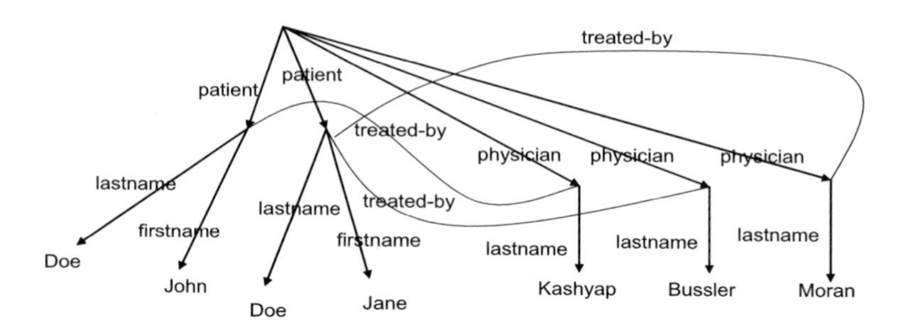

Fig. 4.3. Representation of IDREF attributes in an XML graph

We now continue with a description of the basic RDF data model [56], which consists of three object types:

Resources: All things being described by RDF expressions are called resources. A resource may be an entire Web page such as the HTML document "http://www.w3.org/Overview.html". A resource may be a part of a Web page, e.g., a specific HTML or XML element within the document source. A resource may furthermore also be a whole collection of pages, e.g., an entire Web site. Finally, a resource may be an object that is not directly accessible via the Web, e.g., a printed book. Resources are always named by URIs plus optional anchor IDs. Anything can have a URI; the extensibility of URIs allows the introduction of identifiers for any entity imaginable.

Properties: A property is a specific aspect, characteristic, attribute, or relation used to describe a resource. Each property has a specific meaning, and defines its permitted values, the types of resources it can describe, and its relationship with other properties.

Statements: A specific resource together with a named property plus the value of that property for that resource is an RDF statement. These three individual parts of a statement are called, respectively, the subject, the predicate, and the object. The object of a statement (i.e., the property value) can be another resource or it can be a literal, i.e., a resource (specified by a URI) or a simple string or other primitive datatype defined by XML.

Resources and properties are identified by URIs or uniform resource identifiers. For the purposes of this section, properties will be referred to by a simple name. Consider the simple sentence based on the example presented earlier: *The name of "patient1" is "John Doe"*. This sentence has the following parts:

- Subject (Resource): http://www.hospital.org/Patients/patient1
- Predicate (Property): Name
- Object (literal): "John Doe"

and is illustrated diagrammatically as follows:

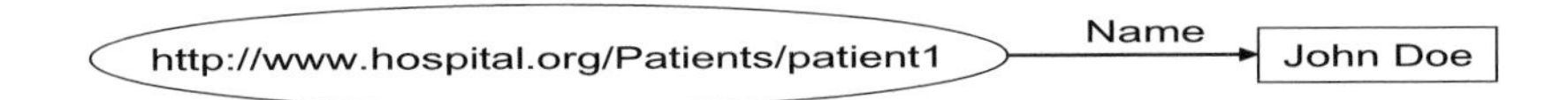

The direction of the arrow is important. The arc always starts at the subject and points to the object of the statement. The simple diagram above may also be read "http://www.hospital.org/Patients/patient1 has name John Doe", or in general "`<subject>` HAS `<predicate>` `<object>`". It is important to note that the RDF representation of a patient requires all resources (patients in this case) to be assigned a URI.

Now, consider the case where we want to say something about the physician of this resource. *The individual whose name is Kashyap, email <kashyap@hospital.org>, treats (or the patient is treated by) http://www.hospital.org/Patients/patient1*. The intention of this sentence is to make the value of the treated-by property a structured entity. In RDF such an entity is represented as another resource. The sentence above does not give a name to that resource; it is anonymous, so in the diagram, it is represented as an anonymous or empty node:

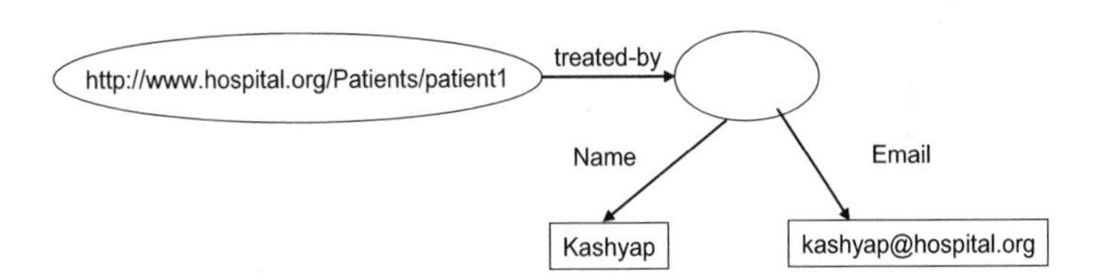

This diagram could be read "http://www.hospital.org/Patients/patient1" is treated by *someone* and *someone* has name Kashyap and email "kashyap@hospital.org". The structured entity above can also be assigned a unique URI leading to the following two sentences. This can be expressed using two sentences: *The individual referred to by http://www.hospital.org/Physicians/physician1 is named Kashyap and has the email address kashyap@hospital.org. The patient http://www.hospital.org/Patients/patient1 was treated by this individual.* The RDF model for these sentences is:

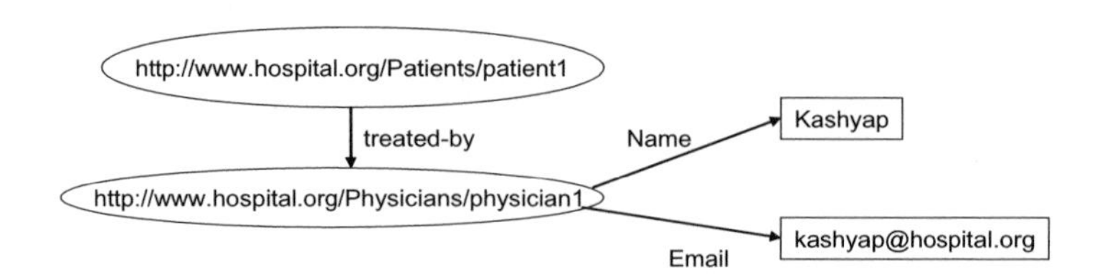

An important design consideration that emerges here is that it might be advantageous to assign URIs to structured entities as opposed to modeling them as anonymous or empty nodes. In the context of data integration, there URIs can be instrumental in identifying similar (sameAs) nodes and merging RDF graphs. This would not be possible with anonymous nodes. RDF defines three types of container objects:

Bag: An unordered list of resources or literals. Bags are used to declare that a property has multiple values and that there is no significance to the order in which the values are given. Bag might be used to give a list of patient medications, where the order of medications does not matter. Duplicate values are permitted.

Sequence: An ordered list of resources or literals. Sequence is used to declare that a property has multiple values and that the order of the values is significant. Sequence might be used, for example, to show the trend of blood pressure values of a patient over time. Duplicate values are permitted.

Alternative: A list of resources or literals that represent alternatives for the (single) value of a property. Alterntives might be use for example to provide the An application using a property whose value is an Alternative collection is aware that it can choose any one of the items in the list as appropriate.

RDF uses the type property, defined below, is used to make a declaration that the resource is one of the container object types defined above. The membership relation between this container resource and the resources that belong in the collection is defined by a set of properties defined expressly for this purpose. These membership properties are named simply "_1", "_2", "_3", etc. Container resources may have other properties in addition to the membership properties and the type property. For example, to represent the sentence *The medical conditions which patient1 suffers from are "Diabetes Mellitus", "Coronary Artery Disease" and "Hypertensions",* the RDF model is

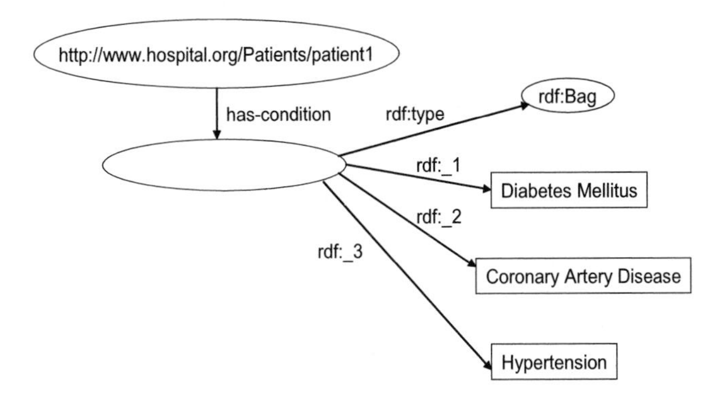

RDF can also be used for making statements about other RDF statements. In order to make a statement about another statement, an RDF model of the original statement is built; this model is a new resource to which we can attach additional properties. Consider the following sentence *Kashyap says that John Doe suffers from Diabetes Mellitus.* Nothing has been said about the patient, instead, a fact about a statement the physician Kashyap makes about the patient has made has been expressed. The original statement is modeled as a resource with four properties: subject, object, predicate and type. This process is formally called *reification* and the model of a statement is called a *reified statement*. To model the example above, we could attach another property to the reified statement (say, "attributedTo") with an appropriate value (in this case, "Kashyap"). The RDF model associated with this statement is presented below:

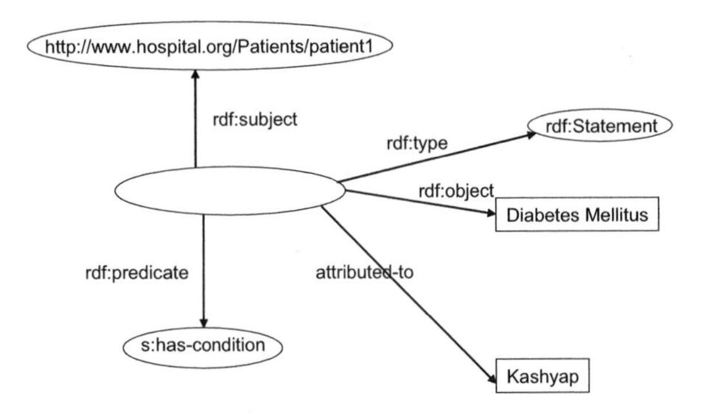

The RDF Data Model is formally defined as follows:

- There is a set called *Resources*.
- There is a set called *Literals*.
- There is a subset of *Resources* called *Properties*, for e.g., *rdf:type*

- There is a set called *Statements*, each element of which is a triple of the form {*pred, sub, obj*}, where *pred* is a *Property*; *sub* is a *Resource*; and *obj* could be either a *Resource* or a *Literal*.
- The set of collections are identified by specially designated resources such as *rdf:Bag*, *rdf:Sequence* and *rdf:Alt*.

Reification may be formally specified as follows:

- *rdf:Statement* is a *Resources*
- *rdf:Predicate, rdf:Subject* and *rdf:Object* are properties used in the reified model.
- Reification of a triple r, {*pred, sub, obj*} consists of elements *s1*, *s2*, *s3*, and *s4* of *Statements* such that
- *s1*: {*rdf:Predicate, r, pred*}
- *s2:* {*rdf:Subject, r, subj*}
- *s3:* {*rdf:Object, r, obj*}
- *s4:* {*rdf:Type, r, rdf:Statement*}

There are interesting similarities and differences between the XML and RDF data models [55] [57]:

- The XML data model structures provide a tree representation (with a horizontal overlay of IDs and IDREFs) of the XML document and can be accessed by applications via a functional interface.
- Parsing RDF documents results in an RDF graph structure, similar to the XML data model, but with a graph instead of a tree. Many RDFS constructs, such as its subclass property, result in constraints in the semantics, such as requiring that certain kinds of relationships are transitive. These semantics are missing from the XML data model.
- The XML data model is strong in capturing positional collections of data since the children of an XML element are textually ordered. It is weak in capturing non-positional collections since these would suggest arc labels, absent in XML, which indicate the role each of the unordered components is playing in the collection.
- The XML data model doesn't support the distinction of various types of tags into classes and properties, as in the case of the RDF data model where there is a clear distinction between resources, classes, properties and values.

OWL is a vocabulary extension of RDF [53]. Thus any RDF graph forms an OWL Full ontology. OWL Full ontologies can thus include arbitrary RDF content, which is treated in a manner consistent with its treatment by RDF. OWL assigns an additional meaning to certain RDF triples. On the other hand, The exchange syntax for OWL is RDF/XML [58], as specified in the OWL Reference Description [59]. Further, the meaning of an OWL ontology in RDF/XML is determined only from the RDF graph/ that results from the RDF parsing of the RDF/XML document. Thus the OWL data model has been defined to be the same as the RDF data model, with OWL offering a vocabulary that provides enhanced semantic interpretation on an RDF Graph. We now present a discussion on the role of model theory in expressing the various interpretations of a data model using mathematical structures; and the entailments supported by a language.

4.2.2 Multiple Syntaxes for RDF: A Short Note

It should be noted that the RDF data model can be serialized using multiple syntaxes. The two possible syntaxes are the XML based syntax and the triples based syntax. Consider the RDF Graph discussed earlier.

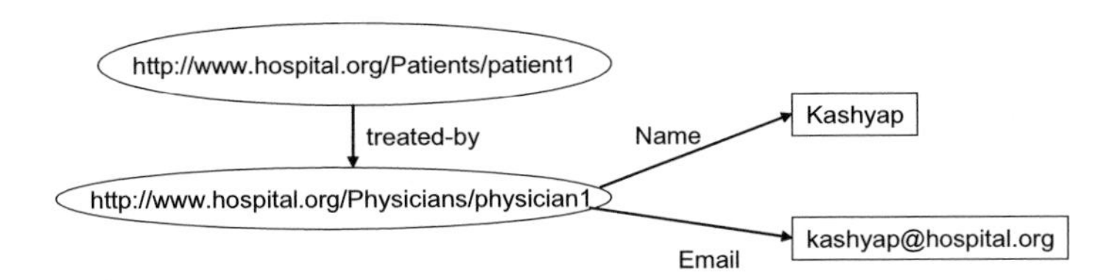

The XML Serialization of the above graph is as follows.

```
<?xml version="1.0"?>
<rdf:RDF xmlns:rdf="http://www.w3.org/1999/02/22-rdf-syntax-ns#">
<rdf:Description
   rdf:about="http://www.hospital.org/Patients/pateint1">
  <treated-by
    rdf:resource="http://www.hospital.org/Physicians/physican1"/>
</rdf:Description>
<rdf:Description
   rdf:about="http://www.hospital.org/Physicians/physician1">
      <Name>Kashyap</Name>
      <Email>kashyap@hospital.org</Email
</rdf:Description>
</rdf:RDF>
```

The triples based serialization of the above graph is as follows.

```
<http://www.hospital.org/Patients/patient1> treated-by
            <http://www.hospital.org/Physicians/physician1> .
<http://www.hospital.org/Physicians/physician1> Name "Kashyap";
                                          Email kashyap@hospital.org .
```

We will be adopting the triples syntax for RDF related discussion in the rest of this chapter.

4.2.3 Model-Theoretic Semantics

Model theory assumes that the language refers to a "world", and describes the minimal conditions that a world must satisfy in order to assign an appropriate meaning for every expression in the language. A particular world is called an *interpretation,* so model theory might be better called 'interpretation theory'. The idea is to provide an abstract, mathematical account of the properties that any such interpretation must have, making as few assumptions as possible about its actual nature or intrinsic structure, thereby retaining as much generality as possible. The chief utility of a formal semantic theory is not to provide any deep analysis of the nature of the things being described by the language or to suggest any particular processing model, but rather to provide a technical way to determine when inference processes are valid, i.e., when they preserve truth. This provides the maximal freedom for implementations while preserving a globally coherent notion of meaning. In a model-theoretic semantics, there is no one single interpretation or model, but instead a collection of interpretations or models. These models can be thought of as the different ways that the world can be and still be compatible with the information in the data or document. We begin with an introductory discussion on the notion of interpretations and entailment in the context of RDF presented in [53].

An interpretation provides just enough information about a possible way the world might be - a 'possible world' - in order to fix the truth-value (true or false) of any ground RDF triple. RDF uses two kinds of referring expressions, URI references and literals, both of which are treated as logical constants, i.e., as expressions having a single value. URI references are treated as simply denoting resources and no further assumptions are made about those resources. An interpretation assigns meanings to symbols in a particular vocabulary of URI references. Some interpretations may assign special meanings to symbols in a particular namespace, called a *reserved* vocabulary. Examples of reserved vocabularies are those associated with RDF, RDF Schema and OWL. A discussion of interpretations and entailments for ground RDF graphs is now presented. The interested reader may refer to [53] for a more detailed exposition.

All interpretations are constructed as being relative to a set of URI references called the vocabulary of the interpretation. A simple interpretation *I* of a vocabulary *V* is defined by:

1. A global, non-empty set *LV* of literal values.
2. A mapping *XL* from the set of literals to *LV.*
3. A non-empty set *IR* of resources, called the domain or universe of *I.*
4. A subset *IP* of *IR* , called the set of properties of *I.*
5. A mapping *IEXT: IP => P(IR x (IR U LV)),* where P stands for powerset. IEXT(x) is a set of pairs, i.e., a binary relational extension called the extension of x.
6. A mapping *IS: V => IR.*

The denotation of a ground RDF graph in *I* is given recursively by the following rules, which extend the interpretation mapping I from names to ground graphs.

- If E is a literal, then *I(E) = XL(E)*
- If E is a URI Reference, then *I(E) = IS(E)*
- If E is an asserted triple `s p o.` then *I(E) = true* if *<I(s), I(o)>* is in *IEXT(I(p))*, otherwise *I(E) = false.*
- If E is a ground RDF graph then *I(E) = false* if *I(E') = false* for some asserted triple E' in E, otherwise *I(E) =true.*

Consider a small vocabulary {a, b, c}. An interpretation of this vocabulary can be specified as follows and is illustrated in Figure 4.4.

IR = {Thing1, Thing2}, IP = {Thing1},
where Thing1, Thing2 are things in the universe.
IEXT = Thing1 => {<Thing1, Thing2>, <Thing2, Thing1>}
IS = a => Thing1, b => Thing1, c => Thing2

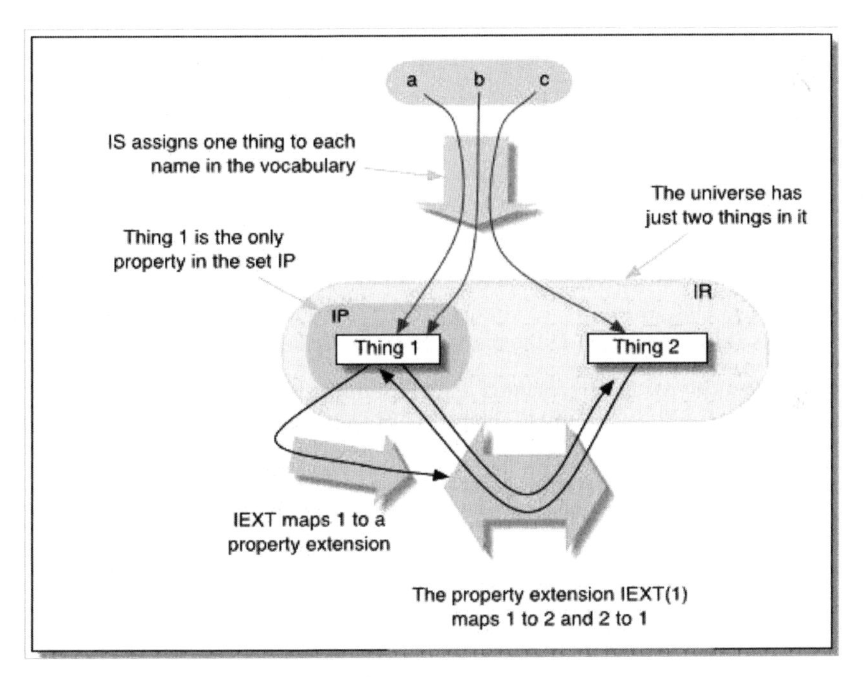

Fig. 4.4. An example of an Interpretation

Based on the above interpretation, the following triples are true:

```
a b c .
c a a .
c b a .
```

I(a b c .) = true if <I(a),I(c)> is in IEXT(I(b)), i.e. if <Thing1,Thing2> is in IEXT(Thing1), which is {<Thing1,Thing2>,<Thing2,Thing1>}

So I(a b c .) is true)

The following triples are false:

```
a c b .
a b b .
c a c .
```

I(a c b .) = true if <I(a),I(b)>, i.e.<Thing1,Thing2>, is in IEXT(I(c)); but I(c) = Thing2 and IEXT is not defined on Thing2

The condition fails and I(a c b .) = false.)

We say that I satisfies E if I(E)=true, and that a set S of expressions (simply) entails E if every interpretation which satisfies every member of S also satisfies E. If {E} entails E' we will say that E entails E'. Any process or technique which constructs a graph E from some other graphs S is said to be (simply) valid iff S always entails E, otherwise invalid. It may be noted that being an invalid process does not mean that the conclusion is false, and being valid does not guarantee truth. However, validity represents the best guarantee that any assertional language can offer: if given true inputs, it will never draw a false conclusion from them. Simple entailment can be recognized by relatively simple syntactic comparisons. The two basic forms of simply valid proof step in RDF are, in logical terms, the inference from (P and Q) to P, and the inference from (foo baz) to (exists (?x) (foo ?x)) . Some examples of entailments are:

- If E is ground, then I satisfies E iff it satisfies every triple in E
- If E and E' are ground, then E entails E' iff E' is a subgraph of E
- The merge of a set of RDF graphs S is entailed by S, and entails every member of S
- S entails E iff a subgraph of the merge of S is an instance of E.

In this section, we have presented a small introduction to the notion of model-theoretic interpretations and entailments. These techniques provide a mathematical basis for specifying the semantics of various metadata representation languages and can be used by tool developers and implementers to validate and their implementations. There has been significant work done for specifying the semantics for various semantic web languages. The interested reader can peruse the following efforts:

- A detailed exposition of model-theoretic semantics for RDF and RDF(S) is presented in [53].
- A detailed presentation of an RDF compatible Model-Theoretic sematnics fow OWL is presented in [60]. This also includes intepretation for constructors for OWL classes that and relationships that hold between classes (e.g., subclass).
- An interesting approach presented in [52] that presents model-theoretic interpretations of XML data. In particular, this apporach deals with ordering information, which is not usually part of RDF semantics, but it is important to capture

document order in XML documents. Additional conditions for an interpretation to be an RDF interpretation are also specified.
- A detailed discussion of the interpretation of RDF Schema (presented in Chapter 5) is presented in [53].

4.3 Query Languages

In the previous section, we discussed data models and semantics for various XML/RDF- and OWL-based specifications that are being proposed and to a limited extent in use for representing metadata descriptions. We now discuss query languages that have been proposed for manipulating these metadata descriptions.

4.3.1 Query Languages for XML Data

There have been various query language proposals for manipulating XML data, such as Lorel [62], XML-QL [55], XML-GL [63], XQBE [386], XSLT [65], XQL [66] and XQuery [64]. We first discuss these proposals in the context of a comparative survey presented in [61] and then present XQuery, the W3C Recommendation in more detail.

XML Query Languages: A Historical Perspective and Comparative Analysis

We present a historical perspective on the development of query languages for semi-structured data. For the reader interested in understanding Semantic Web technologies and specifications and their application to real-world problems, an understanding of the development and evolution of query languages for XML (and later RDF) is important. We begin by first discussing some examples of query languages:

- **Lorel:** Lorel is a SQL/OQL style language for querying semi-structured data, extended to XML data. It supports type coercion and powerful path expressions.
- **XML-QL:** XML-QL supports specification of both queries and transformations. It extends SQL with an explicit CONSTRUCT clause and uses element patterns built on top of XML syntax for matching data in an XML document.
- **XML-GL:** XML-GL is a graphical query language relying on a graphical representation of XML documents and is aimed at supporting a user-friendly interface similar to Query By Example (QBE).
- **XSLT:** An XSLT style sheet consists of a collection of template rules; each template rule has two parts: a pattern which is matched against nodes in the source tree and a template which is instantiated to form part of the result tree. XSLT makes use of the expression language defined by XPath [67].

- **XQL:** XQL is a notation for selecting and extracting XML data, and is designed with the goal of being syntactically simple and compact with reduced expressive power.
- **XBQE:** This is an enhanced version of XML-GL specifically targeted to be a suitable visual interface for XQuery.
- **XQuery:** XQuery is a W3C Candidate Recommendation for querying XML data. It assembles many features from previously defined languages such as: (a) syntax for navigation in hierarchical documents from XPath and XQL, (b) variable bindings from XML-QL, (c) combination of clauses from SQL, and (d) a functional language from OQL.

We now present a summary of the comparison of the query languages discussed above. A more detailed analysis is available in [61]:

Data Model Representation: All the languages assume a tree-based data model. Lorel, XML-QL, XML-GL and XQuery assume cross-links, whereas XBQE also assumes explicit joins between attribute values. Lorel and XQuery differentially interpret IDREFs as strings and references, whereas the others interpret IDREFs as strings only. XQuery and XBQE are highly compatible with existing W3C standards such as XQuery and XPath data model and XML Schema.

Basic Query Abstractions: All the query languages support document selection. Full fledged joins including intra-document, inter-document, inner and outer joins are supported by all languages, except for XQL which supports it partially. All the languages except XML-QL and XBQE support an explicit operator or function to dereference IDREFs. XBQE supports IDREF dereferencing via specification of explicit joins. Only Lorel supports the ability to define views on the data.

Path Expressions: All languages, except XML-GL support partially specified path expressions which makes it easy to query semi-structured data, especially when the extract structure is not known. XML-GL has partial support for this feature. Only Lorel, XML-GL and XBQE support matching of partially specified path expressions with cylic data.

Quantification, Negation, Reduction and Filtering: All languages support existential quantification, whereas universal quantification is supported only by Lorel, XQL and XQuery. Negations are supported by all languages except XML-QL and XBQE, which supports it partially. Reduction or the ability to prune elements from the final result are supported only by XSLT and XBQE. Ad hoc operators for filtering, i.e., the ability to retain specific elements while preserving hierarchy and sequence are supported by XBQE and XQuery, and partially by XSLT.

Restructuring Abstractions: Construction mechanisms to build new XML data are supported by all languages excpet XQL. Grouping constructs are supported by Lorel and XQL and partially by XSLT. XBQE and XQuery support grouping partially through restructuring. Skolem functions that generate unique OIDs are supported by Lorel, XML-QL and XQuery. XBQE and XML-GL partially support skolem functions.

Aggregation, Nesting and Set Operations: Aggregation functions that a scalar value out of a multiset of values are supported by all languages except XML-QL. XSLT and XQL partially support aggregation functions. Nested queries are supported by all languages except XML-GL and XBQE that supports it partially. Set operations are supported by Lorel, XSLT, XQL and XQuery. Set operations are partiallyby XML-QL and XBQE supports only intersection explicitly.

Order Management: Support for ordering of element instances is supported by all languages excetp XQL. All languages support the preserving the order of the result based on the order they appear in the original document. The ability query the order of elements is supported by all languages except XML-GL and XBQE. The ability to ask for numbered instances of XML elements is supported by Lorel, XSLT, XQL and XQuery.

Typing and Extensibility: The ability to embed specialized types in an XML query language is supported only XQuery and to a partial extent, Lorel. Type coercion and the ability to compare values with different type constructors is supported by Lorel, XQL and XQuery. Support for built-in, user-defined and external functions is partially available in all languages except Lorel which lacks support.

Integration with XML: None of the languages provide support for RDF, OWL, XLink and XPointer. All languages excetp Lorel, XML-GL and XBQE support XML namespaces. XML-QL has partial support for XML namespaces. The ability to query tag names rather than tag content are supported by all languages except XML-GL, XQL and XBQE.

Update Support: The ability to update XML elements and attributes is supoted by all languages except XML-QL, XQL and XQuery.

The query languages can be organized into the following three classes:

1. **Core Query Languages:** XQL and XML-QL are representative of this class, playing the same role as core SQL standards and languages in the relational world. Their expressive power is included within thst of XSLT.
2. **Graphic Query Interfaces:** XML-GL and XQBE are representative of this class, playing the same role as QBE in the relational world. It can suitably be adopted as a front-end to any of these query languages to express a comprehensive class of queries (a subset of them in the case of more powerful languages).
3. **Expressive XML Query Languages:** XQuery, Lorel and XSLT are representative of this class, playing the same role as high-level SQL standards and languages (e.g., SQL2 in the relational world). Lorel is strongly object-oriented while XQuery can be considered value-oriented. XQuery is a promising expressive query language that realizes its potentiality by incorporating the experience of XPath and XQL on the one hand, and of SQL/ OQL and XML-QL on the other. The third language of this class, XSLT, covers a lower position in the taxonomy, being less powerful than the previous two. It is a style sheet language with a fairly procedural tendency as opposed to Lorel, which can be considered completely declarative, and to XQuery, which blends the declarative and procedural flavor.

The XQuery Query Language for XML Data

XQuery is a W3C Recommendation [64] designed by the XML Query Working Group. XQuery is a functional language comprised of several kinds of expressions that can be nested and composed with full generality. It is based on the type system of XML Schema and is designed to be compatible with other XML-related standards. The design of XQuery has been subject to a number of influences, the most important perhaps being compatibility with existing W3C standards, including XML Schema, XSLT, XPath, and XML itself. XPath, in particular, is so important and so closely related that XQuery is defined as a superset of XPath. The overall design of XQuery is based on a language proposal called Quilt [71], which in turn was influenced by the functional approach of Object Query Language (OQL) [72], by the syntax of Structured Query Language (SQL) [73], and by previous XML query language proposals including XQL [66], XML-QL [55], and Lorel [62]. We now discuss the XQuery language based on the overview and examples presented in [70].

Consider a small XML database that that contains data from an Electronic Health Record (EHR) system. The database consists of two XML documents named `patients.xml` and `lab-results.xml`. The `patients.xml` document contains a root element named `patients`, which in turn contains a `patient` element for each patient in the EHR system. Each `patient` element has a `status` attribute and subelements named `patient-id`, `name`, `sex`, and `age`. The `lab-results.xml` document contains a root element named `lab-results`, which in turn contains a lab-result element for each laboratory result that has been received from a patient. Each `lab-result` element has subelements named `patient-id`, `lab-test`, and `sequenced-gene`. The XML Graph for the sample data is illustrated in Figure 4.5

Expressions

XQuery is a functional language, which means that it is made up of *expressions* that return values and do not have side effects. XQuery has several kinds of expressions, most of which are composed from lower-level expressions, combined by *operators* or *keywords*. XQuery expressions are fully composable, that is, where an expression is expected, any kind of expression may be used. The value of an expression, in general, is a heterogeneous sequence of nodes and atomic values.

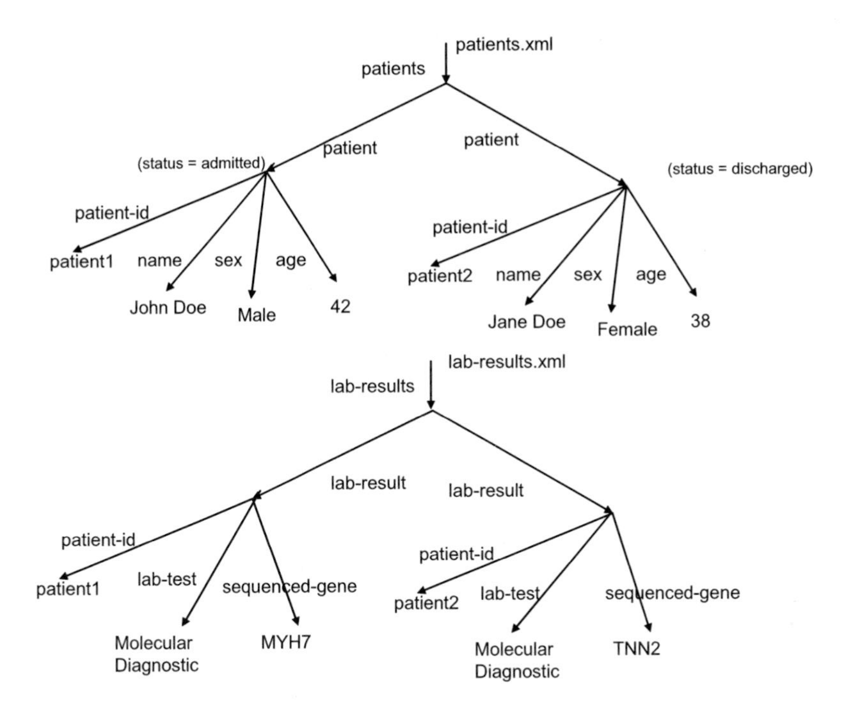

Fig. 4.5. XML Graph corresponding to example Data

Primary Expressions. The simplest kind of XQuery expression is a *literal*, which represents an atomic value which may be of type `integer`, `decimal`, `real`, `double` and `string`. Atomic values of other types may be created by calling constructors. A *constructor* is a function that creates a value of a particular type from a string containing a lexical representation of the desired type. For example in order to create a value of type date, one may invoke the constructor `date("2008-5-31")`. A *variable* in XQuery is a name that begins with a dollar sign. A variable may be bound to a value and used in an expression to represent that value. One way of binding a value to a variable is by using the *LET expression,* for e.g., `let $start := 1`. Parentheses may be used to enforce a particular evaluation order in expressions that contain multiple operators. For example, the expression `(2 + 4) * 5` evaluates to thirty, since the parenthesized expression `(2 + 4)` is evaluated first and its result is multiplied by five. A *context item expression* is a XQuery expreession that evaluates to the context item, which may be either a node (as in the expression `fn:doc("patients.xml")/patients/patient[fn:count(./age)=1]`) or an atomic value (as in the expression `(1 to 100)[. mod 5 eq 0]`).Another simple form of XQuery expression is a *function call.* For example, the core library function substring to extract the first six characters from a string can be invoked as `substring("Martha Washington", 1, 6)`.

Path Expressions. A path expression, used to locate nodes in a document, consists of a series of steps, separated by "`/`" or "`//`", and optionally beginning with a "`/`" or "`//`". An initial "`/`" or "`//`" is an abbreviation for one or more initial steps that are implicitly added to the beginning of the path expression. A step generates a sequence of items, and then filters the sequence with zero or more predicates. A step may either be an axis step or a *filter expression*. An axis step returns a sequence of nodes reachable from a context node along an axis (e.g., `child/descandant` or `parent/ancestor`) and satisfies a node test (e.g., type of node such as `text` or `element`). The value of the path expression is the node sequence that results from the last step in the path.

In XQuery, a *predicate* is an expression, enclosed in square brackets, that is used to filter a sequence of values. Predicates are often used in the steps of a path expression. For example, in the step `patient[name = "John Doe"]`, the phrase `name = "John Doe"` is a predicate that is used to select certain item nodes and discard others. For each item in the input sequence, the result of the predicate expression is coerced to an `xs:boolean` value, called the *predicate truth value*. Those items for which the predicate truth value is `true` are retained, and those for which the predicate truth value is `false` are discarded. If the predicate expression evaluates to a number, the candidate item is selected if its ordinal position in the list of candidate items is equal to the number. For example `patient[2]` selects the second child of a given node. Predicates can be used to test the existence of nodes. For example, `patient[height]` selects `patient` nodes that have a `height` child node, regardless of its value. Common *comparison operators* used in predicates are discussed later in this section. Consider the following examples of path expressions:

- *List the age of all the patients with the name "John Doe".*
  ```
  fn:doc("patients.xml")/*/patient[name = "John Doe"]/age
  ```
  Q1 illustrates a four-step path expression using abbreviated syntax. The first step invokes the built-in document function, which returns the document node for the document named `patients.xml`. The second step is an axis step that finds all children of the document node ("`*`" selects all the children of a node, which in this case is only a single element node named items). The third step finds all child elements at the next level that are named `patient` and that in turn have a child called `name` with the value "`John Doe`". The result of the third step is a sequence of `patient` element nodes. Each of these `patient` nodes is used in turn as the context node for the fourth step, which finds the `age` elements that are children of the given item.
- *List all name elements found in the document patients.xml.*
  ```
  fn:doc("patients.xml")//name
  ```
  When two steps are separated by a double slash rather than by a single slash, it means that the second step may traverse multiple levels of the hierarchy, using the `descendants` axis rather than the single-level child axis. For example, Q2 searches for `name` elements that are descendants of the root node of a given document. The result of Q2 is a sequence of element nodes that could, in principle, have been found at various levels of the node hierarchy.

- *Find the status attribute of the patient that is the parent of a given name.*
  `$name/../@status`
  Within a path expression, a single dot (" . ") refers to the context node, and two consecutive dots ("..") refer to the parent of the context node. This is illustrated by Q3, which begins at the node that is bound to the variable `$name`, traverses to the parent `patient` node, and then traverses the attribute to find an attribute named `status`. The result of Q3 is a single attribute node.

Sequence Expressions. XQuery supports operators to construct, filter, and combine sequences of items. Sequences in XQuery are never nested as nested sequences are flattened out. Sequences can be constructed using the *comma operator* (e.g., `(1, 2, 3)`) and *range expressions* (e.g., `1 to 3`). A *filter expression* consists of a primary expression followed by one or more predicates (e.g., `patient[age > 65]`). XQuery provides the union, intersect and except operators for combining sequences of nodes. The *union* operator takes two node sequences as operands and return a sequence containing all the nodes that occur in either of the operands. The *intersect* operator takes two node sequences as operands and returns a sequence containing all the nodes that occur in both operands. The *except* operator takes two node sequences as operands and returns a sequence containing all the nodes that occur in the first operand but not in the second operand. Duplicate nodes are eliminated based on node identity. Consider the following query: *Construct a new element named elderly male patients, containing copies of all the patient elements in the document patients.xml that have age > 65 and sex as Male.*

```
<elderly-male-patients>
   fn:doc("patients.xml")
       /*/patient[age > 65]
   intersect
   fn:doc("patients.xml")
       /*/patient[sex = "Male"]
</elderly-male-patients>
```

Arithmetic Expressions. XQuery provides arithmetic operators for addition, subtraction, multiplication, division, and modulus, in their usual binary and unary forms. A subtraction operator must be preceded by white space if it could otherwise be interpreted as part of the previous token. For example, `a-b` will be interpreted as a name, but `a - b` and `a -b` will be interpreted as arithmetic expressions. XQuery supports two division operators named `div` and `idiv`. Each of these operators accepts two operands of any numeric type. For example, `$arg1 idiv $arg2` is equivalent to `($arg1 div $arg2) cast as xs:integer?` except for error cases. An example of a query using an arithmetic expression is as follows.
Given a sequence of patient elements, replace the height and weight subelements with a new body-mass-index element containing the value of height divided by weight squared.

```
for $p in $patients
return
```

```
<patient>
   {
    $p/name,
          $p/sex,
          $p/age,
    <body-mass-index>
          {$p/height div ($p/weight * $p/weight)}
    </body-mass-index>
   }
</patient>
```

For those patients whose height or weight is missing (`$p/height` or `$p/weight` evaluates to an empty sequence), the generated `body-mass-index` element will be empty.

Comparison Expressions. Comparison expressions allow two values to be compared. XQuery provides three kinds of comparison expressions, called *value comparisons*, *general comparisons*, and *node comparisons*.

- *Value comparison operators:* `eq, ne, lt, le, gt, ge`. These operators can compare two scalar values, but they raise an error if either operand is a sequence of length greater than one. If either operand is a node, the value comparison operator extracts its value before performing the comparison. For example, `patient[age gt 65]` selects a `patient` node if it has exactly one `age` child node whose value is greater than 65.
- *General comparison operators:* `=, !=, >, >=, <, <=`. These operators can deal with operands that are sequences, providing implicit "existential" semantics for both operands. Like the value comparison operators, the general comparison operators automatically extract values from the nodes. For example, `patient[age > 65]` selects a `patient` node if it has at least one `age` child node whose value is greater than 65.
- *Node comparison operators:* `is, <<, >>`. These operators compare the identities of two nodes. For example, `$node1 is $node2` is true if the variables `$node1` and `$node2` are bound to the same node (that is, the node identity is the same for both variables). A comparison with the `<<` operator returns `true` if the left operand node precedes the right operand node in document order; otherwise it returns `false`. A comparison with the `>>` operator returns `true` if the left operand node follows the right operand node in document order; otherwise it returns `false`.

Logical Expressions. A logical expression is either an *and-expression* or an *or-expression*. If a logical expression does not raise an error, its value is always one of the boolean values `true` or `false`. In addition to and- and or-expressions, XQuery provides a function named `fn:not` that takes a general sequence as parameter and returns true if the effective boolean value of its parameter is false, and false if the effective boolean value of its parameter is true. For example, the following predicate selects patient nodes that have exactly one name child element with the value "John Doe" and also have at least one age child element with any value:

`patient[name eq "John Doe" and age]`. The following step uses the not function with an existence test to find `patient` nodes that have no `age` child element: `patient[fn:not(age)]`.

Constructors. XQuery provides constructors that can create XML structures within a query. Constructors are provided for element, attribute, document, text, comment, and processing instruction nodes. Two kinds of constructors are provided: *direct constructors*, which use an XML-like notation, and *computed constructors*, which use a notation based on enclosed expressions. For example, the following expression constructs an element named `male-patient` containing one attribute named `status` and two child elements named `patient-id` and `sex`:

```
<male-patient status = "admitted">
   <patient-id>patient1</patient-id>
   <sex>Male</sex>
</male-patient>
```

In some cases, the values are evaluated by some expression, which is enclosed in curly braces. The evaluation is done by the constructor. For example, the values of status and the sex of the patient are computed and inserted at run time.

```
<male-patient status = "{$s}">
   <patient-id>patient1</patient-id>
   <sex> {$sex} </sex>
</male-patient>
```

FLWOR Expressions. XQuery provides a feature called a FLWOR expression that supports iteration and binding of variables to intermediate results. This kind of expression is often useful for computing joins between two or more documents and for restructuring data. The name FLWOR, pronounced "flower", is suggested by the keywords `for`, `let`, `where`, `order by`, and `return`. The `for` and `let` clauses in a FLWOR expression generate an ordered sequence of tuples of bound variables, called the tuple stream. The optional `where` clause serves to filter the tuple stream, retaining some tuples and discarding others. The optional `order by` clause can be used to reorder the tuple stream. The `return` clause constructs the result of the FLWOR expression. The `return` clause is evaluated once for every tuple in the tuple stream, after filtering by the `where` clause, using the variable bindings in the respective tuples. The result of the FLWOR expression is an ordered sequence containing the results of these evaluations, concatenated as if by the comma operator.

The following example of a FLWOR expression includes all of the possible clauses. The `for` clause iterates over all the patients in an input document, binding the variable `$p` to each department number in turn. For each binding of `$p`, the `let` clause binds variable `$lr` to all the lab results of the given patient, selected from another input document. The result of the `for` and `let` clauses is a tuple stream in which each tuple contains a pair of bindings for `$p` and `$lr` (`$p` is bound to a patient and `$lr` is bound to a set of lab results of that patient). The `where` clause filters the tuple stream by keeping only those binding pairs that represent patients having at

least two lab results. The `order by` clause orders the surviving tuples in descending order by the number of lab results of the patient. The `return` clause constructs a new `patient-genes` element for each surviving tuple, containing the patient information, number of genes and the list of genes sequenced for the patient.
For each patient from whom more than two genes were sequenced, generate a patient-genes element containing name, sex, age and gene count ordered by the number of genes sequenced.

```
for $p in fn:doc("patients.xml")/*/patient
let $lr := fn:doc("lab-results.xml")/*/lab-result[patient-id = $p/
patient-id]
where fn:count ($lr) > 2
order by fn:count ($lr)
return
 <patient-genes>
   {
     $p,
    <gene-count> {count ($lr)} </gene-count>
    for $sg in fn:distinct-values($lr/sequenced-gene)
        return <gene> {$sg} </gene>
   }
</patient-genes>
```

Conditional Expressions. XQuery supports a *conditional expression* based on the keywords `if`, `then`, and `else`. The expression following the `if` keyword is called the *test expression*, and the expressions following the `then` and `else` keywords are called the *then-expression* and *else-expression*, respectively. If the effective boolean value of the test expression is `true`, the value of the then-expression is returned. If the effective boolean value of the test expression is `false`, the value of the else-expression is returned. Here are some examples of conditional expressions:

```
if ($widget1/unit-cost < $widget2/unit-cost)
  then $widget1 else $widget2
```

In the above example, the test expression is a comparison expression.

```
if ($part/@discounted)
  then $part/wholesale else $part/retail
```

In the above example, the test expression tests for the existence of an attribute named `discounted`, independently of its value.

Quantified Expressions. These expressions support existential and universal quantification and are always `true` or `false`. A *quantified expression* begins with a *quantifier*, which is the keyword `some` or `every`, followed by one or more in-clauses that are used to bind variables, followed by the keyword `satisfies` and a test expression. The value of the quantified expression is defined by the following rules:

1. If the quantifier is `some`, the quantified expression is `true` if at least one evaluation of the test expression has the effective boolean value `true`; otherwise the quantified expression is `false`. This rule implies that, if the in-clauses generate zero binding tuples, the value of the quantified expression is `false`.
2. If the quantifier is `every`, the quantified expression is `true` if every evaluation of the test expression has the effective boolean value `true`; otherwise the quantified expression is `false`. This rule implies that, if the in-clauses generate zero binding tuples, the value of the quantified expression is `true`.

Some examples of quantified expressions are as follows.

```
every $patient in /patients/patient satisfies $patient/@status
```

This expression is `true` if every `patient` element has a `status` attribute (regardless of the values of these attributes):

```
some $patient in /patients/patient satisfies
  ($patient/weight > $patient/height)
```

This expression is `true` if at least one `patient` element has `weight` greater than `height`.

Types

The type system of XQuery consistes of *schema types* and *sequence types*. Sequence types are used to refer to a type in an XQuery expression and describe the type of an XQuery value, which is always a sequence. A schema type is a type that is defined using XML Schema facilities. A schema type can be used as a type annotation on an `element` or `attribute` node (unless it is a non-instantiable type in which case its derived types can be so used). Every schema type is either a *complex type* or a *simple type*; simple types are further subdivided into *list types*, *union types*, and *atomic types*. Atomic types represent the intersection between the categories of sequence type and schema type. An atomic type, such as `xs:integer` or `my:hatsize`, is both a sequence type and a schema type.

Predefined Schema Types. These include the built-in schema types in the namespace `http://www.w3.org/2001/XMLSchema`, which has the predefined namespace prefix `xs`. The schema types in this namespace are defined in XML Schema and augmented by additional types defined in the XQuery/XPath Data Model (XDM).

Sequence Types. A sequence type (except the special type `empty-sequence()`) consists of an *item type* that constrains the type of each item in the sequence, and a *cardinality* that constrains the number of items in the sequence. Apart from the item type `item()`, which permits any kind of item, item types divide into *node types* (such as `element()`) and *atomic types* (such as `xs:integer`). Item types representing element and attribute nodes may specify the required type annotations of those nodes, in the form of a schema type. Thus the item type `element(*,` `us:address)` denotes any element node whose type annotation is (or is derived from) the schema type named `us:address`.

4.3.2 Query Languages for RDF Data

There have been various query language proposals for manipulating RDF data, such as RQL [75], SeRQL [76], TRIPLE [77], RDQL [78], N3 [79] and Versa [80]. We first discuss these proposals in the context of a comparative survey presented in [81] and then present SPARQL, the W3C Proposed Recommendation in more detail.

RDF Query Languages: A Historical Perspective and Comparative Analysis

This section is a continuation of the historical perspective on the development of query languages for semi-structured data, presented in Section 4.3.1. For the reader interested in understanding Semantic Web technologies and specifications and their application to real-world problems, an understanding of the development and evolution of query languages for RDF is important. We begin with a brief description of various query language proposals.

- **RQL:** RQL is a typed language following a functional approach with an OQL-like syntax, which supports generalized path expressions featuring variables on both nodes and edges of the RDF graph. The novelty of RQL lies in its ability to smoothly combine schema and data querying while exploiting the taxonomies of labels and multiple classification of resources.
- **SeRQL:** SeRQL is a light weight yet expressive query and transformation language that seeks to address practical concerns. It is loosely based on several existing languages, most notably RQL, RDQL and N3.
- **TRIPLE:** Triple denotes both a query and rules language and the actual runtime system. The language is derived from F-Logic [82]. RDF triples `(S,P,O)` are represented as F-Logic expressions `S[P->O]`, which can be nested. Triple does not distinguish between rules and queries, which are simply headless rules, where the results are bindings of free variables in the query.
- **RDQL:** RDQL, a W3C member submission, follows a SQL-like select pattern, where a from clause is omitted. For example, `select ?p where (?p,<rdfs:label>,"foo")` collects all resources with label “foo” in the free variable p.
- **N3:** Notation3 (N3) provides a text-based syntax for RDF. Therefore the data model of N3 conforms to the RDF data model. Additionally, N3 allows us to define rules that may be used for specifying queries, for example: `?y rdfs:label "foo" => ?y a :QueryResult.`
- **Versa:** The main building block of Versa is a list of RDF resources. Traversal operations, which have the form `ListExpr - ListExpr -> BoolExpr.` return a

list of all objects of matching triples. For instance, the traversal expression `all() - rdfs:label -> *` would return a list containing all labels.

We now present a comparison of these query languages. A detailed discussion is presented in [81]:

Support for the RDF Data Model: The underlying structure of any RDF document is a collection of triples, called the RDF graph. Since the RDF data model is independent of a concrete serialization syntax, query languages usually do not provide features for query serialization-specific features, e.g., order of serialization. RQL is the only language that provides for this support. RDF has a formal semantics which provides a dependable basis for reasoning and entailment about the meaning of an RDF graph. RQL is partialy compatible with the formal semantics of RDF, and N3 and TRIPLE support RDF semantics and entailment via custom rules. Some implementations of SPARQL, e.g., Jena support entailments via rules engines and ontology-based inferences.

Query Language Properties: The closure property requires that the results of an operation be elements of the data model, i.e., if the language operates on the graph data model, the query results would again have to be graphs. SeRQL and N3 have support for closure, whereas SPARQL provides the CONSTRUCT operator for this purpose. Adequacy is the dual of closure, which requires that all the concepts of the underlying data model be used. SPARQL has support for adequacy. The mapping of the TRIPLE model to the RDF data model is not lossless and hence has partial adequacy. Orthogonality requires that all operations be used independently of the usage context. Of all languages, RDQL is the one which doesn't support orthogonality. Safety requires that a syntactically correct query return a finite set of results on a finite data-set. Typical concepts that cause query languages to be unsafe are recursion, negation and built-in functions. RDQL, N3 and Versa are safe query languages.

Path Expressions: Path expressions are offered in various syntactic forms by all RDF query languages. Some RDF query languages such as SeRQL and VERSA support means to deal with irregularities and incomplete information. SPARQL has support irregularity and incompleteness using the OPTIONAL construct and RQL has partial support for optionality of variable bindings.

Basic Algebraic Operations: In the relational data model several basic algebraic operations are considered, i.e., (i) selection, (ii) projection, (iii) cartesian product, (iv) set difference and (v) set union. The three basic algebraic operations, selection, projection, and product, are supported by all the proposed languages being discussed. Union operations are supported by all languages except RDQL. Difference is supported by RQL, SeRQL, SPARQL and partially supported by Versa.

Quantification: An existential predicate over a set of resources is satisfied if at least one of the values satisfies the predicate. Analogously, a universal predicate is satisfied if all the values satisfy the predicate. Existential quantification is supported by all languages as any selection predicate implicitly has existential bindings. Universal quantification is supported by RQL, SeRQL, SPARQL and partially by TRIPLE.

Aggregation and Grouping: Aggregate functions compute a scalar value from a multiset of values. A special case of aggregation is counting the number of elements in a set which is supported only by RQL, N3 and Versa. Grouping additionally allows aggregates to be computed on groups of values. None of the compared query languages allows us to group values as the SQL GROUP BY clause.

Recursion: Recursive queries typically occur in scenarios where the underlying relationship is transitive in nature. The denotation of a property has to be given in queries as symmetric or transitive relationships are not natively supported in RDF. There is support for transitivity and recursion in TRIPLE and N3 via rules and in Versa via the "traverse" operator.

Reification: Reification is a unique feature of RDF, which allows treating RDF statements as resources themselves, such that statements can be made about statements. Reification is supported to some extent by all languages except N3 and SPARQL?

Collections and Containers: RDF allows us to define groups of entities using collections (a closed group of entities) and containers. All the proposed languages being investigated partially support the ability to retrieve these collection and their elements along with order information.

Namespaces: Given a set of resources, it might be interesting to query all values of properties from a certain namespace or a namespace with a certain pattern. Pattern matching on namespaces is particularly useful for versioned RDF data, as many versioning schemes rely on the namespace to encode version information. All query languages other than TRIPLE and Versa support this feature. RDQL has partial support for this feature.

Literals and Datatypes: RDF supports the type system of XML Schema to create typed literals. An RDF query language should support XML Schema datatypes. A datatype consists of a lexical space, a value space and lexical-to-value mapping. All languages have support for the lexical space, whereas only RQL, SeRQL have support for the value space. RDQL has partial support for the value space.

Entailment: The RDF Schema vocabulary supports the entailment of implicit information such as use of subclass, domain and range relationships. All languages except Versa, support entailment to varying degrees.

Desired Properties: Many of the existing proposals support very little functionality for grouping and aggregation. Surprisingly, except for Versa and SPARQL, no language is capable to do sorting and ordering on the output. Due to the semi-structured nature of RDF, support for optional matches is crucial in any RDF query language and should be supported with a dedicated syntax as in SPARQL. Overall, the languages' support for RDF-specific features like containers, collections, XML Schema datatypes, and reification is quite poor. Since these are features of the data model, the degree of adequacy among the languages is low.

The SPARQL Query Language for RDF Data

SPARQL is a W3C Recommendation [83] which consists of three separate specifications which describe a query language [83], a data access protocol which uses WSDL 2.0 to define simple HTTP and SOAP protcols for querying remote databases [85], and the XML format [84] in which query results will be returned. In this Section, we discuss the SPARQL query language, which consists of the following components:

Graph Patterns: These patterns consists of combinations of triple patterns (one or more of subject, predicate, and object replaced by variables) that help identify a subset of the RDF graphs which need to be retrieved.

RDF Dataset: Many RDF data stores hold multiple RDF graphs, and record information about each graph, allowing an application to make queries that involve information from more than one graph. A SPARQL query is executed against an RDF dataset which represents such a collection of graphs. There is one graph, the default graph, which does not have a name, and zero or more named graphs.

Solution Modifiers: Query patterns generate an unordered collection of solutions, which are then treated as a sequence, initially in no specific order. Sequence modifiers are then applied to create other sequences as appropriate.

Query Form: The last sequence generated after application of the solution modifiers is used to generate a SPARQL result form based on the Query Form such as `SELECT`, `CONSTRUCT` or `ASK`.

Consider two RDF data sources, one from an Electronic Medical Record (EMR) system and another from a Laboratory Information Management System (LIMS). The RDF graphs view of these data sources are illustrated in Figure 4.6. The EMR data illustrated on the left hand side consists of a `Patient` associated to a `Person` via the `isRelatedTo` edge. The `Patient` has an associated `FamilyHistory` which linked by the edge labeled `hasFamilyHistory`. The `FamilyHistory` is associated with a `Person` through `associatedRelative` edge and a `Disease` via the `problem` edge. The `isRelatedTo` edge is reified and the characteristics of the relation such as the `type` and `degree` of the relation are linked by edges with those labels. The LIMS data illustrated on the right hand side consists of a `Patient` and the associated `MolecularDiagnosticTestResult` linked by the edge labeled `hasStructuredTestResult`. The test result sequences a `Mutation` which linked by the edge labeled `identifiesMutation`. In some cases, when a gene mutation is found, this could be indicative of a `Disease` linked by the edge labeled `indicatesDisease`. The `indicatesDisease` edge is reified and the strength of evidence (linked by edges labeled `evidence1` and `evidence2`) by which the patient is inferred to be suffering from that disease. We will use this example data to illustrate various aspects of the SPARQL query language.

Graph Patterns

Graph patterns match against the default graph of an RDF dataset, except for the RDF Dataset Graph Pattern. The result of a query is the set of all pattern solutions that match the query pattern, giving all the ways a query can match the graph being queried. Each result is one solution to the query and there may be zero, one or multiple results to a query.

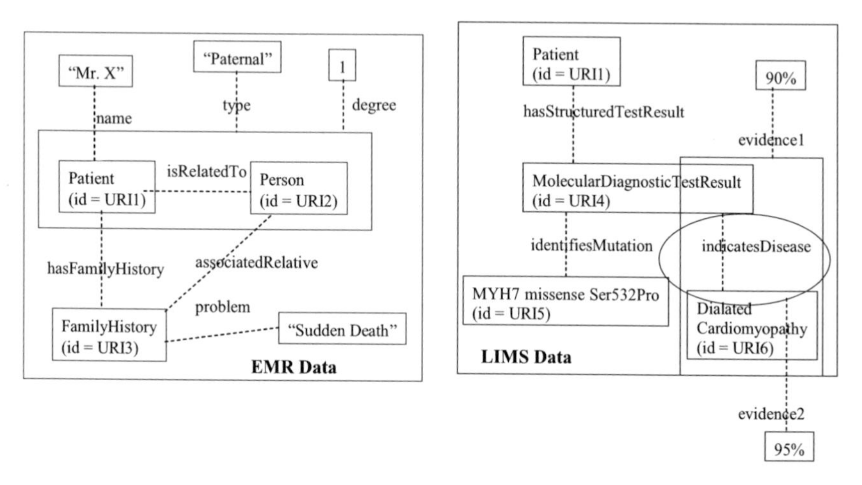

Fig. 4.6. RDF Graphs corresponding to example data

Basic Graph Pattern. A graph pattern is a set of triple patterns, which are created by introducing variables in triples in place of any of subject, predicate and object positions. The following triple pattern has a subject variable (the variable `?patient`), a predicate `name` and an object variable (the variable `?name`).

```
?patient name ?name
```

The following query contains a basic graph pattern of two triple patterns, each of which must match with the same solution for the graph pattern to match. It retrieves the structured test results for a particular patient "John Doe".

```
SELECT ?testResult
WHERE {
        ?patient name "John Doe" .
        ?patient hasSttructuredTestResult ?testResult
     }
```

In the SPARQL syntax, basic graph patterns are sequences of triple patterns mixed with value constraints (discussed later). Other graph patterns separate basic patterns. An example is illustrated below, with a FILTER construct specifying the value constraint.

```
  {  ?patient age ?age .
   FILTER (?age > 65) .
   ?patient height ?height
}
```

Furthermore, RDF literals can also participate in the pattern-matching process. The example below illustrates matching integers.

```
SELECT ?patient
WHERE {
      ?patient ?property 42
     }
```

The example below illustrates matching a user defined datatype.

```
SELECT ?patient
WHERE {
  ?patient  ?property  "abc"^^<http://example.org/datatype#special-
Datatype>
  }
```

The example below illustrates matching against language tags

```
SELECT ?disease
WHERE {
       ?testResult indicatesDisease "Diabetes"@en
      }
```

Group Graph Pattern. A group graph pattern is a set of graph patterns such that all the patterns match using the same substitution. In a SPARQL query string, group graph patterns are delimited by curly braces {}. An example of a group graph pattern is illustrated below.

```
SELECT ?name ?age ?height ?weight
WHERE  {
                   ?patient name ?name .
                   ?patient age ?age .
                   ?patient height ?height .
                   ?patient weight ?weight .

        }
```

Value Constraints. Graph pattern matching creates bindings of variables. It is possible to further restrict solutions by constraining the allowable bindings of variables to RDF terms. Value constraints take the form of boolean-valued expressions; the language also allows application-specific constraints on the values in a solution. An example of a query using value constraints, which identifies elderly patients is illustrated below.

```
SELECT ?name ?age
WHERE   {
             ?patient age ?age .
             FILTER (?age > 65) .
             ?patient name ?name .
        }
```

There is a set of functions and operators in SPARQL for constraints. In addition, there is an extension mechanism to provide access to functions that are not defined in the SPARQL language. Any potential solution that causes an error condition in a constraint will not form part of the final results, but does not cause the query to fail. When matching RDF literals in graph patterns, the datatype lexical-to-value mapping may be reflected in the underlying RDF graph, leading to additional matches where it is known that two literals have the same value. RDF semantics does not require this of all RDF graphs.

Optional Graph Pattern. Basic graph patterns allow applications to make queries where entire query patterns must match for there to be a solution. For every solution of the query, every variable is bound to an RDF term in a pattern solution. However, regular, complete structures cannot be assumed in all RDF graphs and it is useful to be able to have queries that allow information to be added to the solution where the information is available, but not to have the solution rejected because some part of the query pattern does not match. Optional matching provides this facility; if the optional part does not lead to any solutions, variables can be left unbound. Optional parts of the graph pattern may be specified syntactically with the `OPTIONAL` keyword applied to a graph pattern. This is illustrated in the query below which retrieves names and ages for a patient where available. If the age of the patient is not available, then only names are retrieved.

```
SELECT ?name ?age
WHERE  {
          ?patient name  ?name .
          OPTIONAL { ?patient age ?age }
       }
```

Constraints can be specified in an optional graph pattern as follows. In the query below, if the age of a patient is available, only those ages are retrieved which are greater than 65.

```
SELECT  ?name ?age
WHERE   {
          ?patient name ?name .
          OPTIONAL { ?patient age ?age .
                     FILTER (?age > 65) }
        }
```

Optional patterns can occur inside any group graph pattern, including a group graph pattern which itself is optional, forming a nested pattern. The outer optional graph pattern must match for any nested optional pattern to be matched. In the query below, all patients that have a test result which indicates a disease are identified and the type of test and disease are returned along with the name of the patient. In the case where there is no test result associated with a patient, the inner optional pattern that retrieves the indicated disease is not evaluated.

```
SELECT ?name ?test-type ?disease
WHERE {
        ?patient name ?name .
        OPTIONAL { ?patient hasStructuredTestResult ?result .
                   ?result testType ?test-type
                   OPTIONAL { ?result indicatesDisease ?disease}
                 }
    }
```

Union Graph Pattern. SPARQL provides a means of combining graph patterns so that one of several alternative graph patterns may match. If more than one of the alternatives match, all the possible pattern solutions are found. The `UNION` keyword is the syntax for pattern alternatives. The query below retrieves all patients who are minors or seniors.

```
SELECT ?name
WHERE  {
        { ?patient name ?name .
          ?patient age ?age .
          FILTER (?age < 18) }
         UNION
        { ?patient name ?name .
          ?patient age ?age .
        FILTER (?age > 65) }
    }
```

RDF Dataset Graph Pattern. A SPARQL query is executed against an RDF dataset which represents such a collection of graphs. Different parts of the query may be matched against different graphs. When querying a collection of graphs, the GRAPH keyword is used to match patterns against named graphs. This is by using either an IRI to select a graph or a variable to range over the IRIs naming graphs. The query below matches the graph pattern on each of the named graphs in the dataset and forms solutions which have the src variable bound to IRIs of the graph being matched. The src variable may bind itself to different RDF graphs, where for a given patient, these graphs could contain clinical lab tests such as LDL and HbA1c or could contain the results of molecular diagnostic tests that could identify gene mutations.

```
SELECT ?src ?name ?testResult
WHERE
      {
        GRAPH ?src
                 { ?patient name "John Doe"
                   ?patient hasStructuredTestResult ?testResult
                 }
       }
```

RDF Datasets

Many RDF data stores hold multiple RDF graphs, and record information about each graph, allowing an application to make queries that involve information from more than one graph. There is one graph, the default graph, which does not have a name, and zero or more named graphs, each identified by IRI.
An RDF dataset is a set
$\{ G, (<u_1>, G_1), (<u_2>, G_2), \ldots (<u_n>, G_n) \}$
where G and each Gi are graphs, and each $<u_i>$ is an IRI. Each $<u_i>$ is distinct. G is called the default graph. $(<u_i>, G_i)$ are called named graphs. There may be no named graphs. RDF data can be combined by RDF merge [53] of graphs so that the default graph can be made to include the RDF merge of some or all of the information in the named graphs. In the example below, the named graphs contain the same triples as before. The RDF dataset includes an RDF merge of the named graphs in the default graph, relabeling blank nodes to keep them distinct. In the example below, we present a partial representation of the EMR data as the default graph and a partial representation of the LIMS data as a named graph.

```
# Default graph <http://www.hospital.org/EMR>
URI1 name "X" .
URI1 hasFamilyHistory URI3 .
URI3 problem URI7 .
URI7 name "DialatedCardiomyopathy" .
URI1 isRelatedTo URI2 .
URI3 associatedRelative URI2 .
# Named graph: <http://www.laboratory.com/LIMS>
URI1 hasStructuredTestResult URI4 .
URI4 identifiesMutation URI5 .
URI4 indicatesDisease URI6
URI5 name "MYH7" .
URI6 name "Dialated Cardiomyopathy" .
```

Solution Sequence Modifiers

SPARQL supports the following types of solution sequence modifiers:

Projection modifier. The solution sequence can be transformed into one involving only a subset of the variables. For each solution in the sequence, a new solution is formed using a specified selection of the variables. The following example shows a query to extract just the names of patients.

```
SELECT ?name
WHERE
  { ?patient name ?name }
```

Distinct modifier. The solution sequence can be modified by adding the `DISTINCT` keyword which ensures that every combination of variable bindings (i.e., each solution) in the sequence is unique.

```
SELECT DISTINCT ?patient-name ?gene-name
WHERE {
        ?patient name ?patient-name
        ?patient hasStructuredTestResult ?test-result
        ?test-result identifiesMutation ?mutation
        ?mutation name ?gene-name}
      }
```

Order modifier. The `ORDER BY` clause takes a solution sequence and applies ordering conditions. An ordering condition can be a variable or a function call. The direction of ordering is ascending by default. It can be explicitly set to ascending or descending by enclosing the condition in `ASC()` or `DESC()` respectively. If multiple conditions are given, then they are applied in turn until one gives the indication of the ordering.

```
SELECT ?name
WHERE {
        ?patient name ?name
        ?patient age ?age
      }
ORDER BY DESC(?age)
```

Limit modifier. The `LIMIT` form puts an upper bound on the number of solutions returned. If the number of actual solutions is greater than the limit, then at most the limit number of solutions will be returned.

```
SELECT ?name
WHERE {
        ?patient name ?name
      }
LIMIT 20
```

Offset modifier. `OFFSET` causes the solutions generated to start after the specified number of solutions. An `OFFSET` of zero has no effect.

```
SELECT ?name
WHERE {
        ?patient name ?name
        ?patient age ?age
    }
ORDER BY DESC(?age)
LIMIT   5
OFFSET  10
```

Query Forms

SPARQL supports the following types of Query Forms.

Select Form. The SELECT form of results returns the variables directly. The syntax SELECT * is an abbreviation that selects all of the variables.

```
SELECT ?patient-name ?gene-name
WHERE {
         ?patient name ?patient-name
         ?patient hasStructuredTestResult ?test-result
         ?test-result identifiesMutation ?mutation
         ?mutation name ?mutation-name}
      }
```

Construct Form. The CONSTRUCT result form returns a single RDF graph specified by a graph template. The result is an RDF graph formed by taking each query solution in the solution sequence, substituting for the variables in the graph template and combining the triples into a single RDF graph by set union. If any such instantiation produces a triple containing an unbound variable, or an illegal RDF construct (such as a literal in subject or predicate position), then that triple is not included in the output RDF graph. The graph template can contain ground or explicit triples, that is, triples with no variables, and these also appear in the output RDF graph returned by the CONSTRUCT query form. Using CONSTRUCT it is possible to extract parts or the whole of graphs from the target RDF dataset. The access to the graph can be conditional on other information. The construct query below constructs an RDF graph for the patient John Doe which contains all information about the family history and lab test results retrieved from the EMR and LIMS.

```
CONSTRUCT { ?s ?p ?o }
FROM <http://www.hospital.org/EMR>
FROM NAMED <http://www.laboratory.com/LIMS>
WHERE {
         GRAPH ?src {
                     ?patient ?name "John Doe" .
                     ?patient hasFamilyHistory ?family_history .
                     ?family_history ?x ?y .
                     ?patient hasStructuredTestResult ?test-result .
                     ?test-result ?z ?t .
                    }
          }
```

Ask Form. Applications can use the ASK form to test whether or not a query pattern has a solution. No information is returned about the possible query solutions, just whether the server can find one or not.

```
ASK  {
        ?patient name  "John Doe" .
        ?patient age 42
      }
```

4.3.3 Extending Query Languages with Reasoning and Entailment

SPARQL has been designed so that its graph matching semantics can be extended to an arbitrary entailment regime [53]. We present example queries that illustrate how SPARQL can be extended with different types of entailment based on RDF(S) and OWL semantics. Consider the example RDF data presented in the previous section. That RDF data describes a set of patients associated with molecular diagnostic test results. The following query identifies all the genes that share the same function in Gene Ontology (GO) as MYH7.

```
PREFIX eg: <http://ncbi.nlm.nih.gov/entrezgene#>
PREFIX rdfs: <http://www.w3.org/2000/01/rdf-schema#>
PREFIX rdf: <http://www.w3.org/1999/02/22-rdf-syntax-ns#>
PREFIX go: <http://www.geneontology.org/owl#>
SELECT $gene_symbol $label $evidence
WHERE {
      $input_gene eg:symbol "MYH7" .
      $input_gene eg:label $input_label
      $gene eg:symbol $gene_symbol .
      $gene rdf:type $go .
      $go rdfs:label $label .
      $go rdfs:subClassOf $go1 .
      $go1 eg:label $input_label .
      }
ORDER BY $gene_symbol
```

The highlighted conditions in the SPARQL query above encode expressions involving the RDF(S) vocabulary. The SPARQL query supports the associated semantics and entailment of the ***$go rdfs:subClassOf $go1*** in which all the subclasses of the class corresponding to the input label are retrieved. This enables the identification of genes with functions that are specializations of the functions associated with MYH7 in GO. Most SPARQL query processing engines support RDF(S) natively. An example of a SPARQL query which uses OWL entailments is presented below. This query helps to identify the genes and physiological processes in which the functions of the proteins obtained from the gene are realized by processes that are either subprocesses or specializations of the Signal Transduction process.

```
PREFIX rdfs: <http://www.w3.org/2000/01/rdf-schema#
PREFIX owl: <http://www.w3.org/2002/07/owl#>
SELECT ?genename ?processname
WHERE {
        ?gene rdfs:label ?genename
        ?protein rdfs:subClassOf ?protein_superclass .
        ?protein_superclass owl:equivalentClass ?restriction3 .
        ?restriction3 owl:onProperty gene-product-of
        ?restriction3 owl:hasValue ?gene
        ?protein rdfs:subClassOf ?owlrestriction1
        ?owlrestriction1 owl:onProperty has_function.
```

```
        ?owlrestriction1 owl:someValuesFrom ?owlrestriction2
        ?owlrestruction2 owl:onProperty realized_as.
        ?owlrestriction2 owl:someValuesFrom ?process.
      { ?process part-of SignalTransductionProcess
        UNION ?process rdfs:subClassOf SignalTransductionProcess }
        ?process rdfs:label ?processname
}
```

The SPARQL predicates which encode OWL axioms are highlighted in the query below. The OWL entailment regime is invoked here and can the inferences that are drawn are compatible with the OWL semantics.

There have been various proposals for OWL based query languages. Among them are the ASK queries of DIG protocol [400], nRQL queries and the Racer-Pro system [401]. However, DIG queries are limited to atomic (TBox or RBox or ABox) queries whereas nRQL supports only conjunctive ABox queries. A more recent effort is SPARQL-DL [402] which seeks to combine for OWL-DL that can combine TBox/RBox/ABox queries, the semantics for which is based directly on the OWL-DL entailment relation. An earlier proposal, OWL-QL [86] has been proposed as a standard for query-answering dialogs among Semantic Web agents using knowledge represented in the Ontology Web Language (OWL). However it has not achieved much traction so far.

4.4 Clinical Scenario Revisited

We now revisit the Clinical Scenario discussed in Chapter 2 and discuss how semantic web based metadata specifications can be used in designing a solution for the use case scenario. We present an RDF-based representation of the clinical and genomic data obtained from the EMR and LIMS. We illustrate the use of Semantic Web infrastructure to ground these descriptions in URIs and also link them to controlled vocabularies prevalent in the healthcare and life sciences industry. Some example queries are presented and a discussion on the advantages of using a Semantic Web specification are discussed.

4.4.1 Semantic Web Specifications: LIMS and EMR Data

Consider the RDF graphs illustrated in Figure 4.6, the RDF representations of these graphs using the triples syntax is as follows:

```
# Available at http://www.hospital.org/EMR
PREFIX skos: <http://www.w3.org/2004/02/skos/core#>

URI1 name "X" ;
URI1 isRelatedTo URI2 .

_:stmt1 rdf:type rdf:Statement .
```

```
_:stmt1 rdf:subject URI1 .
_:stmt1 rdf:predicate related_to .
_:stmt1 rdf:object URI2 .
_:stmt1 type "Paternal"@en .
_:stmt1 degree 1 .

URI1 hasFamilyHistory URI3 .
URI3 associatedRelative URI2 .
URI3 problem URI7 .

URI7 rdf:type skos:Concept .
URI7 skos:preflabel "Sudden Death"@en .
URI7 skos:inScheme URI8 .
URI8 rdf:type skos:ConceptScheme
URI8 dc:title "Systematized Nomenclature of Medicine (SNOMED)" .

# Available at http://www.laboratory.com/LIMS
PREFIX skos: <http://www.w3.org/2004/02/skos/core#>

URI1 hasStructuredTestResult URI4 .
URI4 identifiesMutation URI5 .
URI4 indicatesDisease URI6

_:stmt2 rdf:type rdf:Statement .
_:stmt2 rdf:subject URI4 .
_:stmt2 rdf:predicate identifiesMutation .
_:stmt2 rdf:object URI6 .
_:stmt2 evidence1 "90%" .
_:stmt2 evidence2 "95%" .

URI5 rdf:type skos:Concept .
URI5 skos:preflabel "MYH7 missense Ser532Pro"@en .
URI5 skos:inScheme URI9 .
URI9 rdf:type skos:ConceptScheme
URI9 dc:title "Human Genome Nomenclature"@en .

URI6 rdf:type skos:Concept .
URI6 skos:preflabel "Dialated Cardiomyopathy"@en .
URI6 skos:inScheme URI10 .
URI10 rdf:type skos:ConceptScheme .
URI10 dc:title "NCI Thesaurus" .
```

The RDF representations of example EMR and LIMS data above illustrate the following key features that are enabled by Semantic Web specifications:

- As discussed earlier, one of the key aspects of the RDF specification is the ability to uniquely identify resources using URIs which are available. In the example above, the same URI (e.g., `URI1`) is used to identify a patient in the EMR and the LIMS data-set. This is an important feature from the point of view of achieving web-scale data linkage and integration.
- The ability to "reify" an edge in the RDF graph (e.g., `URI1 isRelatedTo URI2`) and attaching additional properties and values (e.g., `type` and `"Paternal"`).

- There are multiple standardized vocabularies in use in the healthcare and life sciences. Some examples referenced in the above example are NCI Thesaurus, SNOMED and Human Genome Nomenclature. The Semantic Web specification through the Simple Knowledge Organization Scheme (SKOS) [403] provides a standardized way to link to concepts from these standardized vocabularies. For e.g., the RDF graph refers to a standardized vocabulary code for "Sudden Death" from the SNOMED controlled vocabulary by using the following RDF triples.

```
URI3 problem URI7 .
URI7 rdf:type skos:Concept .
URI7 skos:preflabel "Sudden Death"@en .
URI7 skos:inScheme URI8 .
URI8 rdf:type skos:ConceptScheme
URI8 dc:title "Systematized Nomenclature of Medicine (SNOMED)" .
```

4.4.2 Linking data from Multiple Data Sources

A critical functionality required as a part of the solution for the clinical scenario is the ability to combine clinical and genomic data related to a patient. We now present RDF constructs and SPARQL queries that can enable this integration. As discussed in the earlier section, uniquely identifying the patient using URIs is a critical part of enabling this data linking and integration.

A key construct supported in the SPARQL specification is the ability to define a graph data-set containing a default graph and other default graphs. For instance the RDF graph describing the EMR data can be specified as a default graph and the RDF graph describing the LIMS data can be specified as the named graph. An interesting capability is that these graphs can be distributed and available at different URIs as illustrated below.

```
# Default graph <http://www.hospital.org/EMR>
/* ... RDF representation as illustrated in Section 4.4.1 above ... */

# Named graph: <http://www.laboratory.com/LIMS>
/* ... RDF representation in Section 4.4.1 above ... */
```

In order for merging and integrating the data, we need to represent mappings across the EMR and LIMS data for a given patient. These mappings will be leveraged either by a graph merge or by a SPARQL query to achieve data linking and integration. Consider the merged RDF graph illustrating the integration of clinical and genomic data as illustrated in Figure 4.7 below.

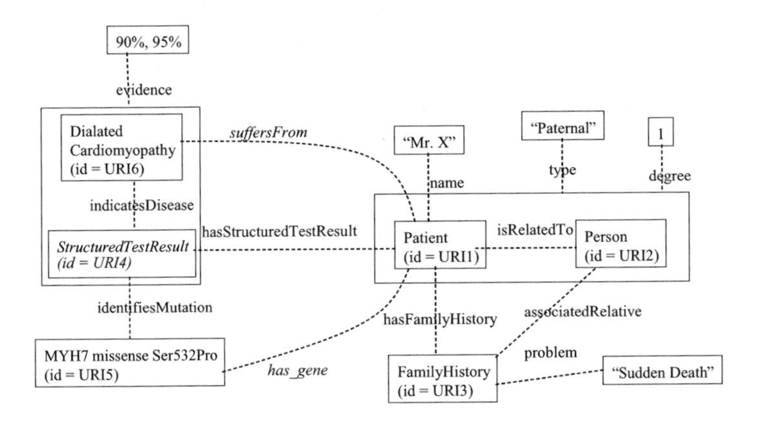

Fig. 4.7. Merged RDF graph combining Clinical and Genomic Data

A default mapping is the ability to match on URIs, a mappping we get for free by using the Semantic Web infrastructure. The merged graph based on ID matching is available based on the GraphMerge operation on the Graph Data-Set illustrated above. However, the merged graph in Figure 4.7 above illustrates two new edges, `suffersFrom` and `hasMutation` which identify the association between the patient and a disease and gene based on the results of the molecular diagnostic test result. The mappings required for enabling this are:

- If a structured test result for a patient indicates a disease with higher then 90% probability, then the patient may suffer from that disease. This can be represented using SPARQL CONSTRUCT expression as follows:

```
CONSTRUCT { ?s suffersFrom ?o}
WHERE {
          ?s hasStructuredTestResult ?result .
          ?result indicatesDisease ?o .
          ?stmt rdf:type rdf:Statement .
          ?stmt rdf:subject ?result .
          ?stmt rdf:predicate indicatesDisease .
          ?stmt rdf:object ?o
          ?stmt evidencel ?evidence_strength
           FILTER (?evidence_strength > 90)
        }
```

- If the structured test result for a patient indicates a mutation, then the patient has the mutation is part of the patient's genome. This can be represented using the following SPARQL CONSTRUCT expression.

```
CONSTRUCT { ?s hasMutation ?o}
 WHERE {
        ?s hasStructuredTestResult ?result .
        ?result identifiesMutation ?o .
      }
```

The merged graph can then be created by appropriately constructing the SPARQL query, where the CONSTRUCT part will contain the new predicates and the WHERE clauses can be appropriately combined.

4.4.3 Advantages and Disadvantages of using Semantic Web Specifications

The solution approach proposed above supports an incremental approach for data integration. Furthermore, the solution leverages the underlying web infrastructure to uniquely identify resources referred to in RDF graphs. This enables easy linking and integration of data across multiple RDF data sources, without the need implement costly data value mapping techniques. The most valuable aspect of this approach is that it enables a flexible approach to ground data representing in RDF graphs to concepts in standardized concepts and vocabularies. The SKOS standard enables the association of semantics with the data in consistent manner. Finally, the key advantage of semantic web specifications is the ability to specify mapping rules at the information level using the SPARQL CONSTRUCT expression. The enables integration and linking at the "information" level in contrast with the current state of art with one of java and perl scripts that implement one-off integration solutions. The externalization and representation of mappings using semantic web specifications enable re-use and configuration of mappings that can be leveraged to implement data linking and integration.

4.5 Summary

In this chapter we presented a detailed discussion of three major metadata frameworks based on the XML, RDF amd OWL specifications. The data models and query languages of these specifications were presented. An introductory discussion on model-theoretic semantics was also presented. A comparison of various query language proposals for XML and RDF data was presented and W3C recommendations, XQuery and SPARQL were discussed in detail. A solution based on semantic web specifications to part of the clinical use case scenario was presented along with a short discusion on the advantages of using semantic web specifications.

5 Ontologies and Schemas

We present a discussion of various ontology and schema-like artifacts used in various fields of activity such as library sciences, relational databases, knowledge representation and medical informatics. We then discuss languages proposed for representation of ontologies and schemas such as XML Schema [87], RDF Schema [88] and OWL [54] and present a comparative evaluation of these languages. The model-theoretic semantics underlying constructs in RDF Schema and OWL are also presented. A discussion of various tools for ontology creation and authoring is presented. This is followed by a discussion of approaches for semi-automatic bootstrapping and generation of ontologies; ontology matching; and ontology management and versioning. Finally, the role of rule-based approaches for representation and reasoning with ontologies will be presented.

5.1 What is an Ontology?

An ontology has been defined [89] as a specification of a conceptualization consisting of a collection of concepts, properties and interrelationships between concepts that can exist for an agent or a community of agents. From our point of view an ontology is a set of terms of interest in a particular information domain and the relationships among them. They can characterize knowledge in an application or domain-specific manner (domain ontologies) or in a domain-independent manner (upper ontologies). This set of terms and interrelationships between them can exist and have been represented in a wide variety of information artifacts such as thesauri, database schemas and UML models to name a few. We view all these artifacts as "ontologies", albeit of varying levels of expressiveness.

A Typology of Terminological Systems

Terminological systems may be viewed as examples of rudimentary forms of ontologies, a typology of which has been proposed in [90]:

- **Terminology:** List of terms referring to concepts in a particular defined domain.
- **Thesaurus:** Terms are ordered, e.g., alphabetically, and concepts may be described by one or more synonymous terms.
- **Vocabulary:** Concepts have definitions, either formal or in free text.

- **Nomenclature:** A set of rules for composing new complex objects or the terminological system resulting from this set of composition rules.
- **Classification:** Concepts are arranged using generic (is_a) relationships.
- **Coding Systems**: Codes designate concepts.

These term lists can be stored in files or database tables and can be used for various purposes such as for metadata annotation, for reference terms for use in various publications and documents and for information exchange between various applications. Some terminologies have been used to systematically record patient data, e.g., choosing a specific code for a patient procedure and for statistical purposes such as estimating the rate of mortality at various hospitals. One of the first terminological systems, the International Classification of Diseases (ICD) [8] was developed in order that "the medical terms reported by physicians, medical examiners and coroners on death certificates can be grouped together for statistical purposes". Based on the typology discussed above, ICD can be typified as a terminology, a thesaurus, and a classification and coding system. Some other applications in which these terminologies have been used in the healthcare context are: clinical documentation in the electronic medical record, clinical decision support, audit, reporting epidemiology and billing. Figure 5.1 illustrates a portion of the ICD.

ICD Version 6

Neoplasms (C00 – D48)
- C00 – C97: Malignant neoplasms
  - C00 – C75: Malignant neoplasms, stated or presumed to be primary, of specified site except lymphoid, haematopoietic and related tissue
    - C00 – C14: Lip, oral cavity and pharynx
    - C15 – C26: Digestive Organs
    - C30 – C39: Respiratory and intrathoracic organs
    - C40 – C41: Bone and articular cartilage
    - C43 – C44: Skin
    - C45 – C49: Mesothelial and soft tissue
    - C50: Breast
    - C51 – C58: Female genital organs
    - C60 – C63: Male genital organs
    - C64 – C68: Urinary tract
    - C69 – C72: Eye, brain and other parts of the central nervous system
    - C73 – C75: Thyroid and other endocrine glands
  - C76 – C80: Malignant neoplasms of ill-defined, secondary and unspecified sites
  - C81 – C96: Malignant neoplasms, stated or presumed to be primary, of lymphoid, haematopoietic and related tissue
  - C96: Malignant neoplasms of independent (primary) multiple sites
- D00 – D09: In situ neoplasms
- D10 – D36: Benign neoplasms
- D37: Neoplasms of uncertain or unknown behavior

Fig. 5.1. ICD: An example of terminology, thesaurus, classification and coding system

Database Schemas

In most applications, database schemas are used to represent implicitly the information model or ontology required by the application. The emphasis of a database schema is more on providing data persistence for an application and the semantics of the data are captured in an implicit manner using primary and foreign keys and functional dependencies. For the purposes of local application requirements, the limited expressiveness provided by a database schema suffices. For most purposes a database schema may be viewed as a rudimentary ontology. An example of a database schema for environmental information is illustrated in Figure 5.2 below. Primary keys in each of the tables are identified and foreign keys are illustrated with arrows between the appropriate columns across multiple tables.

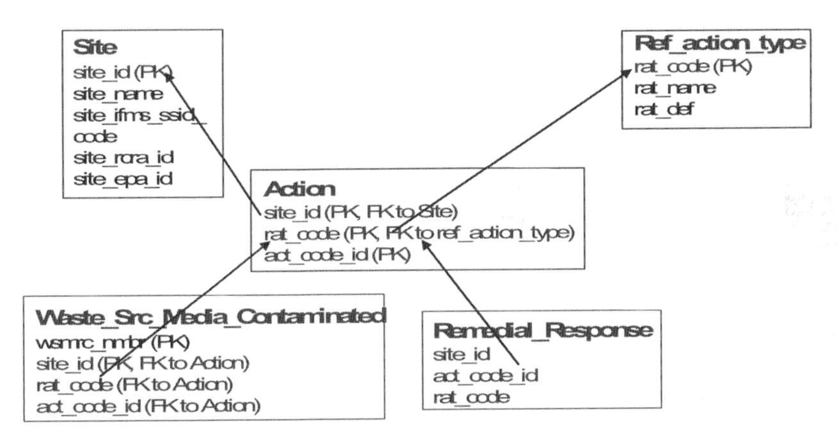

Fig. 5.2. Portion of an environmental ontology represented in a database schema

Entity-Relationship (E-R) Models

Entity-Relationship (E-R) models have been used in information systems literature for representing information models and ontologies. Initial information requirements for an application are typically captured as entity-relationship models before they are mapped to an underlying relational database schema. These models typically have data semantics represented in a more explicit manner than database schemas and are more suitable for representing information models and ontologies. Figure 5.3 illustrates an example of an E-R model representation of the database schema presented in Figure 5.2 above.

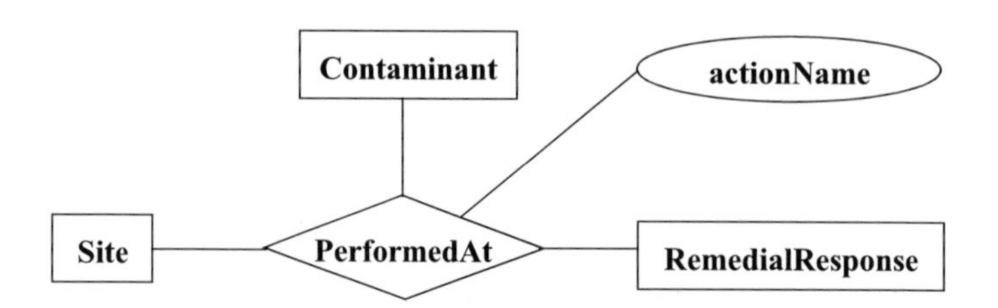

Fig. 5.3. Portion of an environmental ontology represented as an E-R model

UML Models

UML models [40] have been typically used to precisely specify functional and implementation requirements for building software systems. Though not commonly, they have also been used to specify information models and ontologies. An example of a UML representation of the Common Information Model (CIM) proposed by the Distributed Management Task Force is illustrated in Figure 5.4. CIM provides a common definition of management information for systems, networks, applications and services, and allows for vendor extensions. CIM's common definitions enable vendors to exchange semantically rich management information between systems throughout the network.

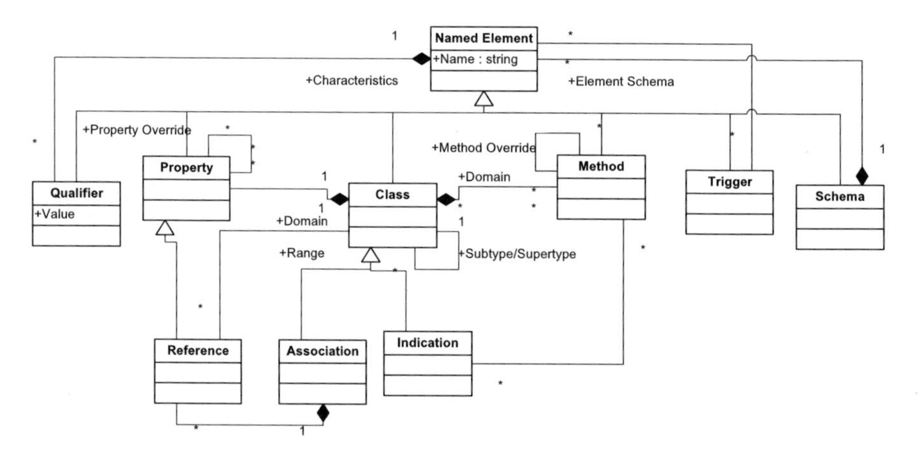

Fig. 5.4. Portion of the Common Information Model represented as a UML model

OWL Ontologies

The Web Ontology Language (OWL) [45] is the W3C Recommendation for representing ontologies on the Web, a detailed description of which is presented later in the chapter. It contains specific constructs to represent the domain and range of properties, subclass and other axioms and constraints on the values that might be assigned to the property of an object. An example of an OWL-based ontology for Parkinson's disease is illustrated in Figure 5.5.

First-Order Logic and Higher Order Theories

Ontologies that capture semantics with a high degree of expressiveness are represented in formalisms that are based on first-order logic. Consider the Bibliographic-data ontology represented using the Knowledge Interchange Format (KIF) [50]:

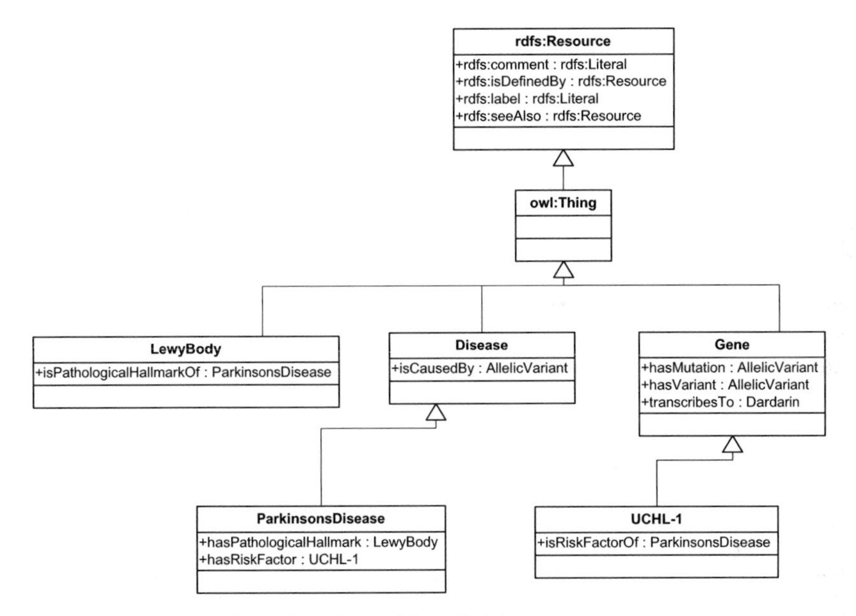

Fig. 5.5. An OWL based ontology for Parkinson's Disease

```
Periodical-Publication Subclass-Partition: Journal, Magazine, Newspa-
per
(<=> (Subclass-Partition ?C ?Class-Partition)
      (And (Class ?C) (Class-Partition ?Class-Partition)
      (Forall (?Subclass)
           (=> (Member ?Subclass ?Class-Partition)
                                  (Subclass-Of ?Subclass ?C)))))
```

The above KIF expression describes the partition of the class `Periodical-Publication` into the subclasses `Journal`, `Magazine` and `Newspaper`. Furthermore it defines the notion of a `Subclass-Partition`: A class has a subclass partition iff all members of the partition are each a subclass of the given class.

Upper Ontologies

An upper ontology (top-level ontology or foundation ontology) is an attempt to create an ontology which describes very general concepts that are the same across all domains. The aim is to have a large number on ontologies accessible under this upper ontology. It is usually a hierarchy of entities and associated rules (both theorems and regulations) that attempts to describe those general entities that do not belong to a specific problem domain. Two examples of upper ontologies are DOLCE [92] and BFO [93].

5.2 Ontology Representation Languages

In the previous Section, we discussed various knowledge artifacts that may be viewed as an ontology. We now discuss the various standards for ontology representation on the Web such as XML Schema, RDF Schema, the Web Ontology Language (OWL), and the Web Services Modeling Language (WSML).

5.2.1 XML Schema

XML Schema is not an Ontology Representation Language in the conventional sense of the term. In a manner similar to the broad range of ontology like artifacts discussed in the previous section, we adopt a broad perspective on an Ontology Representation Language. At the same time from a usage perspective there has been significant work in various communities, e.g., healthcare and finance, where industry standard data and information models have been represented using XML Schema-based specifications. We present a discussion of XML Schema based on the running example in this section. Alternative representations based on RDF Schema, OWL and service oriented representations based on WSML; and comparative pros and cons will be discussed in later sections.

The purpose of a schema is to define a class of XML documents, and so the term "instance document" is often used to describe an XML document that conforms to a particular schema. In fact, neither instances nor schemas need to exist as documents per se they may exist as streams of bytes sent between applications, as fields in a database record, or as collections of XML InfoSet "information items". We now present a discussion of XML Schema based on the examples and exposition in [94]. Consider the example XML document below which shows a patient and the laboratory tests ordered for him or her.

```
<?xml version="1.0"?>
<laboratoryTestOrder orderDate="1999-10-20">
    <recipientAddress country="US">
        <name>Alice Smith</name>
        <street>123 Maple Street</street>
        <city>Mill Valley</city>
```

```
        <state>CA</state>
        <zip>90952</zip>
    </recipientAddress>
    <payorAddress country="US">
        <name>Robert Smith</name>
        <company>CIGNA</company>
        <street>8 Oak Avenue</street>
        <city>Old Town</city>
        <state>PA</state>
        <zip>95819</zip>
    </payorAddress>
    <comment>The bill should be sent to the insurance company and teh
results should be mailed to the patient</comment>
    <panel>
          <test num="872-CL">
             <testName>LDL</testName>
             <quantity>1</quantity>
             <USPrice>148.95</USPrice>
          </test>
          <test num="926-GN">
             <productName>Genes in HCM Panel</productName>
             <quantity>1</quantity>
             <USPrice>500.00</USPrice>
             <shipDate>1999-05-21</shipDate>
          </test>
    </panel>
</laboratoryTestOrder>
```

The laboratory order consists of a main element, `laboratoryTestOrder`, and the subelements `recipientAddress`, `payorAddress`, `comment`, and `panel`. These subelements (except `comment`) in turn contain other subelements, and so on, until a subelement such as `USPrice` contains a number rather than any subelements. Elements that contain subelements or carry attributes are said to have complex types, whereas elements that contain numbers (or strings or dates) but do not contain any subelements are said to have simple types. Some elements have attributes; attributes always have simple types. The complex types in the instance document, and some of the simple types, are defined in the schema for purchase orders. The other simple types are defined as part of XML Schema's repertoire of built-in simple types. The laboratory test order schema is presented below.

```
<xsd:schema xmlns:xsd="http://www.w3.org/2001/XMLSchema">
    <xsd:annotation>
        <xsd:documentation xml:lang="en">
            Laboratory test order schema for Lab.com.
             Copyright 2000 Lab.com. All rights reserved.
         </xsd:documentation>
    </xsd:annotation>
    <xsd:element name="laboratoryTestOrder"
                 type="LaboratoryTestOrderType"/>
    <xsd:element name="comment" type="xsd:string"/>
```

```
    <xsd:complexType name="LaboratoryTestOrderType">
        <xsd:sequence>
              <xsd:element name="recipientAddress" type="USAddress"/>
               <xsd:element name="payorAddress" type="USAddress"/>
             <xsd:element ref="comment" minOccurs="0"/>
             <xsd:element name="panel"  type="Panel"/>
         </xsd:sequence>
          <xsd:attribute name="orderDate" type="xsd:date"/>
    </xsd:complexType>
    <xsd:complexType name="USAddress">
        <xsd:sequence>
             <xsd:element name="name"   type="xsd:string"/>
              <xsd:element name="company" type="xsd:string" minOccurs
= "0"/>
             <xsd:element name="street" type="xsd:string"/>
             <xsd:element name="city"   type="xsd:string"/>
             <xsd:element name="state"  type="xsd:string"/>
              <xsd:element name="zip"    type="xsd:decimal"/>
        </xsd:sequence>
         <xsd:attribute name="country"
                        type="xsd:NMTOKEN" fixed="US"/>
     </xsd:complexType>
    <xsd:complexType name="Panel">
        <xsd:sequence>
             <xsd:element name="test" minOccurs="0"
                                      maxOccurs="unbounded">
             <xsd:complexType>
             <xsd:sequence>
                  <xsd:element name="testName"
                               type="xsd:string"/>
                  <xsd:element name="quantity">
                 <xsd:simpleType>
                     <xsd:restriction
                                base="xsd:positiveInteger">
                          <xsd:maxExclusive value="100"/>
                     </xsd:restriction>
                   </xsd:simpleType>
            </xsd:element>
            <xsd:element name="USPrice"  type="xsd:decimal"/>
            <xsd:element ref="comment"   minOccurs="0"/>
             <xsd:element name="shipDate"
                          type="xsd:date" minOccurs="0"/>
           </xsd:sequence>
           <xsd:attribute name="num"
                          type="LOINC" use="required"/>
       </xsd:complexType>
      </xsd:element>
    </xsd:sequence>
   </xsd:complexType>
   <!-- Unique code for identifying lab tests from the LOINC standard
-->
   <xsd:simpleType name="LOINC">
      <xsd:restriction base="xsd:string">
```

```
      <xsd:pattern value="\d{3}-[A-Z]{2}"/>
   </xsd:restriction>
  </xsd:simpleType>
</xsd:schema>
```

Each of the elements in the schema has a prefix `xsd:` which is associated with the XML Schema namespace through the declaration, `xmlns:xsd="http://www.w3.org/2001/XMLSchema"`, that appears in the `schema` element. The prefix `xsd:` is used by convention to denote the XML Schema namespace, although any prefix can be used. The same prefix, and hence the same association, also appears on the names of built-in simple types, e.g., `xsd:string`. The purpose of the association is to identify the elements and simple types as belonging to the vocabulary of the XML Schema language rather than the vocabulary of the schema author.

Complex Type Definitions, Element and Attribute Declarations

Complex types allow elements in their content and may carry attributes, whereas simple types cannot have element content and cannot carry attributes. New complex types are defined using the `complexType` element and such definitions typically contain a set of element declarations, element references, and attribute declarations. The declarations are not themselves types, but rather an association between a name and the constraints which govern the appearance of that name in documents governed by the associated schema. Elements are declared using the `element` element, and attributes are declared using the `attribute` element. For example, `USAddress` is defined as a complex type, and within the definition of `USAddress` we see five element declarations and one attribute declaration:

```
<xsd:complexType name="USAddress">
       <xsd:sequence>
            <xsd:element name="name"   type="xsd:string"/>
            <xsd:element name="street" type="xsd:string"/>
            <xsd:element name="city"   type="xsd:string"/>
            <xsd:element name="state"  type="xsd:string"/>
            <xsd:element name="zip"    type="xsd:decimal"/>
       </xsd:sequence>
        <xsd:attribute name="country"
                        type="xsd:NMTOKEN" fixed="US"/>
</xsd:complexType>
```

Any element appearing in an instance whose type is declared to be `USAddress` (e.g., `recipientAddress`) must consist of five elements and one attribute. These elements must be called `name`, `street`, `city`, `state` and `zip` as specified by the values of the declarations' `name` attributes, and the elements must appear in the same sequence (order) in which they are declared. The first four of these elements will each contain a string, and the fifth will contain a number. The element whose type is declared to be `USAddress` may appear with an attribute called `country` which must contain the string US. On the other hand, the LaboratoryTestOrderType definition contains element declarations involving complex types, e.g., USAddress, although both declarations use the same type attribute to identify the type, regardless of whether the type is simple or complex. It may be noted that in the definition of LaboratoryTestOrderType, two of the element declarations, for recipientAddress and payorAddress, associate different element names with the same complex type, namely USAddress.

```
<xsd:complexType name="LaboratoryTestOrderType">
        <xsd:sequence>
             <xsd:element name="recipientAddress" type="USAddress"/>
             <xsd:element name="payorAddress" type="USAddress"/>
             <xsd:element ref="comment" minOccurs="0"/>
             <xsd:element name="panel"  type="Panel"/>
         </xsd:sequence>
          <xsd:attribute name="orderDate" type="xsd:date"/>
</xsd:complexType>
```

Occurrence Constraints

Consider the following examples that appear in the definitions of Panel, Test and USAddress respectively:

```
<xsd:element name="test" minOccurs="0"
                         maxOccurs="unbounded">
<xsd:attribute name="num" type="SKU" use="required"/>
<xsd:attribute name="country"
                type="xsd:NMTOKEN" fixed="US"/>
```

The maximum and minimum number of times an element may appear is determined by the value of a maxOccurs and minOccurs attribute in its declaration. This value may be a positive integer such as 41, or the term unbounded to indicate there is no maximum number of occurrences. Attributes are declared with a use attribute to indicate whether the attribute is required, optional, or even prohibited. Default values of both attributes and elements are declared using the default attribute. When an attribute is declared with a default value, the value of the attribute is whatever value appears as the attribute's value in an instance document; if the attribute does not appear in the instance document, the schema processor provides the attribute with a value equal to that of the default attribute. When an element is declared with a default value, the value of the element is whatever value appears as the element's content in the instance document; if the element appears without any content, the schema processor provides the element with a value equal to that of the default attribute. Default attribute values apply when attributes are missing, and default element values apply when elements are empty. The fixed attribute is used in both attribute and element declarations to ensure that the attributes and elements are set to particular values.

Simple Types

Simple types can either be built-in XML Schema datatypes such as `string` or `decimal`; or can be derived from built-in simple or derived datatypes, for example the `LOINC` datatype defined in the XML Schema. It is derived (by restriction) from the simple type `string`. Furthermore, we constrain the values of LOINC using a facet called `pattern` in conjunction with the regular expression `"\d{3}-[A-Z]{2}"` that is read "three digits followed by a hyphen followed by two upper-case ASCII letters":

```
<xsd:simpleType name="LOINC">
   <xsd:restriction base="xsd:string">
      <xsd:pattern value="\d{3}-[A-Z]{2}"/>
   </xsd:restriction>
</xsd:simpleType>
```

Suppose we wish to create a new type of integer called `myInteger` whose range of values is between 10000 and 99999 (inclusive). We base our definition on the built-in simple type integer, whose range of values also includes integers less than 10000 and greater than 99999. To define myInteger, we restrict the range of the integer base type by employing two facets called `minInclusive` and `maxInclusive`:

```
<xsd:simpleType name="myInteger">
  <xsd:restriction base="xsd:integer">
    <xsd:minInclusive value="10000"/>
    <xsd:maxInclusive value="99999"/>
  </xsd:restriction>
</xsd:simpleType>
```

XML Schema defines twelve facets, among which the enumeration facet is particularly useful. The enumeration facet limits a simple type to a set of distinct values. For example, we can use the enumeration facet to define a new simple type called `USState`, derived from string, whose value must be one of the standard US state abbreviations:

```
<xsd:simpleType name="USState">
  <xsd:restriction base="xsd:string">
    <xsd:enumeration value="AK"/>
    <xsd:enumeration value="AL"/>
    <xsd:enumeration value="AR"/>
    <!-- and so on ... -->
  </xsd:restriction>
</xsd:simpleType>
```

List Types

XML Schema has three built-in list types, NMTOKENS, IDREFS and ENTITIES, in addition to list types that can be derived from atomic datatypes. An NMTOKEN is a sequence of one or more letters, digits and most punctuation marks. NMTOKENS is a list of NMTOKEN values separated by white spaces. An IDREF attribute allows the creation of links to an ID within an XML document (one-to-one). IDREFS is a list of IDREFS and allows an attribute to specify a link from one element to many other elements (one-to-many). Several facets can be applied to list types: length, minLength, maxLength, pattern, and enumeration. For example, to define a list of exactly six US states (SixUSStates), we first define a new list type called USStateList from USState, and then we derive SixUSStates by restricting USStateList to only six items:

```
<xsd:simpleType name="USStateList">
  <xsd:list itemType="USState"/>
</xsd:simpleType>
<xsd:simpleType name="SixUSStates">
  <xsd:restriction base="USStateList">
    <xsd:length value="6"/>
  </xsd:restriction>
</xsd:simpleType>
```

Union Types

Atomic types and list types enable an element or an attribute value to be one or more instances of one atomic type. In contrast, a union type enables an element or attribute value to be one or more instances of one type drawn from the union of multiple atomic and list types. For example, the zipUnion union type is built from one atomic type and one list type:

```
<xsd:simpleType name="zipUnion">
  <xsd:union memberTypes="USState listOfMyIntType"/>
</xsd:simpleType>
```

When we define a union type, the memberTypes attribute value is a list of all the types in the union.

Building Content Models

XML Schema enables groups of elements to be defined and named, so that the elements can be used to build up the content models of complex types. In the example below, the choice group element allows only one of its children to appear in an instance. One child is an inner group element that references the named group shipAndBill consisting of the element sequence recipientAddress, payorAddress, and the second child is a singleUSAddress. Hence, in an instance document, the laboratoryTestOrder element must contain either a recipientAddress element followed by a payorAddress element or a singleUSAddress element.

```
<xsd:complexType name="LaboratoryTestOrderType">
  <xsd:sequence>
    <xsd:choice>
      <xsd:group ref="shipAndBill"/>
      <xsd:element name="singleUSAddress" type="USAddress"/>
    </xsd:choice>
    <xsd:element ref="comment" minOccurs="0"/>
    <xsd:element name="panel"  type="Panel"/>
  </xsd:sequence>
  <xsd:attribute name="orderDate" type="xsd:date"/>
</xsd:complexType>
<xsd:group id="shipAndBill">
  <xsd:sequence>
    <xsd:element name="recipientAddress" type="USAddress"/>
    <xsd:element name="payorAddress" type="USAddress"/>
  </xsd:sequence>
</xsd:group>
```

Deriving New Types

XML Schema provides the ability to derive new types from pre existing types by various mechanisms such as extension and restriction. An example of creating new types by extension is presented below.

```
<complexType name="Address">
    <sequence>
      <element name="name"   type="string"/>
      <element name="street" type="string"/>
      <element name="city"   type="string"/>
    </sequence>
</complexType>
<complexType name="USAddress">
    <complexContent>
      <extension base="Address">
        <sequence>
          <element name="state" type="USState"/>
          <element name="zip"   type="positiveInteger"/>
        </sequence>
      </extension>
    </complexContent>
</complexType>
```

The Address type contains the basic elements of an address: a name, a street and a city. From this starting point, a new complex type that contains all the elements of the original type plus additional elements that are specific to addresses in the US is created. The new complex type USAddress is created using the complexType element. In addition, we indicate that the content model of the new type is complex, i.e., contains elements, by using the `complexContent` element, and we indicate that we are extending the base type `Address` by the value of the base attribute on the extension element.

It is possible to derive new types by restricting the content models of existing types. A complex type derived by restriction is very similar to its base type, except that its declarations are more limited than the corresponding declarations in the base type. In fact, the values represented by the new type are a subset of the values represented by the base type. For example, we can create a new type, `RestrictedLaboratoryTestOrderType` type derived by restriction from the base type `LaboratoryTestOrderType`, and provide a new (more restrictive) value for the minimum number of comment element occurrences.

```
<complexType name="RestrictedLaboratoryTestOrderType">
  <complexContent>
    <restriction base="LaboratoryTestOrderType">
      <sequence>
        <element name="recipientAddress" type="Address"/>
        <element name="payorAddress" type="Address"/>
        <element ref="comment" minOccurs="1"/>
        <element name="panel"  type="Panel"/>
      </sequence>
    </restriction>
  </complexContent>
</complexType>
```

This change narrows the allowable number of comment elements from a minimum of 0 to a minimum of 1. Note that all `RestrictedLaboratoryTestOrderType` type elements will also be acceptable as `LaboratoryTestOrderType` type elements.

5.2.2 RDF Schema

RDF provides a way to express simple statements about resources, using named properties and values. However, RDF user communities also need the ability to define the vocabularies (terms) they intend to use in those statements, specifically, to indicate that they are describing specific kinds or classes of resources, and will use specific properties in describing those resources. RDF Schema provides the facilities needed to describe such classes and properties, and to indicate which classes and properties are expected to be used together (for example, to say that a particular property will be used to describe instances of a particular class). In other words, RDF Schema provides a type system for RDF. We now present a discussion of RDF Schema based on the exposition in [96]. We begin by revisiting the example discussed in the previous section and represent information about a patient and his laboratory test orders using RDF Schema constructs. We will be using the XML syntax for representing the RDF and RDF Schema representations in this chapter.

```
<?xml version="1.0"?>
```

```
<!DOCTYPE    rdf:RDF    [<!ENTITY    xsd    "http://www.w3.org/2001/
XMLSchema#">]>
<rdf:RDF xmlns:rdf="http://www.w3.org/1999/02/22-rdf-syntax-ns#"
    xmlns:lab="http://www.example.com/schema.rdf#"
   xml:base="http://www.example.com/2008/06/laborders"/>
  <rdf:Description rdf:ID="o123">
     <rdf:type
  resource="http://www.example.com/schema.rdf#LaboratoryTestOrder"/>
      <lab:recipientAddress rdf:ID="a123"/>
      <lab:payorAddress rdf:ID="a234"/>
      <lab:testPanel rdf:ID="p123"/>
      <lab:orderDate rdf:datatype="&xsd;date">
          1999-10-20
      </lab:orderDate>
  </rdf:Description>
  <rdf:Description rdf:ID="a123">
  <rdf:type resource="http://www.example.com/schema.rdf#USAddress"/>
     <lab:name rdf:datatype="&xsd;string">Alice Smith</lab:name>
     <lab:street rdf:datatype="&xsd;string">
                123 Maple Street
     </lab:street>
     <lab:city rdf:datatype="&xsd;string">Mill Valley</lab:city>
     <lab:state rdf:datatype="&xsd;string">CA</lab:state>
     <lab:zip rdf:datatype="&xsd;decimal">90952</lab:zip>
     <lab:country rdf:datatype="&xsd;string">US</lab:country>
  </rdf:Description>
  <rdf:Description rdf:ID="a234">
  <rdf:type resource="http://www.example.com/schema.rdf#USAddress"/>
     <lab:name rdf:datatype="&xsd;string">Robert Smith</lab:name>
     <lab:company rdf:datatype="&xsd;string">CIGNA</lab:company>
     <lab:street rdf:datatype="&xsd;string">8 Oak Avenue</lab:street>
     <lab:city rdf:datatype="&xsd;string">Old Town</lab:city>
     <lab:state rdf:datatype="&xsd;string">PA</lab:state>
     <lab:zip rdf:datatype="&xsd;decimal">95819</lab:zip>
     <lab:country rdf:datatype="&xsd;string">US</lab:country>
  </rdf:Description>
  <rdf:Description rdf:ID="p123">
    <rdf:type resource="http://www.example.com/schema.rdf#Panel"/>
    <lab:test rdf:ID="872-CL"/>
    <lab:test rdf:ID="926-GN"/>
  </rdf:Description>
  <rdf:Description rdf:ID="872-CL">
    <rdf:type resource="http://www.example.com/schema.rdf#Test"/>
    <lab:testName rdf:datatype="&xsd;string">LDL</lab:testName>
    <lab:quantity rdf:datatype="&xsd;integer">1</lab:quantity>
    <lab:usPrice rdf:datatype="xsd;float">148.95</lab:usPrice>
  </rdf:Description>
  <rdf:Description rdf:ID="926-GN">
    <rdf:type resource="http://www.example.com/schema.rdf#Test"/>
    <lab:testName rdf:datatype="&xsd;string">
      Human Genes in HCM Panel
    </lab:testName>
    <lab:quantity rdf:datatype="&xsd;integer">1</lab:quantity>
```

```
    <lab:usPrice rdf:datatype="xsd;float">500.00</lab:usPrice>
    <lab:shipDate rdf:datatype="xsd;date">1999-05-21</lab:shipDate>
  </rdf:Description>
</rdf:RDF>
```

The RDF graph identified by `rdf:ID="o123"` is used to represent an instance of the `LabOrder` class as reference by the `rdf:type` construct. It may be noted that lab order ID is not a URI and `rdf:about` has not been used to identify the appropriate URI. In this example, we assume that the laboratory (`example.com`) has a database for filing the orders for a given patient. The URI for this order represented as an RDF resource is created by appending the URL corresponding to the XML base (`http://www.example.com/2008/06/laborders/`) to the ID. Thus the URI corresponding to this RDF graph is `http://www.example.com/2008/06/laborders/o123`. The class for laboratory orders has properties such as recipientAddress, payorAddress and panel which points to RDF graphs that might be instances of other classes and may have other properties attached to them. The RDF and RDF Schema specifications are specially designed to reuse the basic types specified by the XML schema specification. This is achieved by using the `rdf:datatype` construct. For example, the RDF expression `<lab:testName rdf:datatype="&xsd;string">LDL</lab:testName>` specifies that the datatype corresponding to the `lab:testName` property is the `string` datatype as specified in the XML Schema specification. The RDF Schema for laboratory test orders is presented below.

```
<?xml version="1.0"?>
<rdf:RDF xmlns:rdf="http://www.w3.org/1999/02/22-rdf-syntax-ns#"
          xmlns:rdfs="http://www.w3.org/2000/01/rdf-schema#">
<rdfs:Class rdf:ID="LaboratoryTestOrder">
<rdfs:Class rdf:ID="USAddress"/>
<rdfs:Class rdf:ID="Panel"/>
<rdfs:Class rdf:ID="Test"/>

<rdfs:Datatype rdf:about="&xsd;string"/>
<rdfs:Datatype rdf:about="&xsd;date"/>
<rdfs:Datatype rdf:about="&xsd;decimal"/>
<rdfs:Datatype rdf:about="&xsd;integer"/>
<rdfs:Datatype rdf:about="&xsd;float"/>

<rdf:Property rdf:ID="recipientAddress">
  <rdfs:domain rdf:resource="#LaboratoryTestOrder"/>
  <rdfs:range rdf:resource="#USAddress"/>
</rdf:Property>
<rdf:Property rdf:ID="payorAddress">
  <rdfs:domain rdf:resource="#LaboratoryTestOrder"/>
  <rdfs:range rdf:resource="#USAddress"/>
</rdf:Property>
<rdf:Property rdf:ID="testPanel">
  <rdfs:domain rdf:resource="#LaboratoryTestOrder"/>
  <rdfs:range rdf:resource="#Panel"/>
</rdf:Property>
```

```
<rdf:Property rdf:ID="orderDate">
  <rdfs:domain rdf:resource="#LaboratoryTestOrder"/>
  <rdfs:range rdf:resource="&xsd;date"/>
</rdf:Property>
<rdf:Property rdf:ID="name">
  <rdfs:domain rdf:resource="#USAddress"/>
  <rdfs:range rdf:resource="&xsd;date"/>
</rdf:Property>
<rdf:Property rdf:ID="street">
  <rdfs:domain rdf:resource="#USAddress"/>
  <rdfs:range rdf:resource="&xsd;string"/>
</rdf:Property>
<rdf:Property rdf:ID="city">
  <rdfs:domain rdf:resource="#USAddress"/>
  <rdfs:range rdf:resource="&xsd;string"/>
</rdf:Property>
<rdf:Property rdf:ID="state">
  <rdfs:domain rdf:resource="#USAddress"/>
  <rdfs:range rdf:resource="&xsd;string"/>
</rdf:Property>
<rdf:Property rdf:ID="country">
  <rdfs:domain rdf:resource="#USAddress"/>
  <rdfs:range rdf:resource="&xsd;string"/>
</rdf:Property>
<rdf:Property rdf:ID="zip">
  <rdfs:domain rdf:resource="#USAddress"/>
  <rdfs:range rdf:resource="&xsd;decimal"/>
</rdf:Property>
<rdf:Property rdf:ID="test">
  <rdfs:domain rdf:resource="#Panel"/>
  <rdfs:range rdf:resource="#Test"/>
</rdf:Property>
<rdf:Property rdf:ID="testName">
  <rdfs:domain rdf:resource="#Test"/>
  <rdfs:range rdf:resource="&xsd;string"/>
</rdf:Property>
<rdf:Property rdf:ID="quantity">
  <rdfs:domain rdf:resource="#Test"/>
  <rdfs:range rdf:resource="&xsd;integer"/>
</rdf:Property>
<rdf:Property rdf:ID="usPrice">
  <rdfs:domain rdf:resource="#Test"/>
  <rdfs:range rdf:resource="&xsd;float"/>
</rdf:Property>
<rdf:Property rdf:ID="shipDate">
  <rdfs:domain rdf:resource="#Test"/>
  <rdfs:range rdf:resource="&xsd;date"/>
</rdf:Property>
```

Describing Classes

A basic step in any kind of description process is identifying the various kinds of things to be described. RDF Schema refers to these "kinds of things" as classes. A class in RDF Schema corresponds to the generic concept of a type or category, and can be used to represent almost any category of thing, such as Web pages, people, document types, databases or abstract concepts. Classes are described using the RDF Schema resources `rdfs:Class` and `rdfs:Resource`, and the properties `rdf:type` and `rdfs:subClassOf`. In RDF Schema, a class is any resource having an `rdf:type` property whose value is the resource `rdfs:Class`. So the class would be described by assigning the class a URIref, say `LabOrder` and describing that resource with an `rdf:type` property whose value is the resource `rdfs:Class`. This can be expressed in an RDF statement as follows:

```
<rdf:Description rdf:ID="LaboratoryTestOrder">
  <rdf:type
      rdf:resource="http://www.w3.org/2000/01/rdf-schema#Class"/>
</rdf:Description>
```

Class descriptions such as the above can also be represented as follows.

```
<rdfs:Class rdf:ID="LaboratoryTestOrder"/>
```

Having described `LaboratoryTestOrder` as a class, resource `o123` would be described as a laboratory test order by the RDF statement:

```
<rdf:Description rdf:ID="o123">
  <rdf:type
    rdf:resource=
      "http://www.example.com/rdf.schema#LaboratoryTestOrder"/>
</rdf:Description>
```

The specialization relationship between two classes is described using the predefined `rdfs:subClassOf` property to relate the two classes. For example, to say that `USAdresss` is a specialized kind of `Address` can be represented using the following RDF statement.

```
<rdfs:Class rdf:ID="USAddress">
  <rdfs:subClassOf rdf:resource="#Address">
</rdfs:Class>
```

It may be noted that RDF itself does not define the special meaning of terms from the RDF Schema vocabulary such as `rdfs:subClassOf`. So if an RDF schema defines this `rdfs:subClassOf` relationship between `USAddress` and `Address`, RDF software not written to understand the RDF Schema terms would recognize this as a triple, with predicate `rdfs:subClassOf`, but it would not understand the special significance of `rdfs:subClassOf`.

Describing Properties

In addition to describing the specific *classes* of things they want to describe, user communities also need to be able to describe specific *properties* that characterize those classes of things (such as `test` to describe the set of tests that belong in a test panel). In RDF Schema, properties are described using the RDF class `rdf:Property`, and the RDF Schema properties `rdfs:domain`, `rdfs:range`, and `rdfs:subPropertyOf`. All properties in RDF are described as instances of class `rdf:Property`. So a new property, such as `lab:test`, is described by assigning the property a URIref, and describing that resource with an `rdf:type` property whose value is the resource `rdf:Property`, for example, by writing the RDF statement.

```
<rdf:Description rdf:ID="test">
  <rdf:type
     rdf:resource="http://www.w3.org/2000/01/rdf-schema#Property"/>
</rdf:Description>
```

Property descriptions may also be represented as follows.

```
<rdfs:Property rdf:ID="test"/>
```

RDF Schema also provides vocabulary for describing how properties and classes are intended to be used together in RDF data. The most important information of this kind is supplied by using the RDF Schema properties `rdfs:range` and `rdfs:domain` to further describe application-specific properties. The `rdfs:range` property is used to indicate that the values of a particular property are instances of a designated class. For example the following RDF statements indicate that the property `lab:test` has values that are instances of class `lab:Test`.

```
<rdf:Property rdf:ID="test">
  <rdfs:range rdf:resource="#Test"/>
</rdf:Property>
```

A property, say `lab:test`, can have zero, one, or more than one range properties. If `lab:test` has no range property, then nothing is said about the values of the `lab:test` property. If `lab:test` has more than one range property, say one specifying `lab:ClinicalTest` as its range, and another specifying `lab:MolecularDiagnosticTest` as its range, this says that the values of the `lab:test` property are resources that are instances of *all* of the classes specified as the ranges, i.e., that any value of `lab:test` is *both* a `lab:ClinicalTest` *and* a `lab:MolecularDiagnosticTest`. This can be represented as follows.

```
<rdf:Property rdf:ID="test">
  <rdfs:range rdf:resource="#ClinicalTest"/>
  <rdfs:range rdf:resource="#MolecularDiagnosticTest"/>
</rdf:Property>
```

The `rdfs:domain` property is used to indicate that a particular property applies to a designated class. The RDF statement below indicate that the property `lab:test` applies to instances of class `lab:Panel`.

```
<rdf:Property rdf:ID="test">
  <rdfs:domain rdf:resource="#Panel"/>
</rdf:Property>
```

These statements indicate that the `lab:test` property have instances of `lab:Panel` as subjects. A given property, say `lab:test`, may have zero, one, or more than one domain property. If `lab:test` has no domain property, then nothing is said about the resources that `lab:test` properties may be used with (any resource could have a `lab:test` property). If `lab:test` has more than one domain property, say one specifying `lab:OutPatientPanel` as the domain and another one specifying `lab:InPatientPanel` as the domain, this says that any resource that has a `lab:test` property is an instance of *all* of the classes specified as the domains, i.e., that any resource that has a `lab:test` property is both a `lab:InPatientPanel` *and* a `lab:OutPatientPanel` (illustrating the need for care in specifying domains and ranges). A statement that the property `lab:test` has the two domains `lab:InPatientPanel` and `lab:OutPatientPanel` can be represented in RDF as follows.

```
<rdf:Property rdf:ID="test">
  <rdfs:domain rdf:resource="#InPatientPanel"/>
  <rdfs:domain rdf:resource="#OutPatientPanel"/>
</rdf:Property>
```

RDF Schema Classes and XML Schema Types

There is a clear difference in the objectives targeted by the RDF Schema and XML Schema specifications. XML Schema seeks to describe the actual structure of a given document, whereas RDF Schema seeks to describe the semantics of information. Hence in the case of RDF Schema, there is a clear commitment to the notion of a `class` and a `property`, whereas XML Schema seeks to represent the notion of an `element` which describes different types of elements that could appear in a document. This leads to different interpretations of XML Schema types and RDF Schema classes which can have an impact on the application being developed. Consider the two representations of `USAdress` as presented earlier. The XML Schema representation is given below.

```
<xsd:complexType name="USAddress">
        <xsd:sequence>
             <xsd:element name="name"   type="xsd:string"/>
             <xsd:element name="street" type="xsd:string"/>
             <xsd:element name="city"   type="xsd:string"/>
             <xsd:element name="state"  type="xsd:string"/>
             <xsd:element name="zip"    type="xsd:decimal"/>
```

```
        </xsd:sequence>
         <xsd:attribute name="country"
                        type="xsd:NMTOKEN" fixed="US"/>
</xsd:complexType>
```

The RDF Schema for US Address is given below. We include the class declaration and the property declarations in which `USAddress` is identified as the domain.

```
<rdfs:Class rdf:ID="USAddress"/>
<rdf:Property rdf:ID="name">
  <rdfs:domain rdf:resource="#USAddress"/>
  <rdfs:range rdf:resource="&xsd;string"/>
</rdf:Property>
<rdf:Property rdf:ID="city"/>
<rdf:Property rdf:ID="state"/>
<rdf:Property rdf:ID="country"/>
<rdf:Property rdf:ID="zip"/>
```

The RDF Schema specifications seek to re-use and interoperate with XML Schema specifications by supporting the ability for properties defined in RDF (e.g., `name`) to refer to XML Schema datatypes (e.g., `"&xsd;string"`). This is crucial for supporting interoperability. However given the widespread usage and acceptance of XML Schema, the desire to use XML Schema types interchangeably and in some cases instead of RDF Schema classes and properties could lead to unintended consequences. Important differences in perspectives are discussed as follows.

- The complex type `USAddress` doesn't represent the semantics of US addresses, but how they would appear in an XML document. An application developer seeking to use the XML Schema type definition as characterizing the class of all US addresses, should be careful to impose consistent interpretations of an element as a class or a property. The challenge here is that XML Schema specifications do not provide guidance on this issue and different interpretations are likely in practice.
- The complex type `USAddress` dictates the order in which the various parts of the address appear in a document. This is required when one seeks to describe the structural representation in an XML document, but is extraneous to the information contained in US addresses, leading to the following disadvantages: (a) The type specification is unnecessarily complex, if the aim is to describe and use the information captured in US addresses; and (b) The type specification is not easily generalizable. For instance, a French application may seek to display the data elements corresponding to US addresses in a different order. This is better achieved by using an RDF Schema representation of the class of US addresses and applying an XSLT transformation to achieve the appropriate display order.
- XML Schema views a complex type as having specific elements and attributes , e.g., `USAddress` has ***elements*** `street`, `city`, `state` and `zip` whereas RDF Schema views properties as applying to classes, e.g., ***properties*** `street`, `city`, `state` and `zip` apply to the `USAddress` ***class***. The scope of properties in an RDF

schema are by default global and these properties could apply to a different class say a generic class called `Address`. The biggest advantage of the RDF Schema viewpoint is that it is easier to extend the use of property definitions to situations that might not have been anticipated in the original description. Thus RDF Schema has better support for generalizability and flexibility.

5.2.3 Web Ontology Language

The Web Ontology Language, OWL, is intended to provide a language that can be used to describe the classes and relations between them that are inherent in Web documents and applications. We now present a discussion based on [97], that demonstrates the use of this language to formalize a domain by defining classes, properties about those classes, axioms involving properties and classes, individuals and properties about those individuals. The OWL language provides three increasingly expressive sublanguages designed for use by specific communities of implementors and users.

- **OWL Lite** supports those users primarily needing a classification hierarchy and simple constraint features. For example, while OWL Lite supports cardinality constraints, it only permits cardinality values of 0 or 1. It should be simpler to provide tool support for OWL Lite than its more expressive relatives, and provide a quick migration path for thesauri and other taxonomies.
- **OWL-DL** supports those users who want the maximum expressiveness without losing computational completeness (all entailments are guaranteed to be computed) and decidability (all computations will finish in finite time) of reasoning systems. OWL-DL includes all OWL language constructs with restrictions such as type separation (a class cannot also be an individual or property, a property cannot also be an individual or class). OWL-DL is so named due to its correspondence with the description logics field of research that has studied a particular decidable fragment of first-order logic. OWL-DL was designed to support the existing Description Logics business segment and has desirable computational properties for reasoning systems.
- **OWL Full** is meant for users who want maximum expressiveness and the syntactic freedom of RDF with no computational guarantees. For example, in OWL Full a class can be treated simultaneously as a collection of individuals and as an individual in its own right. Another significant difference from OWL-DL is that a `owl:DatatypeProperty` can be marked as an `owl:InverseFunctionalProperty`. OWL Full allows an ontology to augment the meaning of the pre-defined (RDF or OWL) vocabulary. It is unlikely that any reasoning software will be able to support every feature of OWL Full.

A domain ontology for Translational Medicine

A subset of a domain ontology for Translational Medicine which will also be used for illustrating aspects of the solution to the clinical use case scenario is illustrated in Figure 5.6 below. The classes and relationships modeled in this ontology are:

- As discussed in earlier sections, the class `LaboratoryTestOrder` represents the order for a laboratory test for a patient. The order may be for a panel of tests represented in the class `Panel` which may contain one or more tests represented in the class `Test`. The order may have a `recipientAddress` and a `payorAddress` represented by the class USAddress, representing the set of addresses in the US.
- The class `Patient` is a core concept that characterizes patient state information, such as values of various patient state parameters, the results of diagnostic tests and his family and lineage information. It is related to the class `Person` through the subclass relationship. Information about a patient's relatives is represented using the `is_related` relationship and the `Relative` concept. The class `FamilyHistory` captures information of family members who may have had the disease for which the patient is being evaluated, and is related to the Patient concept via the `hasFamilyHistory` relationship.

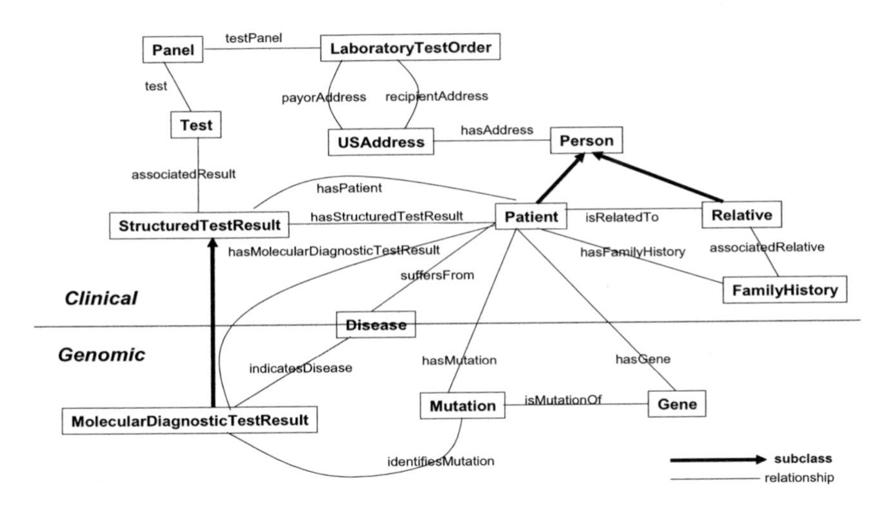

Fig. 5.6. A domain ontology for Translational Medicine

- The `StructuredTestResult` captures results of laboratory tests and is related to the `Patient` class via the `hasStructuredTest` relationship and the `Test` class via the `associatedResult` relationship. Various types of test results such as LDL, AST, ALT, TotalBilirubin, etc. can be represented as subclasses of this class. The `MolecularDiagnosticTestResult` class represents the results of a molecular diagnostic test result, a type of structured test result (represented using the `subclass` relationship). Molecular diagnostics identify mutations (represented using the identifiesMutation relationship) and indicates diseases (represented using the indicatesDisease relationship) in a patient.

- The class Gene represents information about genes and is linked to the Patient class via the hasGene relationship. Genetic variants or mutations of a given gene are represented using the Mutation class which is linked to the Patient class via the hasMutation relationship. The relationship between a gene and mutation is represented using the isMutationOf relationship.
- The Disease class characterizes the disease states which can be diagnosed about a patient, and is related to the Patient class via the suffersFrom relationship and to the molecular diagnostic test results concept via the indicatesDisease relationship.

Simple Classes and Individuals

The most basic concepts in a domain should correspond to classes that are the roots of various taxonomic trees. Every individual in the OWL world is a member of the class owl:Thing. Thus each user-defined class is implicitly a subclass of owl:Thing. Domain-specific root classes are defined by simply declaring a named class. OWL also defines the empty class, owl:Nothing. Based on the ontology illustrated in Figure 5.6, the following classes are created in OWL.

```
<owl:Class rdf:ID="Patient"/>
<owl:Class rdf:ID="LaboratoryTestOrder"/>
<owl:Class rdf:ID="Panel"/>
<owl:Class rdf:ID="Test"/>
<owl:Class rdf:ID="StructuredTestResult"/>
<owl:Class rdf:ID="Disease"/>
<owl:Class rdf:ID="Gene"/>
<owl:Class rdf:ID="Mutation"/>
```

The above OWL statements tell us nothing about these classes other than their existence, despite the use of familiar English terms as labels. And while the classes exist, they may have no members. The fundamental taxonomic constructor for classes is rdfs:subClassOf. It relates a more specific class to a more general class. If X is a subclass of Y, then every instance of X is also an instance of Y.

```
<owl:Class rdf:ID="MolecularDiagnosticTestResult">
   <rdfs:subClassOf rdf:resource="#StructuredTestResult"/>
   ...
</owl:Class>
```

We define MolecularDiagnosticTestResult (laboratory results which require running molecular diagnostic tests on patient samples) is defined as a sublcass of StructuredTestResult. In addition to classes, we want to be able to describe their members. We normally think of these as individuals in our universe of things. An individual is minimally introduced by declaring it to be a member of a class as follows:

```
<MolecularDiagnosticTestResult rdf:ID="result1"/>
```

Simple Properties

A property is a binary relation. Datatype properties relate instances of a class with RDF literals or XML Schema datatypes, whereas object properties relate instances of two classes.

```
<owl:ObjectProperty rdf:ID="hasStructuredTestResult">
    <rdfs:domain rdf:resource="#Patient"/>
    <rdfs:range rdf:resource="#StructuredTestResult"/>
</owl:ObjectProperty>
```

The property `hasStructuredTestResult` has a domain of `Patient` *and* a range of `StructuredTestResult`. That is, it relates instances of the class `Patient` to instances of the class `StructuredTestResult`. Multiple domains mean that the domain of the property is the intersection of the identified classes (and similarly for range). Properties, like classes, can be arranged in a hierarchy.

```
<owl:ObjectProperty rdf:ID="hasMolecularDiagnosticTestResult">
    <rdfs:subPropertyOf rdf:resource="#hasStructuredTestResult" />
    <rdfs:range rdf:resource="#MolecularDiagnosticTestResult" />
    ...
</owl:ObjectProperty>
```

Patients are related to their laboratory test results through the `hasStructuredTestResult` property. `hasMolecularDiagnosticTestResult` is a subproperty with it's range restricted to `MolecularDiagnosticTestResult`, a subclass of `StructuredTestResult`.

Property Characteristics

If a property, P, is specified as transitive, then for any x, y, and z:

```
P(x,y) and P(y,z) -> P(x,z)
```

The property `isRelatedTo` specified below is transitive.

```
<owl:ObjectProperty rdf:ID="isRelatedTo">
    <rdf:type rdf:resource="&owl;TransitiveProperty" />
    <rdfs:domain rdf:resource="#Person"/>
    <rdfs:range rdf:resource="#Person"/>
</owl:ObjectProperty>
```

If a property, P, is tagged as symmetric, then for any x and y:

```
P(x,y) <-> P(y,x)
```

The property `isRelatedTo` defined above is also symmetric. If a property, P, is tagged as functional, then for all x, y, and z:

```
P(x,y) and P(x,z) -> y = z
```

In our running example, `orderDateTime`, i.e., the date and time at which a laboratory test order is placed is functional. A laboratory test order has a unique data and time at which it was ordered. That is, a given `LaboratoryTestOrder` can only be associated with a single date and time using the `orderDateTime` property. It is not a requirement of a `owl:FunctionalProperty` that all elements of the range have values.

```
<owl:DataTypeProperty rdf:ID="orderDateTime">
    <rdf:type rdf:resource="&owl;FunctionalProperty"/>
    <rdfs:domain rdf:resource="#LaboratoryTestOrder"/>
    <rdfs:range  rdf:resource="&xsd;datetime" />
</owl:DataTypeProperty>
```

If a property, P1, is tagged as the `owl:inverseOf` P2, then for all x and y:

```
P1(x,y) iff P2(y,x)
```

Note that the syntax for `owl:inverseOf` takes a property name as an argument.

```
<owl:ObjectProperty rdf:ID="hasStructuredTestResult">
    <owl:inverseOf rdf:resource="#hasPatient"/>
</owl:ObjectProperty>
```

If a property, P, is tagged as InverseFunctional, then for all x, y and z:

```
P(y,x) and P(z,x) implies y = z
```

The inverse of a functional property must be inverse functional.

```
<owl:ObjectProperty rdf:ID="hasStructuredTestResult">
   <rdf:type rdf:resource="&owl;InverseFunctionalProperty"/>
   <owl:inverseOf rdf:resource="#hasPatient"/>
</owl:ObjectProperty>
```

The elements of the range in an inverse functional property define a unique key in the database sense. `owl:InverseFunctional` implies that the elements of the range provide a unique identifier for each element of the domain.

Property Restrictions

The `owl:allValuesFrom` restriction requires that for every instance of the class that has instances of the specified property, the values of the property are all members of the class indicated by the `owl:allValuesFrom` clause.

```
<owl:Class rdf:ID="Patient">
  <rdfs:subClassOf>
    <owl:Restriction>
      <owl:onProperty rdf:resource="#isRelatedTo"/>
      <owl:allValuesFrom rdf:resource="#Relative"/>
    </owl:Restriction>
  </rdfs:subClassOf>
</owl:Class>
```

An patient is related only to persons who are known to be relatives. The `allValuesFrom` restriction is on the `isRelatedTo` property of the `Patient` class and points to the class `Relative`, the instances of which identify those people that are relatives of someone.

The `owl:someValuesFrom` is restriction requires that for every instance of the class that has instanced of the specified property, the values of the property is some member of the class indicated by the `owl:someValuesFrom` clause.

```
<owl:Class rdf:ID="Mutation">
   <rdfs:subClassOf>
    <owl:Restriction>
      <owl:onProperty rdf:resource="#isMutationOf"/>
      <owl:someValuesFrom rdf:resource="#Gene"/>
    </owl:Restriction>
  </rdfs:subClassOf>
</owl:Class>
```

A mutation is mutation of some gene. The `someValuesFrom` restriction is on the `isMutationOf` property of the `Mutation` class and points to the `Gene` class.

The difference between the two formulations is the difference between universal and existential quantification. The first does not require a patient to necessarily have a relative. If it does have a relative, they must all be instances of the class `Relative`. The second requires that there be at least one gene that a mutation is a mutation of.

`owl:cardinality` permits the specification of *exactly* the number of elements in a relation. For example, we specify `LaboratoryTestResult` to be a class with exactly one `Patient`.

```
<owl:Class rdf:ID="StructuredTestResult">
  <rdfs:subClassOf>
    <owl:Restriction>
      <owl:onProperty rdf:resource="#hasPatient"/>
      <owl:cardinality rdf:datatype="&xsd;nonNegativeInteger">
```

```
        1
      </owl:cardinality>
    </owl:Restriction>
  </rdfs:subClassOf>
</owl:Class>
```

Cardinality expressions with values limited to 0 or 1 are part of OWL Lite. This permits the user to indicate 'at least one', 'no more than one', and 'exactly one'. Positive integer values other than 0 and 1 are permitted in OWL-DL. `owl:maxCardinality` can be used to specify an *upper* bound. `owl:minCardinality` can be used to specify a *lower* bound. In combination, the two can be used to limit the property's cardinality to a numeric interval.

`owl:hasValue` allows us to specify classes based on the existence of *particular* property values. Hence, an individual will be a member of such a class whenever at least *one* of its property values is equal to the hasValue resource.

```
<owl:Class rdf:ID="PatientWithMYH7Gene">
  ...
  <rdfs:subClassOf>
    <owl:Restriction>
      <owl:onProperty rdf:resource="#hasGene"/>
      <owl:hasValue rdf:resource="#MYH7"/>
    </owl:Restriction>
  </rdfs:subClassOf>
</owl:Class>
```

Here we specify that the `hasGene` property must have at least one value that is equal to `MYH7`. As for `allValuesFrom` and `someValuesFrom`, this is a local restriction. It holds for `hasGene` as applied to `Patient`.

Ontology Mapping Constructs

It is frequently useful to be able to indicate that a particular class or property in one ontology is equivalent to a class or property in a second ontology. This capability must be used with care. If the combined ontologies are contradictory (all A's are B's vs. all A's are not B's) there will be no extension (no individuals and relations) that satisfies the resulting combination. The property `owl:equivalentClass` is used to indicate that two classes have precisely the same instances. Class expressions can also be the target of `owl:equivalentClass`, eliminating the need to contrive names for every class expression and providing a powerful definitional capability based on satisfaction of a property.

```
<owl:Class rdf:ID="MYH7Mutation">
  <owl:equivalentClass>
    <owl:Restriction>
      <owl:onProperty rdf:resource="#isMutationOf" />
      <owl:hasValue rdf:resource="#MYH7" />
    </owl:Restriction>
```

```
  </owl:equivalentClass>
</owl:Class>
```

MYH7 mutations are *exactly* those things who have the value `MYH7` for the `isMutationOf` property.

The mechanism to specify identity between individuals is similar to that for classes, but declares two individuals to be identical. An example would be:

```
<Gene rdf:ID="GeneImplicatedInHCM">
  <owl:sameAs rdf:resource="#MYH7"/>
</Gene>
```

It may be noted that using `sameAs` to equate two classes is not the same as equating them with `equivalentClass`; instead, it causes the the classes to be interpreted as individuals, and is therefore sufficient to categorize an ontology as OWL Full. In OWL Full `sameAs` may be used to equate anything a class and an individual, a property and a class, etc.,and causes both arguments to be interpreted as individuals.

There also exists a mechanism to specify that two individuals are different from each other as follows.

```
<Gene rdf:ID="MYH7">
    <owl:differentFrom rdf:resource="#TNNT2"/>
</Allergen>
```

A more convenient mechanism exists to define a set of mutually distinct individuals.

```
<owl:AllDifferent>
  <owl:distinctMembers rdf:parseType="Collection">
    <Allergen rdf:ID="#MYH7"/>
    <Allergen rdf:ID="#TNNT2"/>
    <Allergen rdf:ID="#MYL3"/>
  </owl:distinctMembers>
</owl:AllDifferent>
```

Note that `owl:distinctMembers` can only be used in combination with `owl:AllDifferent.`

Complex Class Descriptions

The following example demonstrates the use of the *intersectionOf* construct.

```
<owl:Class rdf:ID="DiabeticPatient">
  <owl:intersectionOf rdf:parseType="Collection">
    <owl:Class rdf:about="#Patient"/>
    <owl:Restriction>
      <owl:onProperty rdf:resource="#suffersFrom"/>
```

```
      <owl:someValuesFrom rdf:resource="#Diabetes"/>
    </owl:Restriction>
  </owl:intersectionOf>
</owl:Class>
```

Classes constructed using the set operations are more like definitions than anything we have seen to date. The members of the class are completely specified by the set operation. The construction above states that `DiabeticPatient` is *exactly* the intersection of the class `Patient` and the set of things that suffer from some instance of the class `Diabetes`. This means that if something is a patient and suffers from some instance of `Diabetes`, then it is an instance of `DiabeticPatient`.

The following example demonstrates the use of the *unionOf* construct. It is used exactly like the *intersectionOf* construct:

```
<owl:Class rdf:ID="StructuredTestResult">
  <owl:unionOf rdf:parseType="Collection">
    <owl:Class rdf:about="#AbnormalStructuredTestResult"/>
    <owl:Class rdf:about="#NormalStructuredTestResult"/>
  </owl:unionOf>
</owl:Class>
```

The class of structured test results includes *both* the extension of abnormal and normal values of the LDL results.

The *complementOf* construct selects all individuals from the domain of discourse that do not belong to a certain class. Usually this refers to a very large set of individuals:

```
<owl:Class rdf:ID="NonDiabeticPatient">
  <owl:complementOf rdf:resource="#DiabeticPatient"/>
</owl:Class>
```

The class of `NonDiabeticPatient` includes as its members all individuals that do not belong to the extension of `DiabeticPatient`.

OWL provides the means to specify a class via a direct enumeration of its members. This is done using the *oneOf* construct. Notably, this definition completely specifies the class extension, so that no other individual can be declared to belong to the class. The following defines a class `Gender` whose members are the individuals `Male` and `Female`.

```
<owl:Class rdf:ID="Gender">
    <owl:oneOf rdf:parseType="Collection">
       <owl:Thing rdf:about="#Male"/>
       <owl:Thing rdf:about="#Female"/>
  </owl:oneOf>
</owl:Class>
```

No other individuals can be a valid `Gender` since the class has been defined by enumeration. Each element of the `oneOf` construct must be a validly declared individual. An individual has to belong to some class. In the above example, each individual was referenced by name. We used `owl:Thing` as a simple construct to introduce the reference.

The disjointness of a set of classes can be expressed using the `owl:disjointWith` constructor. It guarantees that an individual that is a member of one class cannot simultaneously be an instance of a specified other class.

```
<owl:Class rdf:ID="MolecularDiagnosticTestResult">
  <rdfs:subClassOf rdf:resource="#LaboratoryTestResult"/>
  <owl:disjointWith rdf:resource="#ClinicalTestResult"/>
  <owl:disjointWith rdf:resource="#PhysicalExamTestResult"/>
</owl:Class>
```

The above specification only asserts that `MolecularDiagnosticTestResult` is disjoint from all of these other classes. It does not assert, for example, that `ClinicalTestResult` and `PhysicalExamTestResult` are disjoint.

RDF Schema and OWL

We present a discussion on the limitations of the expressive power of RDF Schema and how OWL expressions can be used to overcome some of the limitations:

- In RDF Schema, `rdfs:range` defines the range of a property, e.g., `isAllergicTo` for all classes. It is not possible to declare range restrictions that apply to some classes only. For example, it is not possible to specify that some patients, e.g., `DiabeticPatients` suffer from a particular type of disease, e.g., `Diabetes`, whereas others might suffer from other diseases as well. This can be specified in OWL using the `allValuesFrom` property restriction on the `suffersFrom` property.
- In RDF Schema, one cannot specify the disjointness of classes, e.g., `MolecularDiagnosticTestResult` and `ClinicalTestResult`. This can be specified in OWL using the `disjointWith` construct.
- In RDF Schema, one cannot build new classes by using boolean combinations involving, union, intersection and complement. This can be specified in OWL using the `intersectionOf`, `unionOf`, `oneOf` and `complementOf` constructs.
- RDF Schema doesn't support the specification of cardinality constraints on properties, e.g., a laboratory test result is associated with exactly 1 patient. This can be specified in OWL using the `cardinality`, `minCardinality` and `maxCardinality` constructs.
- Unlike RDF Schema, OWL supports the specification of special characteristics of properties such as `transitivity`, `symmetricity` and `reflexivity` (e.g., `isRelatedTo`) and inverse properties (e.g., `hasStructuredTestResult` and `hasPatient`).

The OWL Specification has been designed to be compatible with RDF and RDF Schema. All varieties of OWL use RDF as their syntax. Instances are declared as in

RDF using RDF descriptions. OWL constructs are specializations of their RDF counterparts, for e.g., `owl:Class` is a specialization of `rdfs:Class`; and `owl:ObjectProperty` and `owl:DatatypeProperty` are specializations of `rdf:Property`.

XML Schema and OWL

We discussed the differences in perspective between RDF Schema and XML Schema in Section 5.2.2. The OWL specification adopts a perspective similar to that of RDF Schema. Some differences in perspectives between XML Schema and OWL are as follows.

- The notion of occurrence constraints or cardinalities appear in XML Schema and OWL. Consider the following XML Schema representation of a Panel of tests involving occurrence constraints.

```
<xsd:complexType name="Panel">
  <xsd:sequence>
    <xsd:element name="test" minOccurs="1" maxOccurs="unbounded"/>
          ...
  </xsd:sequence>
</xsd:complexType>
```

  Consider an OWL-based representation of the same information.

```
<owl:Class rdf:ID="Panel">
  <rdfs:subClassOf>
    <owl:Restriction>
      <owl:onProperty rdf:resource="test">
      <owl:minCardinality>1</owl:minCardinality>
    </owl:Restriction>
  </rdfs:subClassOf>
</owl:Class>
```

  There is a crucial difference in the way the two representations are interpreted. The XML Schema representation evaluates the number of occurrences of a laboratory test element in a panel element. On the other hand, the OWL representation evaluates the number of laboratory test instances that are values of the `test` property. If a given XML document doesn't contain at least one occurrence of a laboratory test element, an error will be flagged. However, in the case of the OWL expression, the laboratory test instances could appear in any other document. Furthermore, unless it is explicitly asserted (or inferred) that there are no laboratory test instances associated with a given panel, an inconsistency or contradiction will not be flagged.
- In XML Schema, new types can be defined as extensions of existing types, whereas in OWL, new classes can be defined in terms of existing classes and properties. Consider the following XML Schema definition of USAddress:

```
<complexType name="Address">
  <sequence>
    <element name="name"   type="string"/>
    <element name="street" type="string"/>
    <element name="city"   type="string"/>
  </sequence>
</complexType>
<complexType name="USAddress">
  <complexContent>
    <extension base="Address">
      <sequence>
        <element name="state" type="USState"/>
        <element name="zip"   type="positiveInteger"/>
      </sequence>
    </extension>
  </complexContent>
</complexType>
```

As discussed earlier each of the elements are assumed to be local to the type definition. Thus when a new derived type, USAddress, is created from a base type, Address, the new elements that are being added need to be specified. This is in contrast to OWL (which is similar to RDF Schema) where all properties are global and can be assigned to each class. There is no need to enumerate all the properties when specifying USAddress as the subclass of Address. The restriction on the property state can be specified by a local property restriction. The resulting expression is simpler to understand and use.

```
<owl:Class rdf:ID="USAddress">
  <owl:intersectionOf rdf:parseType="Collection">
    <owl:Class rdf:about="#Address"/>
    <owl:Restriction>
      <owl:onProperty rdf:resource="#state"/>
      <owl:allValuesFrom rdf:resource="#USState"/>
    </owl:Restriction>
  </owl:intersectionOf>
</owl:Class>
```

There is a critical differences in how these expressions are interpreted. For instance, the XML Schema representation specifies that the `USAddress` type is an extension of the `Address` type, however that does not imply that every instance of a `USAddress` type is also an instance of the `Address` type, though this might be true for the most part. On the other hand, the OWL specification clearly implies through the OWL semantics specification, that every instance of the `USAddress` class is also an instance of the `Address` class.

5.2.4 The Web Service Modeling Ontology (WSMO)

WSMO is a meta-ontology that was devised for describing all aspects of Semantic Web Services. It consists of four top level elements: Web Service, Goal, Ontology and Mediator. The Ontology element itself defines the conceptual model used by all WSMO elements. Hence the description of WSMO as a meta-ontology. It is an ontology that defines how other ontologies can be constructed. WSMO's conceptual model is given semantics through a layered family of logical languages, collectively known as the Web Service Modeling Language (WSML). The WSMO home page[1] provides an extensive set of readily accessible resources describing the conceptual model and WSML languages. In addition, WSMO is the subject of a number of journal articles and books e.g [325], [328], [404]. Full descriptions and detailed examples of WSMO are available at these resources. In this section we provide an introductory description of the WSMO Ontology element and the layered family of logical languages that make up WSML. To illustrate this, we continue with our running example from the medical domain. Chapter 11 looks at the the WSMO Web Service element used for the description of Semantic Web Services.

WSML has a frame-like syntax which means that information about a class and its attributes, or a relation and its parameters, or an instance and its atribute values are grouped together in individual syntactic constructs. This is intended to help with readability of WSML in comparison with OWL or RDF which use XML as their primary syntax and where information about a class, relation or axiom can be spread across several constructs (that said, an XML syntax is also available for WSML). Also in contrast with OWL, WSML attributes are generally defined locally to a class. This is the recommended usage but does not always have to be the case. Attribute identifiers are gloablly unique and it is possible, if neccessary to define global attributes using axioms. More information on the syntax and rules of the WSML language are available in [328] and at the online WSML Language Reference[2].

Concepts

Concepts represent things in a domain of interest. They are defined within a subsumption hierarchy. A concept can be a subconcept of zero, one or many super concepts. For example, the following WSML defines the concepts Test and LiverTest, where Test subsumes LiverTest.

```
concept Test
   testId ofType (1) _IRI
   input ofType TestInput
```

1. http://www.wsmo.org (accessed on May 20, 2008).
2. http://www.wsmo.org/TR/d16/d16.1/v0.3/ (accessed on May 20, 2008).

```
   output ofType TestOutput
concept LiverTest subConceptOf Test
   enzymesToTest ofType LiverEnzymes
   secondControlTest symmetric (0 1) impliesType LiverTest
```

We already mentioned that, generally, attributes are defined locally for a concept. In contrast with OWL, WSML attributes can have cardinality and range constraints which can be used for consistency checking. Similar to the use of database constraints, its possible in WSML for an ontology to be verified as being consistent based on whether or not stated constraints hold. In OWL, the use of cardinality and range constraints are used to create additional inferences such as membership or equality of objects in a class. The designers of WSML held this to be counter-intuitive from the perspective of frame-based languages from which WSML essentially derives. Cardinality in WSML is specified using parenthesis after the ofType *keyword* with either one or two space separated values. The first value indicates the minimal cardinality while the second specifies the maximum. Where only one value is present, it specifies both the minimum and maximum. Additionally, the * character is used to specify no upper limit. For example (0 1) means an attribute is optional. (1) mean it must occur exactly once (known as a functional attribute). (1 *) means that the attribute must exist at least once but with no upper limit. Where the cardinality is not specified, then it is assumed there is no cardinality constraint i.e. (0 *).

WSML allows attributes to be defined that impose constraints as well as those that are used for inferring additional information. The former are identified by the *ofType* keyword. These are the most commonly used attributes and constrain the attribute to the specified type. If the constraint does not hold, the ontology will be inconsistent. The latter set of attributes are identified by the *impliesType* keyword. These attributes do not impose constraints but can be used to infer additional information. The *Test* concept in the listing above has a *testId* attribute with a cardinality of exactly one (functional). Each instance of the *Test* concept must have exactly one *testId* attribute value. The *secondControlTest* attribute of the LiverTest concept is used to indicate that optionally a second control test can be specified. If this attribute exists, the defintion says that its type can be inferred to be the *LiverTest* concept.

Where no datatype is specified, keywords are available to specify attributes as *reflexive*, *transitive*, *symmetric* or the *inverse* of another attribute. They can also be defined as being the subattribute of another attribute. Unlike constraints, these keywords are used by WSML inference engines to infer additional information rather than for consistency checking. For example, the *secondControlTest* is specified as *symmetric* which means that the instance of the concept used for its value can be inferred to also have a *secondControlTest* attribute that points back to the first concept instance.

Relations

Relations are used to model interdependencies between concepts. A relation usually has two or more parameters (typed using concepts) and is defined using a logical expression which must evaluate to *true* over specific instances of the relevant concepts for the relation to hold. For example, in our medical example, we define concepts for *Symptom* and *Condition*, and declare a relation between them below. The relation is called *isSymtomaticOf* and can be read as - an instance of the *Symtom* concept *isSymtomaticOf* an instance of a *Condition* concept. In this WSML snippet, the relation is just declared. The order of the parameters is important but the at this point there is no definition, in terms of a logical expression, for how the relation can be evaluated to be *true*. Such a definition can be provided by the *axiom* WSML element described in the next section.

```
concept Symptom
   clinicalName ofType _string
   durationObserved ofType _time
   recommendedTest ofType Test
concept Condition
   conditionName ofType _string
   recommendedTreatment ofType Treatment
relation isSymptomaticOf(ofType Symptom, ofType Condition)
```

Instances

Instances can be either defined explicitly, by specifying values for a concept and its attributes, or by referring to a unique instance identifier. When instances are defined, values are given to each attribute in accordance with that attribute's definition. For example, the listing below is an instance of a CDA Diastolic Blood Pressure concept from the HL7 CDA medical data model. The value codeBlock3 is an instance of a CodeBlock concept. The other attribute values are of simple built-in types.

```
instance CDADiastolicBP_1_4524 memberOf CDADiastolicBP
  codeValues hasValue codeBlock3
  classCode hasValue "OBS"
  moodCode hasValue "EVN"
  valueType hasValue "PQ"
  value hasValue 86
  valueUnit hasValue "mm[Hg]"
```

Instances of relations can also be defined. An instance of our earlier isSymtomatic relation can be defined using the *relationinstance* keyword as:

```
relationinstance isSymtomaticOf(highDiastolicBP, weakenedLiver)
```

Axioms

An axiom is a logical expression in WSML that can be evaluated to *true* or *false*. It is the means by which rules within a WSMO ontology can be expressed formally and unambiguously so that they can be evaluated by an automated reasoning engine. The axiom below defines the rule for when an instance of a LiverPanelResult can be considered abnormal. It says that an adnormal liver result is defined to exist when the one of the attributes *hasALT*, *hasAST*, or *hasCreatinine* has an anbormally high value.

```
axiom isAbnormalLiverPanelResult
    definedBy
        ?liverPanelResult[hasALT hasValue ?alt,
                          hasAST hasValue ?ast,
                          hasCreatinine hasValue ?creatinine]
            memberOf LivePanelResult
        and
           (?alt > NORMAL_ALT or
           (?ast > NORMAL_AST or
           (?creatinine > NORMAL_CREATININE).
```

Annotations and Non-functional Properties

Every WSMO element can have a set of annotations attached to it to specify metadata that may be outside the domain model for the element. For example, the date the element was last modified, its version, author etc. Non-functional properties extend the notion of annotations so that they can include logical expressions. For both types of metadata, the WSMO specification indicates that a starting point is the use of Dublin Core metadata[3]. For example:

```
ontology hl7_cda_bp_archetype
   nfp
       dc#title hasValue "HL7 Ontology for the BP Archetype"
       dc#description hasValue "An ontology that defines
                                 the HL7 archetype for describing
                                 blood pressure conditions
        dc#dateCreate hasValue "18 May 2008"
   endnfp
```

Importing Ontologies

Ontologies can be imported into a WSML defintion and used directly as long as no conflict in the concept definitions arise. This is specified through the *importsontol-*

3. http://dublincore.org/ (accessed on May 20, 2008).

ogy keyword. Our example ontology below, for the CEN 13606 openEHR data model, imports concept definitions for blood pressure from the *ehr-bp-archetype* ontology. It assumes that no mediation is required to use the imported ontology.

```
wsmlVariant _"http://www.wsmo.org/wsml/wsml-syntax/wsml-rule"
nameSpace {_"http://www.example.org/patientHistory#",
           ehr_bp _"http://www.example.org/ehr-bp-archetype#"}
importsOntology {_"http://www.example.org/ehr-bp-archetype}
concept Patient
   personalDetails ofType (1) PatientPersonalDetails
   ...
```

Mediators

Mediators are used to link WSMO elements defined in different ontologies together. WSMO defines four types of mediators: Ontology-to-Ontology (ooMediator), Goal-to-Goal (ggMediator), Web Service-to-Goal (wgMediator), and Web Service-to-Web Service (wwMediator). The ooMediator is, as yet, the most commonly mediator declaration used for WSMO. The WSML definitions for *mediator* and *ooMediator* are:

```
Class mediator sub-Class wsmoElement
      importsOntology type ontology
      hasNonFunctionalProperties type nonFunctionalProperty
      hasSource type {ontology, goal, webService, mediator}
      hasTarget type {ontology, goal, webService, mediator}
      hasMediationService type {goal, webService, wwMediator}
Class ooMediator sub-Class mediator
      hasSource type {ontology, ooMediator}
```

The WSML language also allows any defintion to declare that it uses a particular mediator using the *usesMediator* keyword followed by the URI for the mediator. For example, an ooMediator, called *openEHR-to-HL7CDA*, defined to bridge the heterogeneity between the data models of openEHR and HL7 CDA could be declared for use in a Web Service by including the following statement:

```
usesMediator _"http://example.org/openEHR-to-HL7CDA"
```

WSML Layered Family of Languages

The formal syntax and semantics for WSMO is provided by the Web Service Modeling Language (WSML) [328] which is actually a layered family of formal description languages. Each member of the family corresponds to a different type of logic, which allows for different expressiveness and reasoning capability. The various logic formalisms are Description Logics, First-Order Logic and Logic Programming, all of which are useful for the modeling of Semantic Web Services. There are five variants, illustrated in Figure 5.7 and described briefly below.

- **WSML-Core.** Defined by the intersection of Description Logics and Horn Logic based on Description Logics Programs [329]. It is the least expressive member of the family and consequently has the best computational characteristics.
- **WSML-DL.** An extension of WSML-Core which captures Description Logics SHIQ(D) and is equivalent to OWL-DL.
- **WSML-Flight.** This is based on a logic programming variant of F-Logic which provides convenient object-oriented and frame-based constructs. It extends WSML-Core with support for meta-modeling, constraints and non-monotonic negation.
- **WSML-Rule.** This extends WSML-Flight in the direction of Logic Programming to provide a powerful rule language.
- **WSML-Full.** Unifies WSML-DL and WSML-Rule under first-order logic with extensions for the support of non-monotonic negation. This language is the least specified of the WSML family as it is not yet clear which formalisms are necessary to achieve it

.

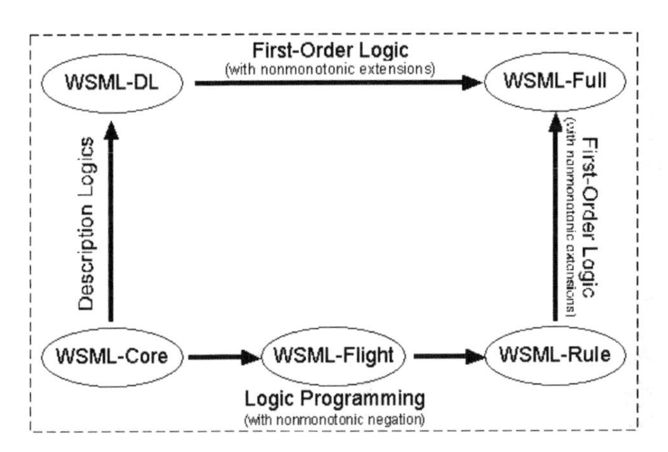

Fig. 5.7. WSML family [328]

In Figure 5.8, the layering of the WSML languages is illustrated with WSML-Core (least expressive) at the bottom and WSML-Full (most expressive) at the top. There are two possible layered combinations:

- WSML-Core + WSML-Flight + WSML-Flight + WSML-Full
- WSML-Core + WSML-DL + WSML-Full + WSML-Full

The two layerings are disjoint to the extent that only the WSML-Core subset provides a common basis. With that restriction, the semantics in the language remain relatively clean in contrast to SWSL-FOL and SWSL-Rule which, as described earlier, share common syntax but not semantics and consequently cannot be used together.

Before leaving the description of WSML, the rationale provided by the authors of WSML for not using OWL as the basis for the language is that OWL was designed as a language for the Semantic Web annotating structures with machine-processable semantics. OWL was not designed as a language rich enough to describe the process models that are part and parcel of the Semantic Web Service conceptual model. This is reflected directly in the OWL-S work where there is no alignment with a framework like MOF and which relies on OWL combined with different notations and semantics for expressing conditions. In some cases this leads to open semantic and decidability issues which are a hindrance to practical usage of the language.

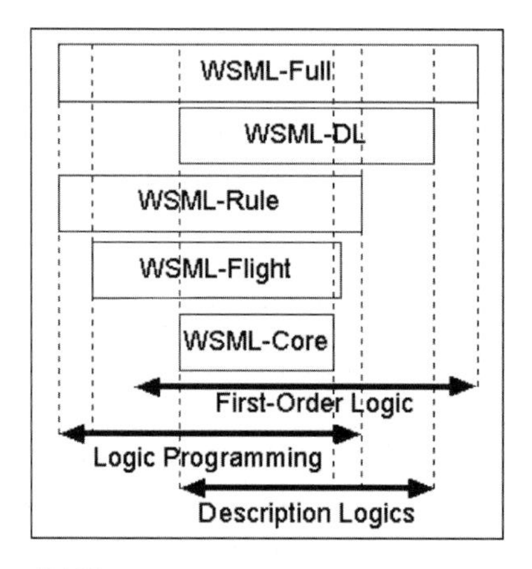

Fig. 5.8. WSML layering [328]

5.2.5 Comparison of Ontology Representation Languages

We now compare XML Schema, RDF Schema, OWL and WSML based on the dimensions and analysis presented in [98]. The dimensions for comparison are as follows:

- **Context:** Modularity is an important consideration, especially in large distributed environments. The same term should be interpreted according to the context in which it is defined. The same term can be defined differently in alternative contexts. It is important that the language be able to express the different contexts in which terms should be interpreted. Namespaces, one of the most elegant features adopted by the XML standard, can be used as a limited form of context. A very subtle but important point is that the URI specified in the tag is not required to point to a schema that defines the terms to be used in that namespace, and so effectively the tag just provides a label to refer to the context. OWL and RDF use XML namespaces to refer to schemas. An important difference from XML is that the namespace URI reference also identifies the location of the RDF schema. Thus, the use of namespaces in RDF seems to be a more clean mechanism to represent contexts.

- **Subclasses and properties:** These express relations between object classes. Subclasses represent "is-a" relations. Properties relate different object classes. Classes and properties are sometimes called concepts and roles, or frames and slots, or objects and attributes, or collections and predicates. In XML Schema, there are no explicit constructs for defining classes and properties. There are only elements and subelements, which can be given an adhoc interpretation as a class/subclass or as a class/property statement. XML allows the definition of new types by extending or restricting existing simple or complex element types. However, the extended and restricted types are not subclasses of the types used in their definition. OWL and RDF Schema provide constructs to define classes, properties, subclasses and subproperties.
- **Primitive datatypes:** A common set of primitive datatypes, such as strings and numbers, that can be used directly or to compose new complex types. XML Schema offers a wide variety of datatypes compatible with database management systems. These datatypes have been adopted by RDF Schema and OWL.
- **Instances:** These objects denote individuals, which can be described in terms of their properties or specified to be members of a class.
- **Property constraints:** These constraints define each property. Properties can be described *with respect to a specific class*, which means that the property can only be described or used for that class. Properties can also be described as *independent* entities, which means that any classes that satisfy the constraints can be related with that property even if this is not indicated in the class definition. XML Schema provides range constraints on elements by the "type" attribute of the element definition. The domain of an element is implicitly the parent element within which it is defined or referred to. A property, and hence its range, is global if the property (or element) is a top-level element (no parent elements); otherwise it is local. Local cardinality constraints on properties can be specified using "minOccurs" and "maxOccurs" attributes while referring to the property inside a parent element. However, it is not possible to define cardinality constraints globally. RDF Schema allows range and domain constraints on properties, which are declared globally. Multiple range and domain statements imply conjunction in both RDF Schema and OWL. RDF Schema does not provide cardinality constraints in its specification. OWL provides for a wide variety of constraints on the ranges of properties, both local and global.
- **Property values:** In addition to range and cardinality, the values that can be assigned to a property can be restricted further. A *default* value can be provided. An *enumeration of possible values* may be given as a set of choices. *Ordered sets* specify the order of the elements, either extensionally or intensionally. Elements are by default ordered in XML. However, we can impose a particular order on the occurrence of elements in XML Schema by the `<xsd:sequence>` tag. In RDF and OWL, we can order a set by using the `<rdf:Seq>` tag.

- **Negation, conjunction and disjunction:** Negated statements of any description allowed in the language are often useful, but the computational cost is steep, and as a result only limited forms of negation are typically supported in a given language. Disjunctive expressions are often used to describe relations among subclasses or property constraints. OWL supports negation and disjunction, through `<owl:complementOf>` (negation) and `<owl:unionOf>` (disjunction). XML Schema also provides a `<union>` tag which gives the disjoint union of various element "types". Conjunction is supported in OWL through the `<owl:intersectionOf>` tag, while multiple `<rdfs:subClassOf>` tags imply conjunction in RDF Schema.
- **Inheritance:** Inheritance indicates that the constraints and values of properties in parent classes are true of the subclasses. Multiple inheritance allows inheritance from multiple parent classes.
- **Definitions**: Definitions indicate whether necessary and sufficient conditions for class membership can be specified. The system can use these definitions to reason about class-subclass relations (subsumption) and to determine whether instances are members of a class (recognition), as in description logic systems. OWL offers the ability to specify necessary and sufficient conditions using the `<owl:equivalentClass>` tag.
- **Logical Consequence:** A key characteristic of a semantic web language is a well defined specification based on a mathematical characterization of the interpretations of the language and the set of logical entailments or consequences supported. A comparison of ontology representation languages.

Table 5.1. A comparison of ontology representation languages

Dimension	Detail	XML Schema	RDF Schema	OWL	WSML
Contexts	Contexts	Yes	Yes	Yes	Yes
Classes	Object Classes and Properties	No, any XML element can be a class or a property	Yes	Yes	Yes
	Inheritance	No, but element types can be extended	Yes	Yes	Yes

Table 5.1. A comparison of ontology representation languages

Dimension	Detail	XML Schema	RDF Schema	OWL	WSML
Property/ Element Constraints	Property/ Element range	Yes, (Global and Local)	Yes (Global only)	Yes (Global and Local)	Yes (Global and Local)
	Property/ Element domain	Yes (implicitly the element under which it is defined)	Yes (Global only)	Yes (Global only)	Yes (implicitly local to the concept to which it belongs but can be explicitly global)
	Property/ Element cardinality	Yes (Local only, min/ max occurs)	No	Yes, (Global and Local)	Yes, (Global and Local)
Datatypes & Instances	Basic Datatypes	Yes	Yes (XSD Datatypes)	Yes (XSD Datatypes)	Yes (XSD Data-types)
	Enumeration of values	Yes	No	Yes	Yes
Datatypes and instances	Instances	Yes	Yes	Yes, using RDF	Yes
Data-Sets	Bounded Lists	No	Yes	Yes	Yes
	Ordered Data-Sets	Yes	Yes	Yes	Yes
Negation, Disjunction & Conjunction	Negation	No	No	Yes	Yes (depending on WSML variant)
	Disjunction	No, but union of element types supported	No	Yes	Yes (depending on WSML variant)
	Conjunction	No	Yes	Yes	Yes
Definitions	Definitions	No	No	Yes	Yes

Table 5.1. A comparison of ontology representation languages

Dimension	Detail	XML Schema	RDF Schema	OWL	WSML
Property types	Inverse	No	No	Yes	Yes
	Transitive	No	No	Yes	Yes
Logical Consequence		No	Limited	Yes	Yes

5.3 Integration of Ontology and Rule Languages

A powerful complementary approach for representing ontologies and other types of knowledge on the Semantic Web is based on declarative rule-based languages such as RuleML and SWRL. Semantic and inferential interoperation across these approaches is critical for enabling the Semantic Web infrastructure. We present and discuss requirements for interoperation, and present a comparison of these two approaches followed by a discussion of some approaches to achieve this integration.

5.3.1 Motivation and Requirements

In this section, we discuss the motivations for combining ontologies and rules and present key requirements for the same. A mapping between ontology and rule languages is important for many aspects of the Semantic Web [160]:

Language Layering: A key requirement for the Semantic Web architecture is to be able to layer rules on top of ontologies in particular to create and reason with rulebases that mention vocabulary specified by ontology-based knowledge bases and to do so in a semantically coherent and powerful manner.

Querying: Rules offer extensive facilities for instance reasoning and querying. Hence, combining description logics with rule paradigms can enable expressive instance queries with respect to terminological knowledge bases represented using ontology representation languages.

Data Integration: The majority of today's structured data resides in relational databases. As the Semantic Web grows in importance, people will probably start publishing their data according to some chosen ontology. This may require loading of data into description logic reasoners, whereas logic programming and rule-based systems can access databases directly through built-in predicates.

Semantic Web Services: Semantic Web Services attempt to describe services in a knowledge-based manner in order to use them for a variety of purposes, including discovery and search; selection, evaluation, negotiation and contracting; composition and planning; execution; and monitoring. Both rules and ontologies are necessary for such service descriptions and play complementary roles. While ontologies are useful for representing hierarchical categorization of services overall and of their inputs and outputs, rules are useful for representing contingent features such as business policies, or the relationships between preconditions and postconditions.

Expressiveness Considerations: There are advanatages and disadvantages of using OWL based representations and rules to represent ontologies and knowledge bases. The language provided by OWL-DL for talking about properties is much weaker. In particular, there is no composition constructor, so it is impossible to capture relationships between a component property and another possibly composite property. The example of the property `uncle` being the composition of the `parent` and `brother` properties cannot be represented in OWL-DL and needs a rule-based representation as follows:

```
uncle(x,y) <= parent(x,z) and brother(z,y)
```

Definite Horn FOL requires that all variables are universally quantified (at the outer level of the rule). This makes it impossible to assert the existence of individuals whose identity might not be known. For example, it is impossible to assert that all persons have a father (known or unknown). This can be expressed as a DL axiom as follows:

```
Class(Person partial restriction(father someValuesFrom(Thing)))
```

5.3.2 Overview of Languages and Approaches

There have been multiple approaches used to integrate rules- and description-logics-based ontologies. Some approaches have been driven by a pragmatic perspective where rules and DLs are viewed as separate components of a system, whereas others have taken an integrated view where the DL and rule specifications are manipulated collectively. The latter approach seeks to represent DLs and variants of rule languages using a First-Order Logic (FOL) framework and identification of the appropriate sublanguage of first-order logic that is some combination of description logics and rule-based languages. A brief summary of the various approaches taken is presented below.

- **Weak Integration of DLs and Rules:** An approach adopted in early description logic systems such as CLASSIC [165] [166] was to include a rule language component. The rules were given a weaker semantic treatment than axioms asserting subclass relationships, were only applied to individuals and did not affect class based inferences.
- **Stronger Integration with Weak DLs and Syntactic Restrictions:** The CARIN system integrated rules and description logics, by using a rather weak description logics and placing severe syntactic restrictions on the occurrence of

description logic terms in the heads of rules, to achieve sound and complete reasoning was still possible [167].

- **Description Logics Programs:** Description Logics Program (DLP) [161] is a sublanguage defined by the intersection of definite equality-free datalog logic programs and description logics. DLP captures a significant fragment of OWL-DL, including the whole OWL-DL fragment of RDF(S), simple frame axioms and more expressive property axioms.
- **OWL Rules Language (ORL):** The OWL Rules Proposal [168] seeks to extend OWL-DL with a form of rules while maintaining maximum backward compatibility with OWL's existing syntax and semantics. This is achieved by adding a new kind of axiom to OWL-DL, namely Horn Rules. The OWL-DL model-theoretic semantics is extended to provide a formal semantics for such an ontology language. This language is more expressive than the DLP discussed above as it extends the expressiveness of OWL-DL as opposed to choosing the intersection of DLs and definite Horn Rules which is necessarily less expressive than either.
- **Semantic Web Rules Language (SWRL):** The proposal for SWRL [312] is a W3C member submission is based on a combination of the OWL-DL and OWL Lite sublanguages of the OWL Web Ontology Language with the Unary/Binary Datalog RuleML sublanguages of the Rule Markup Language [395]. The proposal extends the set of OWL axioms to include Horn-like rules. It thus enables Horn-like rules to be combined with an OWL knowledge base.
- **F-Logic:** F-logic (frame logic) [412], is a knowledge representation and ontology language. It accounts in a declarative fashion for structural aspects of object-oriented and frame-based languages. Features include, among others, object identity, complex objects, inheritance, polymorphism, query methods, encapsulation. F-logic stands in the same relationship to object-oriented programming as classical predicate calculus stands to relational database programming.

5.3.3 Semantic Web Rules Language

The SWRL proposal extends the set of OWL axioms to include Horn-like rules. It thus enables Horn-like rules to be combined with an OWL knowledge base. The proposed rules are of the form of an implication between an antecedent (body) and consequent (head). The intended meaning can be read as: whenever the conditions specified in the antecedent hold, then the conditions specified in the consequent must also hold. Both the antecedent (body) and consequent (head) consist of zero or more atoms. An empty antecedent is treated as trivially true (i.e. satisfied by every interpretation), so the consequent must also be satisfied by every interpretation; an empty consequent is treated as trivially false (i.e., not satisfied by any interpretation), so the antecedent must also not be satisfied by any interpretation. Multiple atoms are treated as a conjunction. Atoms in these rules can be of the form `C(x)`, `P(x,y)`, `sameAs(x,y)` or `differentFrom(x,y)`, where `C` is an OWL description, `P` is an OWL property, and `x,y` are either variables, OWL individuals or OWL data values. We now present examples of rules in the SWRL format which cannot be expressed using OWL.

1. If a patient has a structured test result which is indicative of a particular disease then, the patient suffers from that disease. This rule involves the properties `hasStructuredTestResult`, `indicatesDisease` and `suffersFrom` and specifies that the combination of the first two properties implies the third property. This rule cannot be currently expressed in an OWL axiom. The SWRL representation of this rule is as follows.

```
<ruleml:imp>
  <ruleml:_rlab ruleml:href="#example1"/>
    <ruleml:_body>
      <swrlx:individualPropertyAtom
              swrlx:property="hasStructuredTestResult">
        <ruleml:var>x1</ruleml:var>
        <ruleml:var>x2</ruleml:var>
      </swrlx:individualPropertyAtom>
      <swrlx:individualPropertyAtom
                  swrlx:property="indicatesDisease">
        <ruleml:var>x2</ruleml:var>
        <ruleml:var>x3</ruleml:var>
      </swrlx:individualPropertyAtom>
    </ruleml:_body>
    <ruleml:_head>
      <swrlx:individualPropertyAtom  swrlx:property="suffersFrom">
        <ruleml:var>x1</ruleml:var>
        <ruleml:var>x3</ruleml:var>
      </swrlx:individualPropertyAtom>
    </ruleml:_head>
  </ruleml:imp>
```

2. Another interesting use of SWRL rules, is that one can directly use OWL expressions in the body or head of the rule. This is an interesting approach for integration of rules and ontologies. Consider the rule: If a patient is a diabetic patient, then create an LDL test. This can be represented in SWRL which specifies the OWL description of an allergic patient in the body of the SWRL rule as follows:

```
<ruleml:imp>
  <ruleml:_rlab ruleml:href="#example2"/>
    <ruleml:_body>
      <swrlx:classAtom>
        <owlx:IntersectionOf>
          <owlx:Class owlx:name="Patient"/>
          <owlx:ObjectRestriction owlx:property="suffersFrom">
            <owlx:someValuesFrom owlx:class="Diabetes"/>
          </owlx:ObjectRestriction>
        </owlx:IntersectionOf>
```

```
          <ruleml:var>x1</ruleml:var>
        </swrlx:classAtom>
      </ruleml:_body>
      <ruleml:_head>
        <swrlx:classAtom>
          <owlx:IntersectionOf>
            <owlx:Class owlx:name="Test"/>
            <owlx:ObjectRestriction owlx:property="testName">
              <owlx:hasValue owlx:Individual="LDL"/>
            </owlx:ObjectRestriction>
          </owlx:IntersectionOf>
       <ruleml:var>x2</ruleml:var>
      </ruleml:_head>
    </ruleml:imp>
```

It may be noted that both the head and the body contain OWL expressions (enclosed in the `<owlx:IntersectionOf> </owlx:IntersectionOf>` tags). `Patient`, `isAllergicTo`, `Test` and `testName` are classes and properties defined in an OWL ontology which can be directly referred to from a SWRL Rule. Furthermore, as we will discuss in the context of the use case, the above leads to an interesting differentiation between classification inferences and actionable rules, e.g., creating a new test for checking allergies. We will discuss these issues in more detail in the next section.

5.4 Clinical Scenario Revisited

We now revisit the Clinical Scenario discussed in Chapter 2 and discuss how semantic web based ontology represenation languages can be used to design a solution for the use case scenario. We present an OWL based specification on an ontology for Translational Medicine that spans clinical and genomic domains. An interesting approach for combining OWL-based ontologies with rule-based clinical decision support is also presented. Issues related to decoupling of ontological definitions from clinical decision rules for ordering therapies is presented.

5.4.1 A Domain Ontology for Translational Medicine

We chose OWL as the Ontology Representation Language for designing a solution to the clinical use case. Presented below is an OWL representation of the ontology illustrated in Figure 5.6.

```
<owl:Class rdf:ID="Person"/>
<owl:Class rdf:ID="Patient">
  <rdfs:subClassOf rdf:resource="#Person"/>
</owl:Class>
<owl:Class rdf:ID="Relative">
  <rdfs:subClassOf rdf:resource="#Person"/>
```

```
</owl:Class>
<owl:Class rdf:ID="StructuredTestResult"/>
<owl:Class rdf:ID="MolecularDiagnosticTestResult"/>
<owl:Class rdf:ID="FamilyHistory"/>
<owl:Class rdf:ID="Disease"/>
<owl:Class rdf:ID="Gene"/>
<owl:Class rdf:ID="Mutation"/>
<owl:Class rdf:ID="LaboratoryTestOrder"/>
<owl:Class rdf:ID="Panel"/>
<owl:Class rdf:ID="Test"/>
<owl:Class rdf:ID="USAddress"/>

<owl:ObjectProperty rdf:ID="isRelatedTo">
  <rdf:type rdf:resource="&owl;TransitiveProperty" />
  <rdfs:domain rdf:resource="#Patient"/>
  <rdfs:range rdf:resource="#Relative"/>
</owl:ObjectProperty>
<owl:ObjectProperty rdf:ID="hasFamilyHistory">
  <rdfs:domain rdf:resource="#Patient"/>
  <rdfs:range rdf:resource="#FamilyHistory"/>
</owl:ObjectProperty>
<owl:ObjectProperty rdf:ID="associatedRelative">
  <rdfs:domain rdf:resource="#FamilyHistory"/>
  <rdfs:range rdf:resource="#Relative"/>
</owl:ObjectProperty>
<owl:ObjectProperty rdf:ID="hasStructuredTestResult">
  <rdfs:domain rdf:resource="#Patient"/>
  <rdfs:range rdf:resource="#StructuredTestResult"/>
</owl:ObjectProperty>
<owl:ObjectProperty rdf:ID="hasStructuredTestResult">
   <owl:inverseOf rdf:resource="#hasPatient"/>
</owl:ObjectProperty>
<owl:ObjectProperty rdf:ID="hasMolecularDiagnosticTestResult">
    <rdfs:subPropertyOf rdf:resource="#hasStructuredTestResult" />
    <rdfs:range rdf:resource="#MolecularDiagnosticTestResult" />
</owl:ObjectProperty>
<owl:ObjectProperty rdf:ID="identifiesMutation"/>
  <rdfs:domain rdf:resource="#MolecularDiagnosticTestResult"/>
  <rdfs:range rdf:resource="#Mutation"/>
</owl:ObjectProperty>
<owl:ObjectProperty rdf:ID="indicatesDisease">
  <rdfs:domain rdf:resource="#MolecularDiagnosticTestResult"/>
  <rdfs:range rdf:resource="#Disease"/>
</owl:ObjectProperty>
<owl:ObjectProperty rdf:ID="suffersFrom">
  <rdfs:domain rdf:resource="#Patient"/>
  <rdfs:range rdf:resource="#Disease"/>
</owl:ObjectProperty>
<owl:ObjectProperty rdf:ID="hasMutation">
  <rdfs:domain rdf:resource="#Patient"/>
  <rdfs:range rdf:resource="#Mutation"/>
</owl:ObjectProperty>
```

```
<owl:ObjectProperty rdf:ID="hasGene">
  <rdfs:domain rdf:resource="#Patient"/>
  <rdfs:range rdf:resource="#Gene"/>
</owl:ObjectProperty>
<owl:ObjectProperty rdf:ID="isMutationOf">
  <rdfs:domain rdf:resource="#Mutation"/>
  <rdfs:range rdf:resource="#Gene"/>
</owl:ObjectProperty>
<owl:ObjectProperty> rdf:ID="hasAddress">
  <rdfs:domain rdf:resource="#Person"/>
  <rdfs:range rdf:resource="#USAddress"/>
</owl:ObjectProperty>
<owl:ObjectProperty rdf:ID="recipientAddress">
  <rdfs:domain rdf:resource="#LaboratoryTestOrder"/>
  <rdfs:range rdf:resource="#USAddress"/>
</owl:ObjectProperty>
<owl:ObjectProperty rdf:ID="payorAddress">
  <rdfs:domain rdf:resource="#LaboratoryTestOrder"/>
  <rdfs:range rdf:resource="#USAddress"/>
</owl:ObjectProperty>
<owl:ObjectProperty rdf:ID="testPanel">
  <rdfs:domain rdf:resource="#LaboratoryTestOrder"/>
  <rdfs:range rdf:resource="#Panel"/>
</owl:ObjectProperty>
<owl:ObjectProperty rdf:ID="test">
  <rdfs:domain rdf:resource="#Panel"/>
  <rdfs:range rdf:resource="#Test"/>
</owl:ObjectProperty>
<owl:ObjectProperty rdf:ID="associatedResult">
  <rdfs:domain rdf:resource="#Test"/>
  <rdfs:range rdf:resource="#StructuredTestResult"/>
</owl:ObjectProperty>
<owl:DataTypeProperty rdf:ID="orderDateTime">
    <rdf:type rdf:resource="&owl;FunctionalProperty"/>
    <rdfs:domain rdf:resource="#LaboratoryTestOrder"/>
    <rdfs:range  rdf:resource="&xsd;datetime" />
</owl:DataTypeProperty>
<owl:Class rdf:ID="Patient">
  <rdfs:subClassOf>
    <owl:Restriction>
      <owl:onProperty rdf:resource="#isRelatedTo"/>
      <owl:allValuesFrom rdf:resource="#Relative"/>
    </owl:Restriction>
  </rdfs:subClassOf>
</owl:Class>
<owl:Class rdf:ID="Mutation">
   <rdfs:subClassOf>
    <owl:Restriction>
      <owl:onProperty rdf:resource="#isMutationOf"/>
      <owl:someValuesFrom rdf:resource="#Gene"/>
    </owl:Restriction>
  </rdfs:subClassOf>
</owl:Class>
```

```
<owl:Class rdf:ID="StructuredTestResult">
  <rdfs:subClassOf>
    <owl:Restriction>
      <owl:onProperty rdf:resource="#hasPatient"/>
      <owl:cardinality rdf:datatype="&xsd;nonNegativeInteger">
         1
      </owl:cardinality>
    </owl:Restriction>
  </rdfs:subClassOf>
</owl:Class>
<owl:Class rdf:ID="PatientWithMYH7Gene">
  <rdfs:subClassOf>
    <owl:Restriction>
      <owl:onProperty rdf:resource="#hasGene"/>
      <owl:hasValue rdf:resource="#MYH7"/>
    </owl:Restriction>
  </rdfs:subClassOf>
</owl:Class>
<owl:Class rdf:ID="DiabeticPatient">
  <owl:intersectionOf rdf:parseType="Collection">
    <owl:Class rdf:about="#Patient"/>
    <owl:Restriction>
      <owl:onProperty rdf:resource="#suffersFrom"/>
      <owl:someValuesFrom rdf:resource="#Diabetes"/>
    </owl:Restriction>
  </owl:intersectionOf>
</owl:Class>
<owl:Class rdf:ID="StructuredTestResult">
   <owl:unionOf rdf:parseType="Collection">
     <owl:Class rdf:resource="#NormalStructuredTestResult"/>
     <owl:Class rdf:resource="#AbnormalStructuredTestResult"/>
   </owl:unionOf>
</owl:Class>
<owl:Class rdf:ID="NormalStructuredTestResult">
   <rdfs:subClassOf rdf:resource="#StructuredTestResult"/>
  <owl:disjointWith rdf:resource="#AbnormalStructuredTestResult/>
</owl:Class>
<owl:Class rdf:ID="AbnormalStructuredTestResult">
   <rdfs:subClassOf rdf:resource="#StructuredTestResult"/>
  <owl:disjointWith rdf:resource="#NormalStructuredTestResult/>
</owl:Class>
```

The OWL Specifications above illustrate some of the key features, based on which one may chose OWL as opposed to other alternatives:

- OWL seeks to model the semantics of the information through constructs such as `owl:Class`, `owl:ObjectProperty` and `owl:DatatypeProperty`.
- OWL supports the ability to iteratively add descriptions as more knowledge and information becomes available. For instance, in the ontology below, the declaration of the class `Mutation` could be added first. The property restriction (`<owl:onProperty rdf:resource="#isMutationOf"> <owl:someValuesFrom`

`rdf:resource="#Gene">`) could be added later independently by another domain expert.

- Unlike RDF Schema, OWL supports the ability to locally restrict the values of a particular property, e.g., the values of the property `suffersFrom` are restricted to instances of the class `Diabetes` when applied to instances of the class `DiabetesPatient`. Other classes of patients may be restricted to instances of other diseases.
- In contrast with RDF Schema, OWL supports the ability to support complex classes (`StructuredTestResult` is the union of `NormalStructuredTestResult` and `AbnormalStructuredTestResult`), disjoint classes (`NormalStructuredTestResult` and `AbnormalStructuredTestResult`) and cardinality constraints (e.g., each instance of `StructuredTestResult` has exactly 1 value for the `hasPatient` property).

5.4.2 Integration of Ontologies and Rules for Clinical Decision Support

Consider the following clinical decision rule which determines whether a particular therapy needs to be ordered for a patient.

```
IF the patient has a contraindication to Fibric Acid
THEN Prescribe the Zetia Lipid Management Therapy
```

Consider an extension to the ontology presented in the previous section as follows (Note: this illustrates how an ontology description be iteratively enhanced). The extension is illustrated in below.

```
<owl:ObjectProperty hasLiverPanel>
  <rdfs:subPropertyOf rdf:resource="#hasStructuredTestResult"/>
</owl:ObjectProperty>
<owl:ObjectProperty hasALP>
  <rdfs:subPropertyOf rdf:resource="#hasStructuredTestResult"/>
</owl:ObjectProperty>
<owl:ObjectProperty hasALT>
  <rdfs:subPropertyOf rdf:resource="#hasStructuredTestResult"/>
</owl:ObjectProperty>
<owl:ObjectProperty hasAST>
  <rdfs:subPropertyOf rdf:resource="#hasStructuredTestResult"/>
</owl:ObjectProperty>
<owl:ObjectProperty hasCreatinine>
  <rdfs:subPropertyOf rdf:resource="#hasStructuredTestResult"/>
</owl:ObjectProperty>
<owl:ObjectProperty hasTotalBilirubin>
  <rdfs:subPropertyOf rdf:resource="#hasStructuredTestResult"/>
</owl:ObjectProperty>
<owl:ObjectProperty rdf:ID="isAllergicTo">
  <rdfs:domain rdf:resource="#Patient"/>
  <rdfs:range rdf:resource="#Allergen"/>
</owl:ObjectProperty>
```

```
<owl:ObjectProperty rdf:ID="recommendedTherapy">
  <rdfs:domain rdf:resource="#Patient"/>
  <rdfs:range rdf:resource="#Therapy"/>
</owl:ObjectProperty>
```

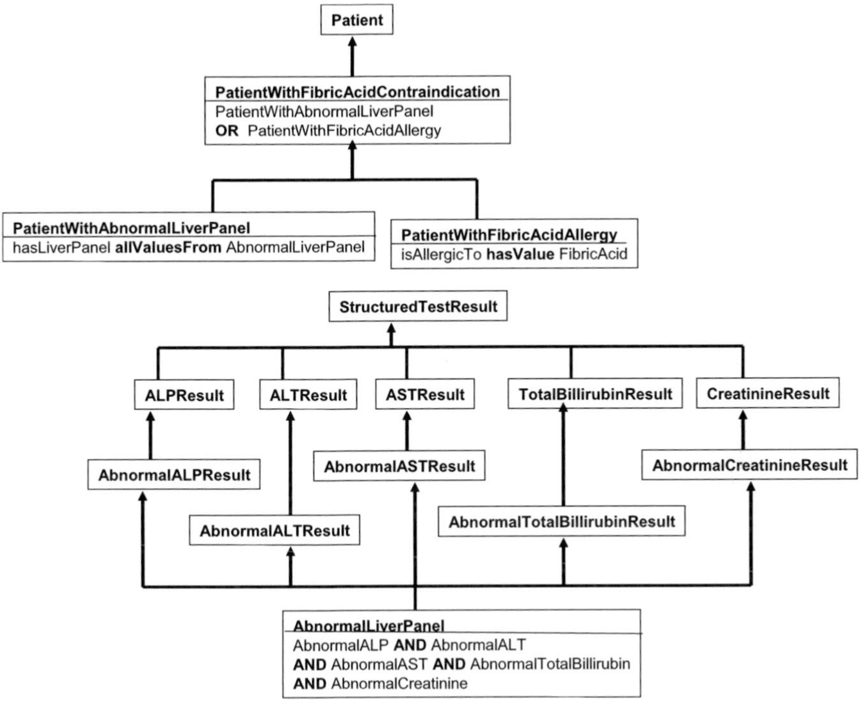

Fig. 5.9. Enhanced ontology to model Contraindication

```
<owl:Class rdf:ID="Allergen"/>
<Allergen rdf:ID="FibricAcid"/>

<owl:Class rdf:ID="Therapy"/>
<Therapy rdf:ID="ZetiaLipidManagementTherapy"/>

<owl:Class rdf:ID="LiverPanelResult>
  <owl:unionOf rdf:parseType="Collection">
    <owl:Class rdf:resource="#NormalLiverPanelResult"/>
    <owl:Class rdf:resource="#AbnormalLiverPanelResult"/>
  </owl:unionOf>
</owl:Class>
<owl:Class rdf:ID="AbnormalLiverPanelResult"/>
  <rdfs:subClassOf rdf:resource="#LiverPanelResult"/>
  <owl:disjointWith rdf:resource="#NormalLiverPanelResult"/>
</owl:Class>
   <owl:Class rdf:ID="NormalLiverPanelResult"/>
  <rdfs:subClassOf rdf:resource="#LiverPanelResult"/>
  <owl:disjointWith rdf:resource="#AbnormalLiverPanelResult"/>
```

```
</owl:Class>
/* Similar definitions for ALPResult, ALTResult, */
/* CreatinineResult and TotalBilirubinResult */

<owl:DatatypeProperty hasALPValue>
  <rdfs:domain rdf:resource="#ALPResult"/>
  <rdfs:range rdf:datatype="&xsd;float"/>
</owl:DatatypeProperty>
/* Similar properties for hasALTValue, hasASTValue, */
/* hasCreatinineValue, hasTotalBilirubinValue */
```

The example clinical decision rule can be specified as follows:

```
<ruleml:imp>
  <ruleml:_rlab ruleml:href="#PropertyBasedRule"/>
    <ruleml:_body>
      <swrlx:individualPropertyAtom swrlx:property="hasALPValue">
        <ruleml:var>x1</ruleml:var>
        <ruleml:var>x3</ruleml:var>
      </swrlx:individualPropertyAtom>
      <swrlx:builtinAtom swrlx:builtin="&swrlb;#greaterThanOrEqual">
           <ruleml:var>x3</ruleml:var>
           <owlx:DataValue owlx:datatype="&xsd;#int">
                ex:NormalALPValue
           </owlx:DataValue>
      </swrlx:builtinAtom>
      ...
       /* Conditions for abnormal ALT, AST, Creatinine and
             Total Bilirubin come here */
       ...
      <swrlx:individualPropertyAtom
                                   swrlx:property="isAllergicTo">
        <ruleml:var>x1</ruleml:var>
        <owlx:Individual>FibricAcid</owlx:Individual>
      </swrlx:individualPropertyAtom>
    </ruleml:_body>
    <ruleml:_head>
      <swrlx:individualPropertyAtom
                         swrlx:property="recommendedTherapy">
        <ruleml:var>x1</ruleml:var>
       <owlx:Individual>ZetiaLipidManagementTherapy</owlx:Individual>
      </swrlx:individualPropertyAtom>
    </ruleml:_head>
</ruleml:imp>
```

The above rule represents the various conditions that correspond to checking whether a patient has a contraindication to Fibric Acid (`<ruleml:_body>`) so that the system can recommend a particular therapy for a patient (`<ruleml:_head>`). This conditions in the rule body also include a set of conditions that indicate an abnormal liver panel; and allergy to fibric acid. An alternative way of representing the above rule can be written by leveraging OWL classes defined in an ontology.

Consider the definition of a FibricAcidContraindication represented using OWL as follows and illustrated in Figure 5.9.

```
<owl:Class rdf:ID="PatientContraindicatedToFibricAcid">
  <owl:unionOf rdf:parseType="Collection">
    <owl:Class rdf:resource="#Patient"/>
    <owl:Restriction>
      <owl:onProperty="#isAllergicTo"/>
      <owl:hasValue="#FibricAcid"/>
    </owl:Restriction>
    <owl:Restriction>
      <owl:onProperty="#hasLiverPanel"/>
      <owl:allValuesFrom="#AbnormalLiverPanelResult"/>
    </owl:Restriction>
  </owl:unionOf>
</owl:Class>
```

The above OWL class defines patients with contraindication to Fibric Acid as patients having an abnormal liver panel and having an allergy to Fibric Acid. Abnormal Live Panel is further defined as:

```
<owl:Class rdf:ID="AbnormalLiverPanel">
 <owl:intersectionOF rdf:parseType="Collection">
    <owl:Restriction>
      <owl:onProperty="#hasALP"/>
      <owl:allValuesFrom="#AbnormalALPResult"/>
    </owl:Restriction>
    <owl:Restriction>
      <owl:onProperty="#hasALT"/>
      <owl:allValuesFrom="#AbnormalALTResult"/>
    </owl:Restriction>
    <owl:Restriction>
      <owl:onProperty="#hasAST"/>
      <owl:allValuesFrom="#AbnormalASTResult"/>
    </owl:Restriction>
    <owl:Restriction>
      <owl:onProperty="#hasCreatinine"/>
      <owl:allValuesFrom="#AbnormalCreatinine"/>
    </owl:Restriction>
    <owl:Restriction>
      <owl:onProperty="#hasTotalBilrubin"/>
      <owl:allValuesFrom="#AbnormalTotalBilirubin"/>
    </owl:Restriction>
  </owl:intersectionOf>
</owl:Class>
```

Based on the above definition, a highly simplified version of the rule can now be specified as:

```
<ruleml:imp>
  <ruleml:_rlab ruleml:href="#DefinedOWLClassBasedRule"/>
```

```
    <ruleml:_body>
      <swrlx:classAtom>
        <owlx:Class owlx:name="PatientContraindicatedToFibricAcid"/>
        <ruleml:var>x1</ruleml:var>
      </swrlx:classAtom>
    </ruleml:_body>
    <ruleml:_head>
    <swrlx:individualPropertyAtom
                          swrlx:property="recommendedTherapy">
       <ruleml:var>x1</ruleml:var>
       <owlx:Individual>ZetiaLipidManagementTherapy</owlx:Individual>
    </swrlx:individualPropertyAtom>
  </ruleml:_head>
</ruleml:imp>
```

The class `Patient` and properties `isAllergicTo, hasLiverPanel` and others provide a framework for describing the patient. The class `PatientContraindicatedToFibricAcid` is a subclass of all patients that are known to have contraindication to fibric acid. This is expressed using an OWL axiom. The class `Allergen` represents various diseases allergens of interest including `FibricAcid`. The classes `AbnormalALPResult, AbnormalALTResult, AbnormalASTResult, AbnormalTotalBilirubinResult` and `AbnormalCreatinineResult` represent ranges of values of abnormal ALP, ALT, AST, total bilirubin and creatinine results respectively. Custom datatypes based on the OWL specifications provide the ability to map XML Schema datatypes to OWL classes. The class `AbnormalLiverPanel` is defined using an axiom to characterize the collection of abnormal values of various component test results (e.g., ALP, ALT, AST) that belong to a liver panel.

The representation of an axiom specifying the definition of `PatientContraindicatedToFibricAcid` enables the knowledge engineer to simplify the rule base significantly. The classification of a patient as being contraindicated to fibric acid is now performed by the ontology engine. The separation of definitions from actions and their implementation in an ontology engine reduces the complexity of the rule base maintenance significantly. It may be noted that the conditions that comprise a definition may appear multiple times in multiple rules in a rule base. Our approach enables the encapsulation of these conditions in a definition, e.g., `PatientContraindicatedToFibricAcid`. Thus all rules can now reference the class `PatientContraindicatedToFibricAcid` which is defined and maintained in the ontology engine. Whenever the definition of `PatientContraindicatedToFibricAcid` changes, the changes can be isolated within the ontology engine and the rules that reference this definition can be easily identified.

Since SWRL is an emerging standard and there is a lack of robust implementations, a commercial rules engine implementation can be used to implement the rules portion. In Chapter 13, we will revisit this discussion and illusrate an implementation approach using the ILOG [405] Business Rules Management System (BRMS).

5.4.3 Advanatages and Disadvantages of using Semantic Web Specifications

The key advantages of using Semantic Web specifications for representing ontologies and schema like artifacts can be characterized along the following dimensions:

- **Representation of semantics as opposed to structure**. Semantic web specifications such as RDF Schema and OWL focus on representation of semantics of the information as opposed to the structure of the document in which the information is captured, as in XML Schema. This results in the creation and authoring of simpler ontologies and schemas since structural details, e.g., sequence, of the document is not needed and not modeled. The commitment to basic semantic constructs, e.g., classes and properties, results in more precise and unambiguous specifications. Finally, RDF Schema and OWL specifications are generizable across different types of data representation formats as they seek to model the information and knowledge contained in various types of data and documents.
- **Incremental specifications of Descriptions**. The perspective of global properties adopted by RDF Schema and OWL enables incremental and distributed creation of ontology and knowledge specifications as different properties along with associated property restrictions can be can associated with a class at different times by different authors. Also since OWL has been engineered to be serialized into RDF and use the RDF Schema Vocabulary, one could imagine incrementally increasing the expressivity of these specifications based on application requirements.
- **Expressiveness Considerations.** As discussed earlier, one can adopt increasingly expressive specifications when the need arises. For example, OWL provides constructs for localizing property ranges for particular classes, cardinality constraints, complex classes involving and boolean operations; and disjointness constraints. Furthermore, OWL specifications can be integrated with rule-based specifications as proposed in the SWRL specifications.
- **Scalability and Robustness.** There are concerns related to the scalability and robustness of semantic web tools such as OWL reasonser and rules engines. However industrial strength Business Rules Management Systmes are known to scale well and open source OWL reasoners such as Pellet and Racer.

5.5 Summary

In this chapter, we presented a discussion on ontology and schema frameworks proposed in the context of Semantic Web specifications. A discussion of different types of ontologies and schemas used in different domains and verticals was presented. The continuum of ontology and schema representation languages XML Schema, RDF Schema, OWL and WSML in the context of Semantic Web standards was presented along with examples from the clinical scenario. In conclusion, approaches for integrating ontology and rule languages were also discussed.

6 Ontology Authoring and Management

Ontologies are a critical component of the Semantic Web architecture. In this chapter, we present a discussion on various aspects of ontology authoring and management. We discuss a collection of ontology building tools and present an evaluation across various dimensions. A brief discussion on techniques for boostrapping of ontologies is presented along with a discussion on techniques of integration, merging and versioning of ontologies.

6.1 Ontology Building Tools

We begin with a survey of ontology building and editing tools that are in use today [99] [100]. The tools may be useful for building ontology schemas (terminological components) alone or together with instance data. Ontology browsers without an editing focus and other types of ontology building tools are not included. Concise descriptions of each software tool are presented and compared according to different criteria which are presented below.

Software architecture and tool evolution, which includes information about the tool architecture (standalone, client/server, n-tier application), how the tool can be extended with other functionalities/modules, how ontologies are stored (databases, text files, etc.) and if there is any backup management system.

Interoperability with other ontology development tools and languages, which includes information about the interoperability capabilities of the tool. We will review the tool's interoperability with other ontology tools (for merge, annotation, storage, inferencing, etc.), as well as translations to and from ontology languages.

Knowledge representation. We will present the KR paradigm underlying the knowledge model of the tool. It is very relevant in order for us to know what and how knowledge can be modeled in the tool. We will also analyze if the tool provides any language for building axioms.

Inference services attached to the tool. We will analyze whether the tool has a built-in inference engine or it can use other external inference engines. We will also analyze if the tool performs constraint/consistency checking, if it can automatically classify concepts in a concept taxonomy and if it is able to manage exceptions in taxonomies.

Usability. We will analyze the existence of graphical editors for the creation of concept taxonomies and relations, the ability to prune these graphs and the possibility to perform zooms of parts of it. We will also analyze if the tool allows some kind of collaborative working and if it provides libraries of ontologies.

6.1.1 Ontology Editors: Brief Descriptions

A brief description of some of the prominent ontology editors in use today is presented below.

Apollo: Apollo [101] is a user-friendly ontology development application, motivated by requirements of industrial users who wished to use knowledge-modeling techniques, but require a syntax and an environment that is easy to use. The application is implemented in Java and supports all the basic primitives of knowledge modeling: ontologies, classes, instances, functions and relations. Full consistency checking is done while editing, for example, detecting the use of undefined classes. Apollo has its own internal language for storing the ontologies, but can also export the ontology into different representation languages, as required by the user.

LinKFactory: Link Factory [102] LinKFactory® is a formal ontology management system developed by Language & Computing NV, designed to build and manage very large and complex language-independent formal ontologies. The LinKFactory system consists of 2 major java-based components: the LinKFactory® Server, and the LinKFactory Workbench (client side component). The LinkFactory Workbench allows the user to browse and model several ontologies and align them. From the knowledge representation and underlying reasoning point of view, LinkFactory has the following characteristics and possibilities: fixed built-in ISA (formal subsumption), DISJOINT, and SAME-AS relationships, definable relationship hierarchy (multiple hierarchies), specification of necessary and sufficient conditions for individual concept definitions, several constraint-checking methods, autoclassification of new concepts on the basis of natural language terms as well as formal definitions, mechanisms to map and/or merge various ontologies; and automatic text analysis to assign links to the ontology.

OntoStudio: OntoStudio [103], the successor of OntoEdit, is a engineering environment for ontology creation and maintenance. OntoStudio is built on top of a powerful internal ontology model. This paradigm supports representation-language-neutral modeling as much as possible for concepts, relations and axioms. Several graphical views on structures in the ontology support representation of different phases of the ontology engineering cycle. The tool allows the user to edit a hierarchy of concepts or classes. A concept may have several names, which essentially is a way to define synonyms for that concept. The tool for reorganization of concepts within the hierarchy is based on a “copy-and-paste” like functionality. The tool is based on a flexible plug-in framework, allowing easy extension and customization. Some available plug-ins are: (a) inferencing for consistency checking, classification and execution of rules; (b) collaborative engineering of ontologies; and (c) an ontology server for administration, collaborative sharing of and persistent storage for ontologies.

Ontolingua Server: The Ontolingua Server is a set of tools and services that support the building of shared ontologies between distributed groups, and that have been developed by the Knowledge Systems Laboratory (KSL) at Stanford University. The ontology server architecture provides access to a library of ontologies, translators to languages (Prolog, CORBA IDL, CLIPS, Loom, etc.) and an editor to create and browse ontologies. Remote editors can browse and edit ontologies, and remote or local applications can access any of the ontologies in the ontology library using the OKBC (Open Knowledge-Based Connectivity) protocol.

Ontosaurus: Ontosaurus [105], developed by the Information Sciences Institute (ISI) at the University of South California, consists of two modules: an ontology server, which uses Loom as its knowledge representation language, and an ontology browser server that dynamically creates HTML pages (including image and textual documentation) that displays the ontology hierarchy. The ontology can be edited by HTML forms, and translators exist for translation from LOOM to Ontolingua, KIF, KRSS and C++.

Protege: Protégé [106] is an open source ontology editor that has seen wide usage from modeling cancer protocol guidelines to nuclear power stations. Protégé provides a graphical and interactive ontology design and knowledge base development environment. Tree controls allow quick and simple navigation through a class hierarchy. Protégé uses forms as the interface for filling in slot values. The knowledge model of Protégé is OKBC compatible, and includes support for classes and the class hierarchy with multiple inheritance; template and own slots; specification of pre defined and arbitrary facets for slots, which include allowed values, cardinality restrictions, default values, and inverse slots; and metaclasses and metaclass hierarchy. The Protege architecture supports a database backend and caching mechanism. Its component-based architecture enables system builders to add new functionality by creating appropriate plug-ins, which fall into one of the three categories: (1) backends for import and export of knowledge bases in various formats; (2) slot widgets for display and edit of slot values in domain and task specific ways; and (3) tab plug-ins which are typically applications tightly linked with Protege knowledge bases. Current back-end plug-ins include export and import from-RDF Schema, XML Schema and OWL files. Currently, tabs that enable advanced visualization, ontology merging, version management and inferences are available.

WebODE: WebODE [107] [108] is engineering workbench that provides services for the ontology development process. Ontologies are represented using a very expressive knowledge model, based on the reference set of intermediate representations of the METHONTOLOGY methodology [109], which includes ontology components such as concepts (with instance and class attributes), partitions, ad hoc binary relations, predefined relations (taxonomic and part-of), instances, axioms, rules, constants and bibliographic references. It also allows the import of terms from other ontologies. Ontologies in WebODE are stored in a relational database underlying a well-defined service-oriented API for ontology access that makes easy integration with other systems. Ontologies built with WebODE can be easily integrated with other systems by using its automatic export/import services from and into XML, and translation services into and from various ontology specification languages (currently, RDF(S), OWL, CARIN and FLogic) and systems such as Java and Jess. Authoring is aided both by form-based and graphical user interfaces, a user-defined views manager, a consistency checker, an inference engine, an axiom builder and the documentation service. Two interesting and novel features of WebODE are: instance set for instantiating the same conceptual model for different scenarios, and conceptual views from the same conceptual model. The graphical user interface allows browsing all the relationships defined on the ontology as well as pruning these views with respect to selected types of relationships. WebODE also supports collaborative authoring of ontologies. Constraint-checking capabilities are also provided for type constraints, numerical values constraints, cardinality constraints and taxonomic consistency verification.

WebOnto: WebOnto [110] is a tool developed by the Knowledge Media Institute (KMi) of the Open University (England). It supports the collaborative browsing, creation and editing of ontologies, which are represented in the knowledge-modeling language OCML. Its main features are: management of ontologies using a graphical interface; the automatic generation of instance editing forms from class definitions; support for Problem Solving Methods (PSMs) and task modeling; inspection of elements incorporating inheritance of properties and consistency checking; a full tell and ask interface, and support for collaborative work; by means of broadcast/receive; and making annotations.

ICOM: ICOM [111] supports the conceptual design phase of an information system. An Extended Entity-Relationship (EER) conceptual data model, enriched with multidimensional aggregations and inter-schema constraints, is used. ICOM is fully integrated with a very powerful description logic reasoning server which acts as a background inference engine. The ICOM modeling language can express: (a) the standard E-R data model, enriched with IsA links (i.e., inclusion dependencies), disjoint and covering constraints, full-cardinality constraints, and definitions attached to entities and relations by means of view expressions over other entities and relationships in the schema; (b) aggregated entities together with their multiply hierarchically organized dimensions; and (c) a rich class of (inter-schema) integrity constraints, as inclusion and equivalence dependencies between view expressions involving entities and relationships possibly belonging to different schemas. ICOM reasons with (multiple) diagrams by encoding them in a single description logic knowledge base, and shows the result of any deductions such as inferred links, new stricter constraints, and inconsistent entities or relationships. The DLR description logics is used to encode the schemas and to express the views and the constraints. The Java-based tool allows for the creation, the editing, the managing, and the storing of several interconnected conceptual schemas, with a user-friendly graphical interface (including an auto-layout facility).

IODE: OntologyWorks IODE [112] is a data and information modeling tool for creating high-definition ontologies, specifically designed for supporting ontology development for database and application development. IODE incorporates a library of vetted and comprehensive domain-independent content that knits together and guides the development of ontologies, ensuring interoperability across domains and across time. It uses a powerful logic (SCL) for representation of ontologies and verifies the consistency of these ontologies, both internally and with respect to domain-independent content. IODE supports existing W3C standards such as RDF and OWL, and supports management of ontology versions in a transactional environment.

Visual Ontology Modeler: The Visual Ontology Modeler [113] by Sandpiper Software is a visual application for building component-based ontologies. It is a UML-based modeling tool that enables ontology development and management for use in collaborative applications and interoperability solutions. Some key features are: (a) A multi-user, network-based environment for ontology development in a rich, graphical notation; (b) Automated import/export facilities in XML schema, RDF, OWL, DAML, and MOF formats; (c) A feature-rich set of ontology authoring wizards that create and maintain the required UML model elements for the user, saving time and substantially reducing construction errors and inconsistencies. The Visual Ontology Modeler implements Sandpiper's UML Profile for Knowledge Representation, which extends UML to enable modeling of frame-based knowledge representation concepts such as class, relation, function and individual frames, as well as the slots and facets that constrain those frames. It also includes a library of ontologies, including the IEEE Standard Upper Ontology (SUO), concepts relevant to XML schema, RDF, and DAML generation, and other basic concepts to develop rich ontologies. The framework supports analysis, alignment, development, merging, and evolution, with consistency checking and validation for OWL-DL, first order, and production rules-related applications.

Semtalk: Semtalk [114] is a Microsoft Visio based graphical modeling tool, which is used for business process modeling, product configuration and visual glossaries. Since it is based on an open extensible meta-model, new modeling tools can be created with reasonable effort. Most of these solutions make use of SemTalk's ability to represent ontologies or at least taxonomies in a visual way using Microsoft Visio. The native modeling language supported by the SemTalk consistency engine is a mixture of RDF(S) and OWL. It supports multiple inheritance, instances, and object and datatype properties. A UML-based graphical representation is adopted for the graphical representation of ontologies. Semtalk provides export and import interfaces to RDF(S), OWL, DAML and F-Logic.

COBra: COBra [115] is an ontology browser and editor for GO and OBO ontologies. It has been specifically designed to be usable by biologists to create links between ontologies, and supports: (a) drag-and-drop editing of GO ontologies; (b) mapping between two ontologies; and (c) translation to OWL and other Semantic Web languages. COBra supports manual creation of links between terms in two ontologies, e.g.,links or mappings between tissues in an anatomy and the cell types of the tissues can be recorded and stored. COBra supports import/export of ontologies from and into GO and GO XML/RDF/RDF(S), DAG Edit and OWL.

Generic Knowledge Base (GKB) Editor: The GKB Editor [116] is a tool for graphically browsing and editing knowledge bases across multiple frame representation systems (FRSs) in a uniform manner. It offers an intuitive user interface in which objects and data items are represented as nodes in a graph, with the relationships between them forming the edges. A sophisticated incremental browsing facility allows the user to selectively display only that region of a KB that is currently of interest, even as that region changes. The GKB Editor consists of three main modules: a graphical interactive display based on Grasper-CL, and a library of generic knowledge-base functions, and corresponding libraries of frame-representation-specific methods, based on Open Knowledge Base Connectivity (OKBC).

SWOOP: Most existing ontology development toolkits provide an integrated environment to build and edit ontologies, check for errors and inconsistencies (using a reasoner), browse multiple ontologies, and share and reuse existing data by establishing mappings among different ontological entities. However, their UI design (look and feel) and usage style are inspired by traditional KR-based paradigms, whose constrained and methodical framework have steep learning curves, making them cumbersome to use for the average Web user. SWOOP [117] is a hypermedia-inspired ontology editor that employs a Web browser metaphor for its design and usage. Such a tool would be more effective (in terms of acceptance and use) for the average web user by presenting a simpler, consistent and familiar framework for dealing with entities on the Semantic Web. Some features of SWOOP are: (a) Web-browser-like look and feel including URI-based access and hyperlink-based navigation; (b) inline editing with HTML renderer; (c) browsing, comparison and mapping of multiple ontologies; (d) ontology partitioning and explanation; (e) collaborative annotation support; and (e) sound and complete conjunct Abox queries.

WSMT: The Web Services Modeling Toolkit [377] is an open source graphical development environment for all elements of the Web Service Modeling Ontology (WSMO). It is built as a set of plug-ins for the Eclipse development environment and allows ontology engineers to graphically build their ontologies. Particular focus is placed on visualization of large ontologies with zoom-in and zoom-out capabilities. Verification and consistency checking are provided through a plugged-in WSML reasoner. Another plug-in allows the creation of mappings between ontologies that can be used in a WSMO execution environment.

WSMO Studio: WSMO Studio [378] also provides an open source Eclipse-based development environment for building WSMO ontologies. Similarly to WSMT, reasoning support can be plugged in for different WSML language variants. It also directly supports WSMO annotations of existing WSDL Web Service descriptions via the W3C SAWSDL. Another difference with WSMT is that WSMO Studio does not focus on graphical visualization of ontologies.

TopBraid Composer: TopBraid Composer is an enterprise-class modeling environment for developing Semantic Web ontologies and building semantic applications. Fully compliant with W3C standards, Composer offers comprehensive support for developing, managing and testing configurations of knowledge models and their instance knowledge bases. Composer incorporates a flexible and extensible framework with a published API for developing semantic client/server or browser-based solutions, that can integrate disparate applications and data sources. Implemented as an Eclipse plug-in, Composer is used to develop ontology models, configure data source integration as well as to customize dynamic forms and reports. Other than W3C standards, there is support for importing UML models, XML Schemas and relational databases. It supports integration with leading RDF data stores such as Jena, Pellet and Racer.

Neon Toolkit: The NeOn toolkit [406], based on the OntoStudion Editor core, is an extensible Ontology Engineering Environment. It contains plugins for ontology management and visualization. The core features of the Neon toolkit include support for basic schema editing operations, visualization and browsing of ontologies, the ability to import and export ontologies in various representation languages such as F-Logic, subsets of RDF(S) and OWL. It is designed around an open and modular architecture, which includes infrastructure services such as registry and repository, and supports distributed components for ontology management, reasoning and collaboration in networked ontologies. Building on the Eclipse platform, the Toolkit provides an open framework for plug-in developers. A number of commercial plugins are available that extent the toolkit by various functionalities including support for rules, development and interpretation of mappings, ability to access databases and import database schemas and specify queries in a Query-Editor.

6.1.2 Ontology Editors: A Comparative Evaluation

We present a comparative evaluation based on the dimensions identified earlier.

Table 6.1. Ontology editing tools: Architecture

Tool	SW Architecture	Extensibility	Ontology Storage	Backup Management
Apollo	Standalone	Plug-ins	Files	No
LinKFactory	3-tier	Plug-ins	DBMS	No
OntoStudio	Client Server	Plug-ins	DBMS	No
Ontolingua Server	Client Server	No	Files	No
Ontosaurus	Client Server	No	Files	No
Protege	Standalone	Plug-ins	DBMS	No
WebODE	3-tier	Plug-ins	DBMS	Yes
WebOnto	Client Server	No	Files	Yes
ICOM	Client Server	No	XML Files	
IODE	Standalone	No	Deductive DBMS	Yes

Table 6.1. Ontology editing tools: Architecture

Tool	SW Architecture	Extensibility	Ontology Storage	Backup Management
Visual Ontology Modeler	Plug-in to Rational Rose	Yes	As UML class diagram	Yes
Semtalk	Plug-in for Microsoft Visio	No	Visio files	No
COBra	Standalone	No	Flat files (multiple formats)	No
GKB	Standalone and client server	No	Yes	No
SWOOP	Web-based client server	Yes via plug-ins	As HTML models	No
WSMT	Standalone Eclipse plug-in	Plug-ins	Flat file	No
WSMO Studio	Standalone Eclipse plug-in	Plug-ins	Flat file	No
Topbraid Composer	Standalone Eclipse plug-in	Plug-ins	DBMS	Yes
Neon Toolkit	Standalone Eclipse plug-in	Plug-ins	DBMS	Yes

Table 6.1 presents a comparison of various ontology editors and tools based on its software architecture (standalone, client/server, n-tier application), extensibility, storage of the ontologies (databases, ASCII files, etc.) and backup management. From this perspective, most of the tools are moving toward Java platforms, and most of them are moving to extensible architectures as well. Storage in databases is and backup management are weak points of ontology tools.

Interoperability (Table 6.2) with other ontology development tools, merging tools, information systems and databases, as well as translations to and from some ontology languages, are important for integration of ontologies into applications. Most of the new tools export and import to adhoc XML and other markup languages.

Table 6.2. Ontology editing tools: Knowledge representation and methodological support

Tool	KR Knowledge Model	Axiom Language	Methodological Support
Apollo	Frames (OKBC)	Unrestricted	No
LinKFactory	Frames + First Order Logic	Restricted First Order Logic	Yes
OntoStudio	Frames + First Order Logic	FLogic	OntoKnowledge
Ontolingua Server	Frames + First Order Logic	KIF	No
Ontosaurus	Description Logics	LOOM	No
Protege	Frames + First Order Logic + Metaclasses	PAL	No
WebODE	Frames + First Order Logic	WAB	Methontology
WebOnto	Frames _ First Order Logic	OCML	No
ICOM	Description Logics with extension	DLR	No
IODE	Common Logic, extended with temporal reasoning and quantification over predicates	FOL	Yes
Visual Ontology Modeler	Description Logics	DL	Own - collaborative ontology development
Semtalk	OWL	OWL Full is possible	No
COBra	RDF and OWL	Not used	No
GKB	Multiple Frame Representation Systems	LOOM and others	No
SWOOP	OWL	OWL-DL	No
WSMT	WSML	WSML	No
WSMO Studio	WSML	WSML	No
Topbraid Composer	RDF, OWL and SWRL	OWL-DL	No
Neon Toolkit	F-Logic, RDF and OWL	F-Logic, OWL-DL	Yes, Neon Ontology Dev Process and Lifecycle

From the knowledge representation point of view (Table 6.2), there are two families of tools: description-logic-based tools, and other tools, which allow representation of knowledge following a hybrid approach based on frames and first order logic. Additionally, Protégé provides flexible modeling components like metaclasses. Some ontology building methodologies that are supported are: the OntoKnowledge methodology, GALEN methodology and Methontology. None of the tools provide project management facilities, and provide only a little support for ontology maintenance and evaluation.

Table 6.3. Ontology editing tools: Inference services

Tool	Inbuilt Inference Engine	External Inference Engine	Constraint, Consistency Checking	Automatic Classification	Exception Handling
Apollo	No	No	Yes	No	No
LinKFactory	Yes	Yes	Yes	Yes	No
OntoStudio	Yes (Ontobroker)	No	Yes	No	No
Ontolingua Server	No	ATP	No	No	No
Ontosaurus	Yes	Yes	Yes	Yes	No
Protege	Yes (PAL)	Jess, FLogic, Pellet	Yes	No	No
WebODE	Yes (Prolog)	Jess	Yes	No	No
WebOnto	Yes	No	Yes	No	No
ICOM	Not in GUI	Connect to ICOM server	Yes	No	No
IODE	Yes	No	Yes	No	Yes
Visual Ontology Modeler	No	DL reasoner, rules engines	Yes	No	No
Semtalk	No	No	No	No	No
COBra	No	No	No	No	No
GKB	No	Yes	Yes	No	No

Table 6.3. Ontology editing tools: Inference services

Tool	Inbuilt Inference Engine	External Inference Engine	Constraint, Consistency Checking	Automatic Classification	Exception Handling
SWOOP	No	Yes (Pellet or other engine)	Only with reasoner plug-in	No	No
WSMT	No	Yes, via plug-in	Yes	No	No
WSMO Studio	No	Yes, via plug-in	Yes	No	No
Topbraid Composer	No	Yes, OWLIM, Pellet, Jena, Oracle Rules	Yes	Yes	Yes
Neon Toolkit	Yes, Ontobroker	Yes, KAON-2 Engine	Yes	No	Yes

Before selecting a tool, it is also important to know which inference services are attached to it (Table 6.3). This includes built-in and other inference engines, constraint and consistency-checking mechanisms, automatic classifications and exception handling, among others. LinkFactory has its own inference engine, OntoStudio uses OntoBroker, Ontolingua uses ATP, Ontosaurus uses the Loom classifier, Protégé uses PAL and can also be linked to DL reasoners, WebODE uses Ciao Prolog and WebOnto uses the OCML inference engine. Besides, WebODE and Ontosaurus provide evaluation facilities. LinkFactory performs automatic classification. Finally, none of the tools provide exception-handling mechanisms.

Table 6.4. Ontology editing tools: Usability

Tool	Graphical Taxonomy	Graphical prunes	Zoom	Collaboration	Ontology Libraries
Apollo	Yes	Yes	No	Yes	Yes
LinKFactory	Yes	Yes	Yes	Yes	Yes
OntoStudio	No*	No	No*	Yes	Yes
Ontolingua Server	Yes	No	No	Yes	Yes
Ontosaurus	Yes	Yes	Yes	Yes	No
Protege	Yes	Yes	Yes	No	Yes

Table 6.4. Ontology editing tools: Usability

Tool	Graphical Taxonomy	Graphical prunes	Zoom	Collaboration	Ontology Libraries
WebODE	Yes	Yes	No	Yes	No
WebOnto	Yes	Yes	No	Yes	Yes
ICOM	Yes	No	No	No	Import and export as XML
IODE	No	No	No	Yes	Yes
Visual Ontology Modeler	Yes	No	No	Yes	Yes
Semtalk	Yes	No	No	Yes	Yes
COBra	Yes	No	No	No	Limited to GO and OBO
GKB	Yes	No	Yes	Yes	No
SWOOP	Yes	No	No	Yes	No
WSMT	Yes	No	Yes	No	No
WSMO Studio	No	No	No	No	Yes
Topbraid Composer	Yes	Yes	Yes	Yes	No
Neon Toolkit	Yes	Yes	Yes	Yes	No

Related to the usability of tools (Table 6.4), WebOnto has the most advanced features related to the cooperative and collaborative construction of ontologies. In general, more features are required in existing tools to ensure the successful collaborative building of ontologies. Finally, other usability aspects related to help system, editing and visualization should be improved in most of the tools.

6.2 Ontology Bootstrapping Approaches

There have been various approaches for semi-automatic generation of ontologies or taxonomies from underlying unstructured text data. These approaches can be broadly characterized as:

- Supervised-machine-learning based approaches, which require a large number of training examples, traditionally generated manually.
- Natural Language Processing (NLP) based approaches applied for generating ontological concepts and relationships. These are based on rules that analyze

patterns based on syntactic categories, which requires significant human involvement, making it expensive and infeasible for large-scale applications.
- Statistical clustering methods have been used to partition data-sets, categorize search results and visualize data. However, they have not focussed on generating labels for clusters and creation of new taxonomies.

Machine learning approaches are for the most part supervised, for which a set of manually generated positive and negative training examples are used. An approach using the concept-forming system COBWEB [118] has been used to perform incremental conceptual clustering on structured instances of concepts extracted from the Web [119]. Experimental and theoretical results on learning the CLASSIC description logic were presented in [120], and were used to construct concept hierarchies. An approach to bootstrapping a classification taxonomy based on a set of structured rules was proposed in [121]. A supervised approach presented in [122] supports semi-automatic and incremental bootstrapping of a domain-specific information extraction system.

Empirical and corpus-based NLP methods to build domain-specific lexicons have been proposed in [123] and used in [124]. Approaches that learn meanings of unknown words based on other word definitions in the surrounding context have been presented in [125] [126]. Case-based methods that match unknown word contexts against previously seen word contexts are described in [127] [128]. Approaches presented in [129] [130] apply shallow parsing, tagging and chunking, along with statistical techniques to extract terminologies or enhance existing ontologies. Full parse tree construction followed by decomposition into elementary dependency trees has been used to create medical ontologies from French text corpora in [131]. In [132], a thesaurus is built by performing clustering according to a similarity measure after having retrieved triples from a parsed corpus.

Linguistic structures such as verbs, appositions and nominal modifications have been used to identify hypernymic propositions in biomedical text [133]. Lexico-syntactic patterns have been investigated for inferring hyponymy from textual data in [134]. Salient words and phrases extracted from the documents are organized hierarchically using subsumption type co-occurrences in [130]. A description of supervised and unsupervised approaches to extract semantic relationships between terms in a text document is presented in [135]. A generalized association rule algorithm proposed in [136] detects non-taxonomic relationships between concepts and also determines the right level of abstraction at which to establish the relationship.

Effectively mining relevant information from a large volume of unstructured documents has received considerable attention in recent years [137] [138] [139]. A survey on the use of clustering in information retrieval is presented in [140]. Document clustering has been used for browsing large document collections in [141], using a "scatter/gather" methodology. These approaches create vector space representations of documents and use Euclidean or cosine-distance-based similarity metrics like the Euclidean ones to extract clusters from groups of documents. Clustering of Web documents to organize search results has been proposed in [142] [143].

There is a realization amongst researchers that one needs to leverage the strengths of a wide variety of techniques across machine learning, natural language processing and statistical approaches to address the difficult problem of ontology generation. Frameworks for hybrid approaches have been proposed:

- The ontology learning framework developed by Maedche and Staab [144].
- The Thematic Mapping System [144].
- The Taxaminer approach, which presents a framework to combine the techniques enumerated above [145].
- A complementary approach that uses the structure and content of HTML-based pages on the Web to generate ontologies [146].

6.3 Ontology Merge and Integration Tools

Ontology merging and integration, including functionalities related to versioning and keeping track of various changes, are very important in the context of the ontology design process. Furthermore, a large number of ontologies are being used to annotate content on the Web. There is a need to be able to reconcile annotations based on multiple ontologies and also support query processing across multiple ontologies. An approach to address the above challenge is to establish mappings between ontologies and to merge them at run time, as proposed in [147] [148]. A large number of ontology mapping and merging tools have appeared to address these issues. We now present a brief survey and a comparative evaluation of various ontology merge and integration tools [99]. The criteria used for the comparative evaluation of these tools are as follows:

Knowledge used during the merge process: The merging process can be much more efficient if additional knowledge can be made to bear on the process. Some examples of these knowledge resources are: electronic dictionaries, thesauri, lexicons, concept definitions and slot values, graph structures, instances of concepts and inputs from the user.

Interoperability: It is important because key activities such as transformation of formats and evaluation can be performed by other non-merging tools. Some important considerations are interoperability with other ontology tools or information systems and whether ontologies expressed in different languages can be merged.

Management of different versions of ontologies: A change in the source ontology results in a change in the merged ontology. Some important considerations are whether the tool takes advantage of the former versions of the ontologies and whether it warns that the merged ontology is not an accurate reflection of the source ontologies.

Components manipulated by the tools: An important consideration is about which components that can be merged by the ontology development tools or about which suggestions can be made by the merging tools. The main components that need to be considered are concepts (including slots, and taxonomic and other relationships), axioms, rules and instances.

Editing and Visualization: This is very important for the usability of the tool. Some important considerations are support for a step-by-step view of the process, a simultaneous view of the source ontologies being merged, graphical prunes (views) of the ontologies being merged, zooming and the ability to hide/show information.

6.3.1 Ontology Merge and Integration Tools: A Brief Description

In this section, we present a brief description of some of the ontology merge and integration tools in use today.

Chimaera: Chimaera [149] is a merging and diagnostic Web-based browser ontology environment. It contains a simple editing environment in the tool and also allows the user to use the full Ontolingua editor/browser environment for more extensive editing. It facilitates merging by allowing users to upload existing ontologies into a new workspace (or into an existing ontology). Chimaera will suggest potential merging candidates based on a number of properties. Chimaera allows the user to choose the level of vigor with which it suggests merging candidates. Higher settings, for example, will look for things like possible acronym expansion. Chimaera also supports a taxonomy resolution mode. It looks for a number of syntactic term relationships, and when attached to a classifier, it can look for semantic subsumption relationships as well. Chimaera includes analysis capability that allows users to run a diagnostic suite of tests selectively or in its entirety. The tests include incompleteness tests, syntactic checks, taxonomic analysis, and semantic checks. Terms that are used but that are not defined, terms that have contradictory ranges, and cycles in ontology definitions are also detected.

PROMPT: PROMPT [150] is a tool for semi-automatic guided ontology merging, and is available as a plug-in for Protege. PROMPT leads the user through the ontology-merging process, identifying possible points of integration, and making suggestions regarding what operations should be done next, what conflicts need to be resolved, and how those conflicts can be resolved. PROMPT's ontology-merging process is interactive. A user makes many of the decisions, and PROMPT either performs additional actions automatically based on the user's choices or creates a new set of suggestions and identifies additional conflicts among the input ontologies. The tool takes into account different features in the source ontologies to make suggestions and to look for conflicts. These features include names of classes and slots, class hierarchy, slot attachment to classes, facets and facet values. Some conflicts identified by PROMPT are: name conflicts, dangling references, redundancy in the class hierarchy (more than one path from a class to a parent other than the root) and slot value restrictions that violate class inheritance.

ODEMerge: ODEMerge [151] is a tool to merge ontologies that is integrated in WebODE [107]. This tool is a partial software support for the methodology for merging ontologies [152], which proposes the following steps: (1) transformation of formats of the ontologies to be merged; (2) evaluation of the ontologies; (3) merging of the ontologies; (4) evaluation of the result; and (5) transformation of the format of the resulting ontology to be adapted to the application where it will be used. WebODE helps in steps (1), (2), (4) and (5) of the merging methodology, and ODEMerge carries out the merge of taxonomies of concepts in step (3). Besides, ODEMerge helps in the merging of attributes and relations, and it incorporates many of the rules identified in the methodology. ODEMerge uses the source ontologies to be merged, and the synonymy, hyponymy and hypernymy relationships between terms across ontologies in the merging process. Customized dictionaries can be added to provide the relationships, and new merging rules can also be defined. ODEMerge supports the merging of ontologies in all the ontology languages supported by the WebODE tool.

6.3.2 Evaluation of Ontology Merge and Integration Tools

We now present a comparative evaluation of the various tools based on the dimensions identified earlier.

Table 6.5. Information used during the merge process

Feature	PROMPT	ODEMerge	Chimaera
Thesauri, Dictionaries	No	No	No
Lexicons	No	No	No
Concept Definitions and Slot Values	Yes	Yes	Yes
Graph Structure	Yes	Yes	No
Instances of concepts	Yes	No	No
User Input	Yes	Yes	Yes

Table 6.5 compares the information used by these tools (electronic dictionaries, lexicons, etc.) during the merge process. The more information a tool uses during this process, the more work it is able to perform without the user's participation. Most of the tools start the merging process by searching for similar concepts of ontologies.

Table 6.6. Interoperability

Feature	PROMPT	ODEMerge	Chimaera
Tools and systems Interoperability	Yes	Yes	Yes
Multiple ontology language support	Yes	Yes	Yes

Interoperability with other ontology tools is also an important aspect (Table 6.6) and is usually determined by the ontology development platform in which the merge tool is integrated. Another important aspect is whether the tool can merge ontologies expressed in different languages. All these tools are able to merge ontologies expressed in different languages (XML, RDFS, OIL, etc.).

Table 6.7. Management of different versions

Feature	**PROMPT**	**ODEMerge**	**Chimaera**
Leveraging previous ontology versions	No	No	No
Notifications of changes in source ontologies	No	No	No

Given that ontologies usually evolve, the management of different ontology versions is also important (Table 6.7). None of these tools takes advantage of former versions of the ontologies to be merged, and none of them warns users about changes in the source ontologies.

Table 6.8. Components of ontologies manipulated by the tools

Feature	**PROMPT**	**ODEMerge**	**Chimaera**
Concepts	Merge, Suggest	Merge	Merge, Suggest
Own Slots	Merge, Suggest	Merge	Merge, Suggest
Template Slots	Merge, Suggest	Merge	Merge, Suggest
Taxonomies	Merge, Suggest	Merge	Merge, Suggest
Concepts	Merge, Suggest	Merge	Merge, Suggest
Relations	Merge	Merge	Not currently supported
Partitions and/or Decompositions	No	Merge	Both supported
Relations or Functions?	Merge, Suggest (relations only)	Merge	Not currently supported
Arity	Merge, Suggest (binary relations)	Merge	No
Sets of axioms	No	No	No
Sets of rules	No	No	No
Instances	Merge, Suggest	No	Not currently supported
of Concepts	Merge, Suggest	No	No

Table 6.8. Components of ontologies manipulated by the tools

Feature	PROMPT	ODEMerge	Chimaera
of Relations	Merge, Suggest	No	No
Claims	No	No	No

We present in Table 6.8 which kinds of components can be merged by the tool and about which kinds of component merging suggestions can be proposed by the tools. All the tools allow merging concepts, taxonomies, relations and instances. However, no tool allows merging axioms and rules. From among all of them, PROMPT is the tool that provides most suggestions to the users.

Table 6.9. Editing and visualization support

Feature	PROMPT	ODEMerge	Chimaera
Step by Step view of Process	Graphical, tabular, hierarchical	Non-graphical	HTML text
Simultaneous view of source ontologies	Yes	No	Yes
Graphical view of source ontologies	Through host tool	Through host tool	No
Zoom	Through host tool	Through host tool	No
Hide/Show	Through host tool	Through host tool	Yes: for subclass/ superclass relationships and child slots

Edition and visualization features (Table 6.9) are strongly influenced by the ontology development platform in which these tools are integrated.

6.4 Ontology Engines and Reasoners

In this section, we present a brief description of ontology reasoners and engines. Whereas most of the ontology-reasoning systems are based on description logics, some reasoners are implemented based on rules and first-order logic theorem provers.

CEL: The system CEL [153] is a description logic system that offers both sound and complete polynomial-time algorithms and expressive means that allow its use in real-world applications. It is based on recent theoretical advances that have shown that the description logics (DL) EL, which allows for conjunction and existential restrictions, and some of its extensions have a polynomial-time subsumption problem even in the presence of concept definitions and so-called general concept inclusions (GCI). The DL EL+ handled by CEL extends EL by so-called role inclusions (RI). On the practical side, it has turned out that the expressive power of EL+ is sufficient to express several large life science ontologies. In particular, the Systematized Nomenclature of Medicine (SNOMED) [7] can be expressed using EL with RIs and acyclic concept definitions. The Gene Ontology (GO) [10] can also be expressed in EL with acyclic concept definitions and one transitive role (which is a special case of an RI). Finally, large parts of the Galen Medical Knowledge Base (GALEN) [154] can be expressed in EL with GCIs and RIs.

CEL is a tractable fragment of OWL 1.1 [407], which is an extension of OWL and is currently a W3C member submission. A W3C group is currently working on creating the next version of the OWL, to be christened OWL 2.0, based on this submission. This tractability is achieved by eliminating the `allValuesFrom` construct and retaining the `someValuesFrom` construct. CEL also supports the role inclusion axioms e.g., `hasStructuredTestResult o indicatesDisease => suffersFrom`. The constructs from OWL 1.1 which cause intractability are cardinality restrictions, union, negation, inverse properties, functional and inverse functional properties.

FaCT++: FaCT++ [155] is the new generation of the well-known FaCT [156] OWL-DL reasoner. FaCT++ uses the established FaCT algorithms, with a different internal architecture, and is implemented using C++ for efficiency and portability. Some interesting features of FaCT are: (a) its expressive logic (in particular the SHIQ reasoner): SHIQ is sufficiently expressive to be used as a reasoner for the DLR logic, and hence to reason with database schemas; (b) its support for reasoning with arbitrary knowledge bases (i.e., those containing general concept inclusion axioms); (c) its optimized tableaux implementation (which has now become the standard for DL systems); and (d) its CORBA- based client server architecture.

fuzzyDL: fuzzyDL [157] is a Description Logics Reasoner supporting Fuzzy Logic reasoning. The fuzzyDL system includes a reasoner for fuzzy SHIF with concrete fuzzy concepts (ALC augmented with transitive roles, a role hierarchy, inverse roles, functional roles, and explicit definition of fuzzy sets). Some interesting features of fuzzyDL are: (a) extension of the classical Description Logics SHIF to the fuzzy case; (b) explicit definitions of fuzzy concepts with left-shoulder, right-shoulder, triangular and trapezoidal membership functions; (c) concept modifiers in terms of linear hedges; (d) support for General Inclusion Axioms; (e) support for "Zadeh semantics" and Lukasiewicz logic; and (f) backward compatibility, i.e. it support for classical description logic reasoning.

KAON2: KAON2 [158] is an infrastructure for managing OWL-DL, SWRL, and F-Logic ontologies and has the following interesting features: (a) an API for programmatic management of OWL-DL, SWRL, and F-Logic ontologies; (b) a stand alone server providing access to ontologies in a distributed manner using RMI; (c) an inference engine for answering conjunctive queries (expressed using SPARQL syntax); (d) a DIG interface, allowing access from tools such as Protege; and (e) a module for extracting ontology instances from relational databases.

Pellet: Pellet [159] is an open source OWL-DL reasoner written in Java, originally developed at the University of Maryland's Mindswap Lab, and funded by a diverse group of organizations. Pellet is based on the tableaux algorithms developed for expressive Description Logics (DLs). It supports the full expressiveness of OWL-DL including reasoning about nominals (enumerated classes). In addition to OWL-DL, as of version 1.4, Pellet supports all the features proposed in OWL 1.1, with the exception of n-ary datatypes. Thus the expressiveness of supported DL is SROIQ(D), which extends the well-known DL SHOIN(D) (the DL that corresponds to OWL-DL) with qualified cardinality restrictions, complex subproperty axioms (between a property and a property chain), local reflexivity restrictions, reflexive, irreflexive, symmetric, and anti-symmetric properties, disjoint properties, and user-defined datatypes. Pellet provides many different reasoning services such as consistency checking, concept satisfiability, classification and realization. It also incorporates various optimization techniques described in the DL literature and contains several novel optimizations for nominals, conjunctive query answering and incremental reasoning.

RacerPro: RacerPro [160] provides a first implementation of the Semantic Web Rules Language (SWRL) in its latest version. It also supports services for OWL ontologies and RDF data descriptions such as: (a) consistency checking for OWL ontologies and a set of data descriptions; (b) inference of implicit subclasses and synonyms for resources (classes or instances); (c) OWL-QL query processing for OWL documents (ontologies and their instances); and (d) incremental query answering for information retrieval tasks. RacerPro implements a highly optimized tableau calculus for a very expressive description logics. It offers reasoning services for multiple T-boxes and for multiple A-boxes as well. The system implements the description logic $ALCQHI_R$, also known as SHIQ. This is the basic logic ALC augmented with qualifying number restrictions, role hierarchies, inverse roles, and transitive roles. In addition to these basic features, RacerPro also provides facilities for algebraic reasoning including concrete domains for dealing with min/max restrictions over the integers; linear polynomial (in)equations over the reals or cardinals with order relations; and equalities and inequalities of strings. RacerPro combines description logic reasoning with, for instance, reasoning about spatial (or temporal) relations within the A-box query language nRQL. Bindings for query variables that are determined by A-box reasoning can be further tested with respect to an associated constraint network of spatial (or temporal) relationships.

Jena: The Jena2 inference subsystem [408], is designed to allow a range of inference engines or reasoners to be plugged into Jena. Such engines are used to derive additional RDF assertions which are entailed from some base RDF together with any optional ontology information and the axioms and rules associated with the reasoner. The primary use of this mechanism is to support the use of languages such as RDFS and OWL which allow additional facts to be inferred from instance data and class descriptions. However, the machinery is designed to be quite general and, in particular, it includes a generic rules engine that can be used for many RDF processing or transformation tasks. Other than the generic rules engine, there are other pre-defined reasoners included in the Jena2 system, such as a transitive reasoner that stores and traverses class and property lattices, a RDFS rule reasoner that implements a configurable set of RDFS entailments, and a set of reasoners for various subsets of OWL.

JESS: Jess [409], is a rules engine and scripting environment written entirely in Sun's Java language at Sandia National Laboratories in Livermore, CA. Using Jess, it is possible to build Java software that has the capacity to perform reasoning using knowledge supplied in the form of declarative rules. Jess includes a full-featured development environment based on the award-winning Eclipse platform. Jess uses an enhanced version of the Rete algorithm [410] to process rules. Jess has many unique features including backward chaining and working memory queries, and Jess can directly manipulate and reason about Java objects.

6.5 Clinical Scenario Revisited

Consider the clinical use case scenario presented in Chapter 2. In particular consider the important issue of knowledge change propagation in this section. Consider the definition in natural language of fibric acid contraindication:

```
A patient is contraindicated for fibric acid if he or she has an
allergy to fibric acid or has an abnormal liver panel.
```

Suppose there is a new (hypothetical) biomarker for fibric acid contraindication for which a new molecular diagnostic test is introduced in the market. This leads to a redefinition of a fibric acid contraindication as follows.

```
The patient is contraindicated for fibric acid if he has an allergy to
fibric acid or has elevated Liver Panel or has a genetic mutation.
```

Let us also assume that there is a change in a clinically normal range of values for the lab test AST which is a part of the liver panel lab test. This leads to a knowledge change and propagation across various knowledge objects that are sub-components and associated with the fibric acid contraindication concept. A diagrammatic representation of the OWL representation of the new fibric contraindication with the changes marked in red ovals is illustrated below. The definition of "fibric acid contraindication" changes, triggered by changes at various levels of granularity.

A potential sequence of change propagation steps are enumerated below:

1. The clinically normal range of values for the AST lab result changes.
2. This leads to a change in the abnormal value ranges for the AST lab result
3. This leads to a change in the definition of an abnormal liver panel.

4. This leads to a change in what it means to be a patient with an abnormal liver panel.
5. The definition of fibric acid contraindication changes due to the following changes:
   (A) The change in the definition of a patient with an abnormal liver panel as enumerated in steps 1-4 above.
   (B) Introduction of a new condition: a patient having a mutation: “Missense: XYZ3@&%” (hypothetical). This is a new condition which could lead to a change in what it means to be a patient with a contraindication to fibric acid.

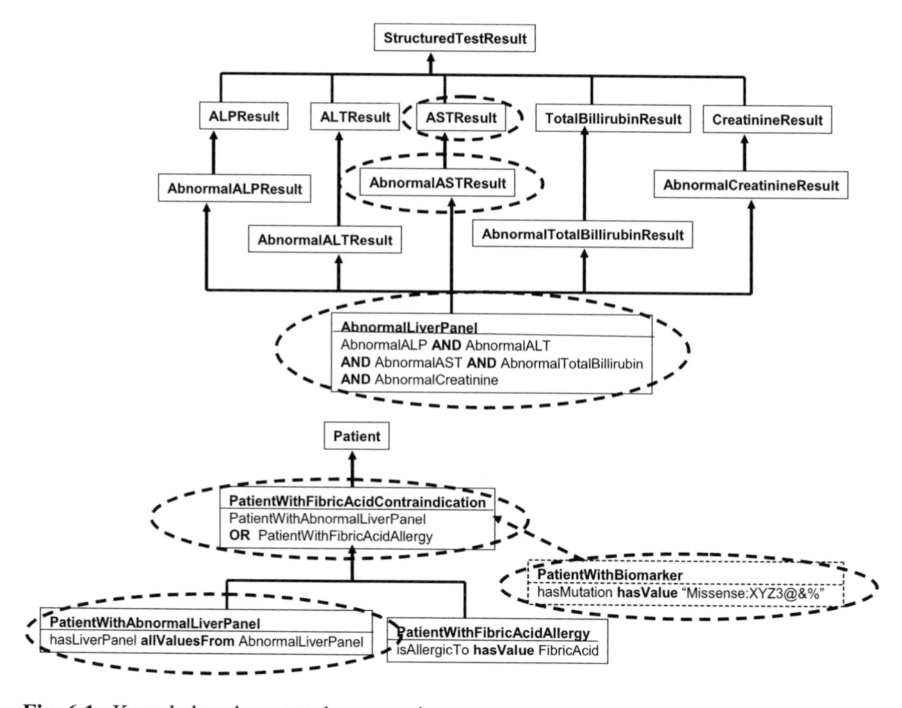

Fig. 6.1. Knowledge change and propagation

It may be noted that in our discussion in Section 6.3.1, none of the ontology editors and tools today support versioning and change management functionality. In our solution approach, we propose to load these ontologies as data into a rules engine and write specialized rules to identify the impacts of a change operation.

6.6 Summary

In this chapter, we presented a discussion on different aspects of ontology authoring, bootstrapping and management. In particular, we present a survey of ontology building tools, ontology-reasoning engines, and techniques for ontology bootstrapping, matching, merging and integration. A more detailed account of ontology authoring and management may be obtained from the Handbook on Ontologies by Staab and Ruder [411]. The clinical use case is revisited and an approach for modeling knowledge change and propagation as ontology versioning and change management is presented.

7 Applications of Metadata and Ontologies

A key value proposition enabled by the use of metadata descriptios based on concepts from domain specific ontologies, is the ability to describe Web and other types of content using semantic descriptions with fine grained abstractions. These descriptions could appear in the form of annotations in the case of unstructured data. Alternatively, in the case of structured data created according to a well defined schema, these descriptions can be created based on a mapping between the schema and a domain specific ontology. These metadata descriptions may also be used to query the underlying structured data as well. Finally, with the mappings between the schema and domain ontologies form a critical component, that enables domain ontolog driven information integration. In this chapter, we discuss:

- Structured and semi-structured metadata annotations of unstructured and semi-structured documents on the Web. Tools and techniques to support metadata annotation are presented in Section 7.1.
- Structured metadata annotations of structured Web resources with a well-defined set of types and schemas. Techniques to support schema and ontology mapping are discussed in Section 7.2.
- Approaches for ontology driven information integration are discussed in Section 7.3.

7.1 Tools and Techniques for Metadata Annotation

Knowledge about documents has traditionally been managed through the use of metadata, which can concern the world around the document, e.g., the author, and often at least part of the content, e.g., keywords. The Semantic Web proposes annotating document content using semantic information from domain ontologies. The result is Web pages with machine interpretable markup that provide the source material with which agents and Semantic Web Services can operate. The goal is to create annotations with well-defined semantics, however those semantics may be defined. This is a crucial requirement for interoperability, as it ensures that the annotator and annotation consumer actually share meaning.

Semantic Web annotations go beyond familiar textual annotations about the content of the documents, such as "clause seven of this contract has been deleted because . . ." and "the test results need to go in here". This kind of informal annotation is common in word processor applications and is intended primarily for use by document creators. Semantic annotation formally identifies concepts and relations between concepts in documents, and is intended primarily for use by machines. For example, a semantic annotation might relate "Paris" in a text to an ontology which both identifies it as the abstract concept "City" and links it to the instance "France" of the abstract concept "Country", thus removing any ambiguity about which "Paris" it refers to. Annotations can be utilized to make the knowledge contained in unstructured sources (medical images such as X-rays) available in a structured form, allowing both accurate and focussed retrieval and knowledge sharing for a given patient's case. Moreover, they can be processed to automatically draft textual reports about the patient, the diagnostic information that is available and the assessments made about the data by the medical team.

In this section, we discuss some requirements identified for semantic annotation and review the systems that currently exist to support annotations of documents presented in [171]. Seven requirements, which are used to assess the capabilities of existing annotation systems, are identified for semantic annotation systems.

7.1.1 Requirements for Metadata Annotation

The metadata task may be considered from four viewpoints: ontologies, documents, annotations that link ontologies to documents, and end users. Each viewpoint suggests one or more requirements, e.g., the need for tools to support multiple, evolving ontologies (ontology viewpoint) and the need to support the reuse and versioning of documents (document viewpoint). Some requirements for metadata annotation are as follows [171]:

1. **Standard Formats**: Using standardized formats and data models is preferable whenever possible. Two types of standard are required, standards for describing ontologies such as OWL [45] and standards for annotations such as tRDF [42].
2. **User-Centered/Collaborative Design:** Since few organizations have the capacity to employ professional annotators, it is crucial to provide knowledge workers with easy to use interfaces that simplify the annotation process and place it in the context of their everyday work. Thre is a need to facilitate collaboration between users, with experts from different fields contributing to and reusing metadata annotations. Other issues for collaboration include implementing systems with access control functionality. For example, in a medical context, physicians might share all information about patients among themselves but only share anonymized information with planners. Issues related to access policies, trust and provenance are important in this context.
3. **Ontology Support (Multiple Ontologies and Evolution):** Metadata annotation tools need to be able to support multiple ontologies. For example, in a medical context, there may be one ontology for general metadata about a patient and other domain-specific ontologies that deal with diagnosis and medications. In addition, systems will have to cope with changes made to ontologies over time, such as incorporating new classes or modifying existing ones. This is a

crucial requirement as in some domains such as the biomedical domain, standardized vocabularies and ontologies are regularly updated. In this case, the problem is ensuring consistency between ontologies and annotations with respect to ontology changes. Some important issues for the design of an annotation environment are to determine how changes should be reflected in the knowledge base of annotated documents and whether changes to ontologies create conflicts with existing annotations. Knowledge workers may require facilities to help them explore and edit the ontologies they are using.

4. **Document Evolution (Document and Annotation Consistency):** Ontologies change sometimes but some documents change many times. What should happen to the annotations on a document when it is revised? Is it even desirable, in general, to transfer annotations to a new version of a document, or do versions of annotations need to be maintained in parallel with document versions. For example, if a contract were prepared for a new client, annotations that referred to a legal ontology could be retained, but annotations which referred to previous clients could be removed. How can this selective transfer of annotations be achieved?
5. **Annotation Storage:** The Semantic Web model assumes that annotations will be stored separately from the original document, whereas the "word processor" model assumes that comments are stored as an integral part of the document, which can be viewed or not as the reader prefers. The Semantic Web model, which decouples content and semantics, works particularly well for the Web environment in which the authors of annotations do not necessarily have any control over the documents they are annotating.
6. **Automation:** Easing the knowledge acquisition bottleneck can be enabled by the provision of facilities for automatic markup of document collections to facilitate the economical annotation of large document collections. To achieve this, the integration of knowledge extraction and natural language processing technologies into the annotation environment is vital.

7.1.2 Tools and Technologies for Metadata Annotation

In this section, we discuss annotation frameworks, tools and environments that produce semantic metadata annotations, i.e., metadata annotations that are based on a vocabulary presented by ontologies.

Metadata Annotation Frameworks

We discuss two frameworks for annotation in the Semantic Web, the W3C annotation project Annotea [172], and CREAM [173], an annotation framework being developed at the University of Karlsruhe. These frameworks can be implemented by multiple tools.

Annotea is a W3C project, which specifies infrastructure for annotation of Web documents, with emphasis on the collaborative use of annotations. The main format for Annotea is RDF and the kinds of documents that can be annotated are limited to HTML or XML-based documents. XPointer is used as the method for locating annotations within a document. The Annotea approach concentrates on a semiformal style of annotation, in which annotations are free text statements about documents. These statements must have metadata (author, creation time, etc.) and may be typed according to user-defined RDF schemas of arbitrary complexity. The storage model proposed is a mixed one with annotations being stored as RDF held either on local machines or on public RDF servers. The Annotea framework has been instantiated in a number of tools including Amaya, Annozilla and Vannotea, which are discussed later in this section.

The CREAM framework looks at the context in which annotations could be made. It specifies components required by an annotation system, including the annotation interface, with automatic support for annotators, the document management system and the annotation inference server. CREAM subscribes to W3C standard formats with annotations made in RDF or OWL and XPointers used to locate annotations in text, which restricts it to Web-native formats such as XML and HTML. The CREAM framework supports annotating the databases from which deep Web pages are generated so that the annotations are generated automatically with the pages. It is supported by a storage model that allows users to choose whether they want to store annotations separately on a server or embedded in a Web page. The CREAM framework allows for relational metadata, defined as "annotations which contain relationship instances". Relational metadata is essential for constructing knowledge bases which can be used to provide semantic services. Examples of tools based on the CREAM framework are S-CREAM and OntoMat-Annotizer, discussed later in the section.

Metadata Annotation Tools

The most basic annotation tools allow users to manually create annotations. They have a great deal in common with purely textual annotation tools but provide some support for ontologies. The W3C Web browser and editor Amaya [174] can markup Web documents in XML or HTML. The user can make annotations in the same tool they use for browsing and for editing text, making Amaya a good example of a single point of access environment. The Annozilla [179] browser aims to make all Amaya annotations readable in the Mozilla browser and to shadow Amaya developments. Teknowledge [180] produces a similar plug-in for Internet Explorer.

The Mangrove system is another example of manual but user-friendly annotation [175]. The annotation tool itself is a straightforward graphical user interface that allows users to associate a selection of tags with text that they highlight. Mangrove has recently been integrated with a semantic email service [176], which supports the initiation of semantic email processes, such as meeting scheduling, via text forms. The COHSE Annotator [188] produces annotations that are compatible with Annotea. The annotator is provided as a plug-in suitable for use in Mozilla or Internet Explorer, giving the user a choice of working environment. The COHSE architecture has been used to support a number of domain applications, including the generation of semantic annotation for visually impaired users [190] and enriching a Java tutorial site [189].

Multimedia annotation is the next phase of development for annotation, expanding the range files types that can be marked up into images, video and audio. Vannotea [177] has been developed by the University of Brisbane for adding metadata to MPEG-2 (video), JPEG 2000 (image) and Direct 3D (mesh) files, with the mesh being used to define regions of images. It has been designed to allow input from distributed users enabling deployment to annotate cultural artifacts in a collaborative annotation exercise involving both museum curators and indigenous groups [177].

Some manual annotation tools provide more sophisticated user support and a degree of semi-automatic or automatic annotation facilities. The OntoMat-Annotizer is a tool for making annotations based on the CREAM framework. A Web browser displays the page being annotated and provides user-friendly function, such as drag-and-drop creation of instances and the ability to markup pages while they are being created. OntoMat has been extended to include support for semi-automatic annotation. The first of these extensions was S-CREAM [181], which uses an information extraction (IE) system (Amilcare [182]). The system learns how to reproduce the user annotation, to be able to suggest annotations for new documents. OntoMat also incorporates methods for deep annotation [183]. M-OntoMat-Annotizer [184] supports manual annotation of image and video data by indexers with little multimedia experience by automatic extraction of low-level features that describe objects in the content.

SHOE Knowledge Annotator [186] was an early system which allowed users to markup HTML pages in SHOE guided by ontologies available locally or via a URL. Users were assisted by being prompted for inputs. Running SHOE took a step toward automated markup by assisting users to build wrappers for Web pages that specify how to extract entities from lists and other pages with regular formats. A recent addition is the RDF annotator SMORE [185] which allows mark-up of images and emails as well as HTML and text. A tool with similar characteristics to SMORE is the Open Ontology Forge (OOF) [187], an ontology editor that supports annotation, taking it a step further toward an integrated environment to handle documents, ontologies and annotations.

Automation can generally be regarded as falling into three categories. The most basic kind uses rules or wrappers written by hand that try to capture known patterns for the annotations. Supervised systems learn from sample annotations marked up by the user. A problem with these methods is that picking enough good examples is a nontrivial and error-prone task. In order to tackle this problem unsupervised systems employ a variety of strategies to learn how to annotate without user supervision, but their accuracy is limited.

Lixto is a Web information extraction system which allows wrappers to be defined for converting unstructured resources into structured ones. The tool allows users to create wrappers interactively and visually by selecting relevant pieces of information [191]. MnM was designed to markup training data for IE tools rather than as an annotation tool per se [192]. It stores marked up documents as tagged versions of the original, rather than in RDF format. It provides an HTML browser to display the document and ontology browser features. MnM provides open APIs to link to ontology servers and for integrating information extraction tools, making it flexible with the formats and methods it uses.

Melita [193] is a user-driven automated semantic annotation tool which makes two main strategies available to the user. It provides an underlying adaptive information extraction system (Amilcare) that learns how to annotate the documents by generalizing on the user annotations. It also provides facilities for rule writing (based on regular expressions) to allow sophisticated users to define their own rules. Documents are not selected based on the expected usefulness, to the IE system, of annotating the document. The Amilcare IE system has been incorporated in K@, a legal KM system with RDF-based semantic capabilities produced by Quinary [194].

CAFETIERE is a rule-based system for generating XML annotations developed as part of the Parmenides project [195]; it has been used to annotate the GENIA biomedical corpus [208]. Text mining techniques supplemented with slot-based constraints are used to suggest annotations to analysts [196]. The Parmenides project also experimented with a clustering approach to suggest concepts and relations to extend ontologies [197].

Armadillo is a system for unsupervised creation of knowledge bases from large repositories (e.g., the Web) as well as for document annotation [198]. It uses the redundancy of the information in repositories to bootstrap learning from a handful of seed examples selected by the user. Seeds are searched in the repository. Then Adaptive IE is used to generalize over the examples and find new facts. Confirmation by several sources (e.g., documents) is then required to check the quality of the newly acquired data. After confirmation, a new round of learning can be initiated. This bootstrapping process can be repeated until the user is satisfied with the quality of the learned information.

KnowItAll [199] automates extraction of large knowledge bases of facts from the Web. The pointwise mutual information (PMI) measure is used. The PMI measure is roughly the ratio between the number of search engine hits obtained by querying with the discriminator phrase (e.g., "Liege is a city") and the number of hits obtained by querying with the extracted fact (e.g., "Liege"). Three extensions to the system (pattern learning, subclass extraction and list extraction) which are shown to improve overall performance have also been provided.

The SmartWeb project is also investigating unsupervised approaches for RDF knowledge base population [200]. Their approach uses class and subclass names from the ontology to construct examples. The context of these examples is then learned. In this way, instances can be identified which have similar contexts, but which may use different terminology from the ontology. SmartWeb is aimed at broadband mobile access.

Another approach to learning annotations which exploits the sheer size of the Web is Pattern-based Annotation through Knowledge On the Web (PANKOW) [201]. PANKOW uses a range of relatively rare, but informative, syntactic patterns to markup candidate phrases in Web pages without having to manually produce an initial set of marked-up Web pages and go through a supervised learning step. AeroSWARM8 is an automatic tool for annotation using OWL ontologies based on the DAML annotator AeroDAML [202]. This has both a client/server version and a Web-enabled demonstrator in which the user enters a URI and the system automatically returns a file of annotations on another Web page.

SemTag is another example of a tool which focusses only on automatic markup [124]. It is based on IBM's text analysis platform Seeker and uses similarity functions to recognize entities which occur in contexts similar to marked-up examples. The key problem of large-scale automatic markup is identified as ambiguity, e.g., identical strings, such as "Niger", which can refer to different things, a river or a country. A Taxonomy-Based Disambiguation (TBD) algorithm is proposed to tackle this problem. SemTag is proposed as a bootstrapping solution to get a semantically tagged collection off the ground. It is intended as a tool for specialists rather than one for knowledge workers.

KIM [203] [204] uses information extraction techniques to build a large knowledge base of annotations. The annotations in KIM are metadata in the form of named entities (people, places, etc.) which are defined in the KIMO ontology and identified mainly from references to extremely large gazetteers. In the Rich News application KIM has been used to help annotate television and radio news by exploiting the fact that Web news stories on the same topic are often published in parallel [207].

The Rainbow project is taking a Web-mining-led approach to automating annotation. Rainbow is in fact a family of independent applications which share a common Webservice front end and upper-level ontology [205]. The applications include text mining from product catalogs as well as more general pattern-matching applications such as pornography recognition in bit map image files. The generated RDF is stored in Sesame databases for semantic retrieval [210].

A traditional approach to information extraction is used by the h-TechSight Knowledge Management Platform, in which the GATE rule-based IE system is used to feed a semantic portal [206]. This work is of particular interest because the automatically generated annotations are monitored to produce metrics describing the "dynamics" of concepts and instances which can be fed back to end users [209]. It is envisaged that dynamics data will be used to inform the manual evolution of ontologies.

Integrated Annotation Environments

In this section, we discuss systems that are aimed at integrating annotation into standard tools and making annotation simultaneous to writing. WiCKOffice [211] demonstrates how writing within a knowledge-aware environment has useful support possibilities, such as automatic assistance for form filling using data extracted from knowledge bases. AktiveDoc [212] enables annotation of documents at three levels: ontology-based content annotation, free text statements and on-demand document enrichment. Support is provided during both editing and reading. Semi-automatic annotation of content is provided via adaptive information extraction from text (using Amilcare). AktiveDoc is designed for knowledge reuse; it is able to monitor editing actions and to provide automatic suggestions about relevant content. Armadillo supports searches of relevant knowledge in large repositories; annotations in the document are used as context for searches. Annotations are saved in a separate database; levels of confidentiality are associated with annotations to ensure confidentiality of knowledge when necessary. AeroDAML can provide automation within authoring environments. For example, the SemanticWord annotator [213] provides graphical-user-interface based tools to help analysts annotate Microsoft Word documents with DAML ontologies as they write.

Two systems discussed next, are not strictly annotation tools, but produce annotation-like services on demand for users browsing unannotated resources. Magpie [214] operates from within a Web browser and does "real-time" annotation of Web resources by highlighting text strings related to an ontology of the user's choice. The Thresher uses wrappers to generate RDF on the fly as users browse deep Web resources [215]. The user can access semantic services for recognized objects. Writing wrappers is a complex task which Thresher tackles by providing facilities for nontechnical users to markup examples of a particular class. These are then used to induce wrappers automatically. Thresher is part of the Haystack semantic browser [216], which enables users to personalize the ontologies they use.

7.1.3 Comparative Evaluation

We now revisit the requirements presented in Section 7.1.1, and discuss a comparative evaluation of the various tool discussed in the previous section with respect to the requirements.

1. **Standard Formats:** The discussion in the previous section shows that the W3C standards, particularly Annotea, are becoming dominant in this area. Systems like CAFETIERE, which use their own XML-based annotation scheme, are rare. This requirement has been fulfilled, although the standards may need to be augmented to tackle inadequacies in the existing standards.
2. **User-Centered/Collaborative Design:** The most common home environment of the tools we have seen is a Web browser, a natural result of the fact that most of them were designed for the Semantic Web. The downside is that it both

focusses development on native Web formats like HTML and XML and tends to divorce the annotation process from the process of document creation. More attention needs to be paid to developing built-in or plug-in semantic annotation facilities in commonly used packages to encourage knowledge workers to view annotation as part of the authoring process, not as an afterthought, and also to support annotation in collaborative environments, as for example in Vannotea. Most of the tools discussed in the previous section did not address issues of provenance or access rights. Standard methods to restrict access to databases or the file system are available. As a result of offering this kind of support for trust, provenance and access policies concerning annotations are important issues which need to be addressed.

3. **Ontology Support (Multiple Ontologies and Evolution):** Annotation tools have adapted rapidly to recent changes in ontology standards for the Web, with many of the more recent tools already supporting OWL. However, support for doing anything more complex than searching and navigating an ontology browser is the exception. Ontology maintenance, which directly affects the maintenance of annotations, is poorly supported, or not supported at all, by the current generation of tools. This perhaps reflects the assumption that knowledge workers will use existing ontologies rather than editing or creating them. However there are signs that annotation systems are giving users more control of ontologies. Melita allows users to split a concept and then view all the instances that have been created for the old concept and reassign them. The COHSE architecture includes a component for maintaining the ontology but this does not appear to be available from the annotator. The Open Ontology Forge supports the creation of new classes from a root class. h-TechSight monitors the dynamics of instances and concepts to assist endusers in manual ontology evolution. Parmenides has gone further and experimented with clustering methods to suggest ontology changes. However there is still a long way to go and we believe that ontology maintenance presents a significant research challenge.
4. **Document Evolution (Document and Annotation Consistency):** Keeping annotations synchronized with changes to documents is challenging and this is one area in which the current annotation standards are inadequate. The Annotea approach adopted by many of the tools stores annotations separately from the document and uses XPointer to locate them in the document. There are strong arguments in favor of separate storage of annotations and documents, but the problem with the XPointer approach is that connections are one-way from annotations to documents and, therefore, too easily broken by edits at the document end. An environment in which documents and annotations are stored separately, but closely coordinated is required. A number of practical fixes have been implemented in OntoMat, including the ability to search for similar documents that have already been annotated, and a proposal to use pattern matching to help relocate annotations in suitable places in the new document. However, a coordinated approach is needed to tackle the issues of versioning annotations as

documents evolve. These include determining who has permission to edit annotations, at which points in the document life cycle is it appropriate to update the annotations, and what automatic interventions are possible to reduce the burden on users.

5. **Annotation Storage:** In the Semantic Web, documents and their annotations are stored separately. This is unavoidable since documents and annotations are likely to be owned by different people or organizations and stored in different places. A variety of approaches to separate storage were seen in the tools discussed. The Annotea approach calls for RDF servers. Web storage technologies that have been used are RDF triplestore (Armadillo and AktiveDoc), Label Bureaus (SemTag) and DLS (COHSE). An alternative model is to store annotations directly in the document. This approach has been used for in Semantic-Word and MnM. Separate storage of annotations results in decoupling of semantics and content and facilitates document reuse because it is possible to set up rules which control and automate which kinds of annotations are transferred to new documents and which are not. It allows information from heterogeneous resources to be queried centrally as a knowledge base. It also makes it easy to produce different views of a document for users with different roles in an organization or different access rights, thus facilitating knowledge sharing and collaboration. The results of the comparative evaluation of the various tools with respect to the above requirements is presented in Table 7.1 below.

Table 7.1. Comparison of metadata tools

Annotation Tool	**User-Centered Design**	**Ontology support**	**Document Evolution**	**Annotation Storage**
Amaya	Web browser, editor	Annotation server	XPointer	Local, annotation server
Mangrove	Graphical annotation tool			RDF database (Jena)
Vannotea	Collaboration support			Annotation server
OntoMat	Drag/drop, create, annotate	Ontobroker annotations inference server	Xpointer, pattern matching	Annotation server, embedded in Web page, separate file
M-OntoMat Annotizer	Extraction of visual descriptors			Annotation server
SHOE Knowledge Annotator	Prompting	Ontology server		Embedded in Web page
SMORE	Web browser, editor	Ontology server, editing		
Open Ontology Forge	Web browser, drag, drop, create, annotate	Local, editable ontologies	Xpointer	Local RDF or XML file
COHSE Annotator	Plug-in for Mozilla and Internet Explorer	Ontology server	Xpointer	Annotation server

Table 7.1. Comparison of metadata tools

Annotation Tool	User-Centered Design	Ontology support	Document Evolution	Annotation Storage
Lixto				
MnM	Web browser	Ontology server	Store annotated page	Embedded inWeb page
Melita	Control IE intrusiveness	Local, editable ontologies	Regular expressions	
Parmenides		Additions based on clustering		
Armadillo				RDF triple store
KnowItAll				
SmartWeb				RDF Knowledge base
PANKOW	CREAM			
Aero-SWARM	Web Services	Local ontologies		
SemTag				Label Bureau (PICS)
KIM	Various plug-in front ends	KIMO		RDF Knowledge-base
Rainbow Project	AmphorA XHTML database	Shared upper-level ontology		RDF repository (Sesame)
h-TechSight	KM Portal	Ontology editor, dynamics metrics		Tagged HTML web server
WiCKOffice	Office application, support for form filling			Annotation server
AktivDoc	Integrated editing environment			RDF triple store
Semantic-Word	Microsoft Word GUIs		Markup tied to text regions	
Magpie	Web browser plug-in			
Thresher	Haystack semantic browser	Ontology personalization		

6. **Automation:** Automation is vital to ease the knowledge acquisition bottleneck, as discussed above. Many of the systems we examined had some kind of automatic and semi-automatic support for annotation. Most of these handled just text, using mainly wrappers, IE and natural language processing although there are some systems, notably M-OntoMat-Annotizer and parts of the Rainbow Project, looking to automate the handling of other media. Language technolo-

gies present usability challenges when deployed for knowledge workers since most are research tools or designed for use by specialists. A first step in addressing these challenges is Melita, where attention has been paid in finding ways to enable a seamless user interaction with the underlying IE system. In addition to the usability challenges there are also research challenges, among which extraction of relations is important for semantic annotation. A comparison of annotation tools for automation is presented in Table 7.2 below.

Table 7.2. A comparison of annotation tools based on automation support

Annotation Tool	Automation	Type of Analysis	Learning
Amaya	No		
Mangrove	No		
Vannotea	No		
OntoMat	Yes	PANKOW, Amilcare (IE)	Supervised learning
M-OntoMat Annotizer	Yes	Extraction of spatial description	Genetic algorithms
SHOE Knowledge Annotator	Yes	Running SHOE (wrappers)	No
SMORE	Yes	Screen scraper	No
Open Ontology Forge	Yes	String matching	No
COHSE Annotator	Yes	Ontology string matching	No
Lixto	Yes	Wrappers	No
MnM	Yes	POS tagging, named entity recognition	Supervised learning
Melita	Yes	String matching, POS tagging, named entity recognition	Supervised learning
Parmenides	Yes	Text mining with constraints	Unsupervised learning
Armadillo	Yes	String matching, POS tagging, named entity recognition	Unsupervised learning
KnowItAll	Yes	String matching, Hearst patterns	Unsupervised learning
SmartWeb	Yes	Shallow linguistic parsing	Unsupervised learning
PANKOW	Yes	Hearst patterns	Unsupervised learning
AeroSWARM	Yes	AeroText	No
SemTag	Yes	Seeker, similarity, TBD	Unsupervised learning
KIM	Yes	String matching, POS tagging. named entity recognition	No

Table 7.2. A comparison of annotation tools based on automation support

Annotation Tool	Automation	Type of Analysis	Learning
Rainbow Project	Yes	Hidden markov models, bit-map classification	Supervised learning
h-TechSight	Yes	Shallow linguistic analysis	No
WiCKOffice	Yes	Named entity recognition	No
AktivDoc	Yes	String matching, POS tagging, named entity recognition	Unsupervised and supervised learning
SemanticWord	Yes	AeroDAML	No
Magpie	Yes	String matching, named entity recognition	No
Thresher	Yes	Screen scraping, wrappers	Supervised learning

7.2 Techniques for Schema/Ontology Mapping

Schema and ontology matching is a critical problem in many application domains, such as Semantic Web, schema/ontology integration, data warehouses, and e-commerce. Many different matching solutions have been proposed so far. In the following we present a discussion of schema and ontology matching techniques based on classifications presented in [217] [218].

7.2.1 A Classification of Schema-matching Approaches

Schema-matching approaches can be classified as follows [217] [218]:

- **Elementary matchers:** These consist of instance-based and schema-based, element- and structure-level, linguistic- and constraint-based matching techniques.
- **Combination of matchers:** These consist of various ways of combining the schema matchers using committee-based or hybrid approaches.

Elementary schema-based matching techniques are classified based on two perspectives (Figure 7.1). These two perspectives are presented as two trees sharing their leaves. The leaves represent classes of elementary matching techniques and their concrete examples, identified as basic techniques in the figure. The two perspectives are discussed below.

Granularity/Input Interpretation: This is based on the granularity of the match, i.e., whether it is at the element or structural level, and how these techniques interpret this information. This perspective is illustrated from the top in a descending manner in Figure 7.1 till it reaches the Basic Techniques Layer. Elementary matchers are further distinguished based on the following criteria:

Element-level vs. structure-level. Element-level matching techniques compute mapping elements by analyzing entities in isolation, ignoring their relations with other entities. Structure-level techniques compute mapping elements by analyzing how entities appear together in a structure.

Syntactic vs. external vs. semantic. The key characteristic of the syntactic techniques is that they interpret the input as a function of its syntactic structure. External techniques exploit auxiliary (external) resources of a domain and common knowledge in order to interpret the input. These resources might be human input or some thesaurus expressing the relationships between terms. The key characteristic of the semantic techniques is that they use some formal semantics (e.g., model-theoretic semantics), possibly with some sort of reasoning to interpret the input and justify their results.

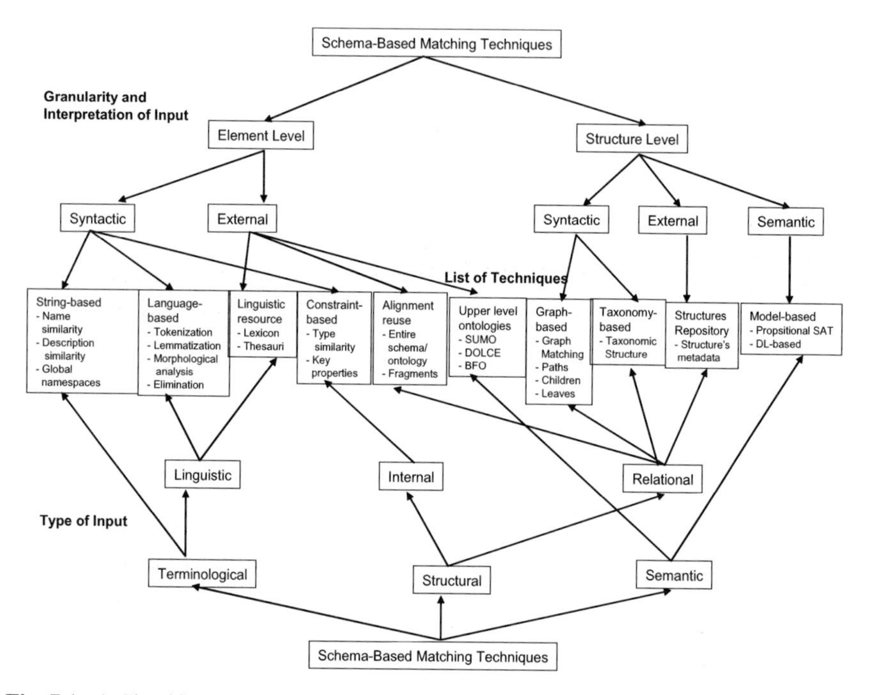

Fig. 7.1. A Classification of schema-based matching approaches

Type of Input. This is based on the type of input used by the elementary matching techniques. This perspective is illustrated from the bottom in an ascending manner in Figure 7.1 till it reaches the Basic Techniques Layer. Elementary matchers are further distinguished based on the following criteria:

- The first level is categorized depending on which kind of data the algorithms work on: string (*terminological*), structure (*structural*) or model (*semantics*). The two first ones are found in the ontology descriptions, the last one requires some semantic interpretation of the ontology and usually uses some semantically-compliant reasoner to deduce the correspondences.

- The second level of this classification decomposes further these categories if necessary: *terminological* methods can be *string-based* (considering the terms as sequences of characters) or based on the interpretation of these terms as linguistic objects (*linguistic*). The structural methods category is split into two types of methods: those which consider the *internal* structure of entities (e.g., attributes and their types) and those which consider the relation of entities with other entities (*relational*).

We discuss below the main classes of the Basic Techniques Layer and the associated matching systems according to the above classification in more detail. Techniques based on upper-level ontologies and DL-based techniques have not been implemented in any matching system yet. However, their use in matching systems seems quite likely in the near future.

Element-level techniques

String-based techniques consider strings as sequences of letters in an alphabet. They assume that the more similar the strings, the more likely they denote the same concepts. A comparison of different string-matching techniques, from distance-like functions to token-based distance functions can be found in [219]. Some examples of string-based techniques which are extensively used in matching systems are *prefix/suffix*, *edit distance*, and *n-gram*.

Prefix/Suffix. Two strings are input and a check of whether the first string starts/ends with the second one is performed. Prefix is efficient in matching cognate strings and similar acronyms (e.g., int and integer). This test can be transformed into a smoother distance by measuring the relative size of the prefix and the strings. These techniques have been used in [225] [231] [232] [233].

Edit distance. This distance takes as input two strings and computes the edit distance between the strings, that is, the number of insertions, deletions, and substitutions of characters required to transform one string into another, normalized by the length of the longest string.

N-gram. This test takes as input two strings and computes the number of common n-grams (i.e., sequences of n characters) between them. These techniques have been used in [225] [231] [234].

Language-based techniques consider names as words in some natural language (e.g., English) and apply Natural Language Processing (NLP) techniques that exploit morphological properties of the input words.

Tokenization. Names of entities are parsed into sequences of tokens by a tokenizer which recognizes punctuation, cases, blank characters, digits, etc. (e.g., see [230]).

Lemmatization. The strings, underlying tokens are morphologically analyzed in order to find all their possible basic forms (e.g., see [230]).

Elimination. The tokens that are articles, prepositions, conjunctions, and so on, are marked to be discarded (e.g., see [232]).

Usually, the above-mentioned techniques are applied to names of entities before running string-based or lexicon-based techniques in order to improve their results. However, language-based techniques may be considered as a separate class of matching techniques, since they can be naturally extended, for example, in a distance computation (by comparing the resulting strings or sets of strings).

Constraint-based techniques are algorithms which deal with the internal constraints being applied to the definitions of entities, such as types, cardinality of attributes, and keys.

Datatype comparison involves comparing the various attributes of a class with regard to the datatypes of their value. Contrary to objects that require interpretation, the datatypes can be considered objectively and it is possible to determine how a datatype is close to another (ideally this can be based on the interpretation of datatypes as sets of values and the set-theoretic comparison of these datatypes). For instance, the datatype `day` can be considered closer to the datatype `workingday` than the datatype `integer`. This technique is used in [228].

Multiplicity comparison attribute values can be collected by a particular construction (set, list, multiset) on which cardinality constraints are applied. It is possible to compare the so constructed datatypes by comparing (i) the datatypes on which they are constructed and (ii) the cardinality constraints that are applied to them. For instance, a set of between two and three children is closer to a set of three people than a set of ten to twelve flowers (if children are people). This technique is used in [228].

Linguistic resources such as common knowledge or domain-specific thesauri are used to match words (in this case names of schema/ontology entities are considered as words of a natural language) based on linguistic relations between them (e.g., synonyms, hyponyms).

Common knowledge thesauri are used to obtain the meaning of terms used in schemas/ontologies. For example, WordNet [237] is an electronic lexical database for English (and other languages), where various senses (possible meanings) of words or expressions are put together into sets of synonyms. Relations between schema/ontology entities can be computed in terms of bindings between WordNet senses; see, for instance [221] [230]. Other matchers exploit thesauri based on their structural properties, e.g., WordNet hierarchies. In particular, hierarchy-based matchers measure the distance, for example, by counting the number of arcs traversed, between two concepts in a given hierarchy.

Domain-specific thesauri usually store some specific domain knowledge, which is not available in common knowledge thesauri (e.g., proper names) as entries with synonym, hypernym and other relations; see, for instance [232].

Alignment reuse techniques exploit alignments of previously matched schemas and ontologies, for instance, when we need to match schema/ontology o and o'', given the alignments between o and o', and between o' and o'' from the external resource, storing previous match operation results. The alignment reuse is motivated by the intuition that many schemas/ontologies to be matched are similar to already-matched schemas/ontologies, especially if they are describing the same application domain. These techniques are particularly promising when dealing with large schemas/ontologies consisting of hundreds and thousands of entities. In these cases, first, large match problems are decomposed into smaller sub-problems, thus generating a set of schema/ontology fragment-matching problems. Then, reusing previous match results can be more effectively applied at the level of schema/ontology fragments compared to entire schemas/ontologies. The approach was first introduced in [217], and later was implemented as two matchers, i.e., reuse of (i) entire schemas/ontologies alignments, or (ii) their fragments; see, for details [220] [225] [235].

Upper-level formal ontologies can be also used as external sources of common knowledge. Examples are the Suggested Upper Merged Ontology (SUMO) [49] and Descriptive Ontology for Linguistic and Cognitive Engineering (DOLCE) [92]. The key characteristic of these ontologies is that they are logic-based systems, and therefore, matching techniques exploiting them can be based on the analysis of interpretations. Even though current matching systems do not use these techniques, it is likely that this will happen in the near future. In fact, the DOLCE ontology aims at providing a formal specification (axiomatic theory) for the top-level part of WordNet. Therefore, systems exploiting WordNet now in their matching process might also consider using DOLCE as a potential extension.

Structure-level techniques

Graph-based techniques view database schemas, taxonomies and ontologies as graph-like structures containing terms and their interrelationships. Usually, the similarity comparison between a pair of nodes from the two schemas/ontologies is based on the analysis of their positions within the graphs. The intuition behind this is that if two nodes from two schemas/ontologies are similar, their neighbors might also be somehow similar.

Graph matching. Matching graphs is a combinatorial problem and is usually solved by approximate methods. In schema/ontology matching, the problem is encoded as an optimization problem (finding the graph matching minimizing some distance like the dissimilarity between matched objects) which is further resolved with the help of a graph-matching algorithm. This optimization problem is solved through a fix-point algorithm (improving gradually an approximate solution until no improvement is made). Examples of such algorithms are [233] and [228].

Children. The (structural) similarity between inner nodes of the graphs is computed based on similarity of their children nodes, that is, two non-leaf schema elements are structurally similar if their immediate children sets are highly similar. A more complex version of this matcher is implemented in [225].

Leaves. The (structural) similarity between inner nodes of the graphs is computed based on similarity of leaf nodes, that is, two non-leaf schema elements are structurally similar if their leaf sets are highly similar, even if their immediate children are not; see, for example [225] [232].

Relations. The similarity computation between nodes can also be based on their relations. For example, if class `Photo` and `Camera` relates to class `NKN` by relation `hasBrand` in one ontology, and if class `DigitalCamera` relates to class `Nikon` by relation `hasMarque` in the other ontology, then knowing that classes `Photo` and `Camera` and `DigitalCamera` are similar, and also relations `hasBrand` and `hasMarque` are similar, we can infer that `NKN` and `Nikon` may be similar.

Taxonomy-based techniques consider only the specialization relation. The intuition behind taxonomic techniques is that is-a links connect terms that are already similar (each being a subset of the other); therefore their neighbors may be also somehow similar.

Bounded path matching. Bounded path matchers take two paths with links between classes defined by the hierarchical relations, compare terms and their positions along these paths, and identify similar terms; see, for instance [234].

Super(sub)-concept rules. These matchers are based on rules capturing the above stated intuition. For example, if super-concepts are the same, the actual concepts are similar to each other. If sub-concepts are the same, the compared concepts are also similar; see, for example [224] [226].

Repository of structures stores schemas/ontologies and their fragments together with pairwise similarities (e.g., coefficients in the [0,1] range) between them. When new structures are to be matched, they are first checked for similarity to the structures which are already available in the repository. The goal is to identify structures which are sufficiently similar to be worth matching in more detail, or to reuse already existing alignments. Obviously, the determination of similarity between structures should be computationally cheaper than matching them in full detail. In order to match two structures, [235] proposes using some metadata describing these structures, such as structure name, root name, number of nodes, maximal path length, etc. Then, these indicators are analyzed and are aggregated into a single coefficient, which estimates the similarity between them.

Model-based algorithms handle the input based on its semantic interpretation (e.g., model-theoretic semantics). Examples are propositional satisfiability (SAT) and description logics (DL) reasoning techniques. As from [221] [229] [230], the approach is to decompose the graph(tree)-matching problem into a set of node-matching problems. Then, each node-matching problem, namely each pair of nodes with possible relations between them, is translated into a propositional formula of form, $Axioms \Rightarrow rel(context_1, context_2)$, and checked for validity. Axioms encode background knowledge (e.g., `HypertrophicCardioMyopathy subClassOf Disease` codifies the fact that Hypertrophic Cardiomyopathy is a kind of disease), which is used as premises to reason about relations rel (e.g., `=`, `subClassOf`, `unsatisfiability`) holding between the nodes $context_1$ and $context_2$. A propositional formula is valid iff its negation is unsatisfiable. The unsatisfiability is checked by using state-of-the-art SAT solvers. Propositional language used for codifying matching problems into propositional unsatisfiability problems is limited in its expressiveness; namely it allows for handling only unary predicates. Thus, it cannot handle, for example, binary predicates, such as properties or roles, which are expressible in OWL and various variants of DLs. The relations (e.g., `=`, `subClassOf`, `unsatisfiability`) can be expressed using subsumption in DLs. In fact, first merging two ontologies (after renaming) and then testing each pair of concepts and roles for subsumption is enough for aligning terms with the same interpretation (or with a subset of the interpretations of the others). Currently, there are no systems supporting DL-based techniques.

7.2.2 Schema-matching Techniques: Overview

We now look at some recent schema-based state-of-the-art matching systems in the context of the classification presented in Figure 7.1. A summary of the various characteristics of these techniques is presented in Table 7.3.

Similarity Flooding. The Similarity Flooding (SF) [233] approach utilizes a hybrid-matching algorithm based on the ideas of similarity propagation. Schemas are presented as directed labeled graphs; the algorithm manipulates them in an iterative fix-point computation to produce an alignment between the nodes of the input graphs. The technique starts with string-based comparison (common prefix and suffix tests) of the vertex labels to obtain an initial alignment which is refined within the fix-point computation. The basic concept behind the SF algorithm is the similarity spreading from similar nodes to the adjacent neighbors through propagation coefficients. From iteration to iteration the spreading depth and the similarity measure increase till the fix-point is reached. The result of this step is a refined alignment which is further filtered to finalize the matching process. SF considers the alignment as a solution to a clearly stated optimization problem.

Artemis. Analysis of Requirements: Tool Environment for Multiple Information Systems (Artemis) [222] was designed as a module of the MOMIS mediator system [238] for creating global views. It performs affinity-based analysis and hierarchical clustering of source schema elements. Affinity-based analysis represents the matching step: in a hybrid manner it calculates the name, structural and global affinity coefficients exploiting a common thesaurus. The common thesaurus is built with the help of Ontology Development Tools, WordNet or manual input. It represents a set of intensional and extensional relationships which depict intra- and inter-schema knowledge about classes and attributes of the input schemas. Based on global affinity coefficients, a hierarchical clustering technique categorizes classes into groups at different levels of affinity. For each cluster it creates a set of global attributes and the global class. The logical correspondence between the attributes of a global class and source schema attributes is determined through a mapping table.

Cupid. Cupid [232] implements a hybrid-matching algorithm comprising linguistic and structural schema-matching techniques, and computes similarity coefficients with the assistance of a domain-specific thesaurus. Input schemas are encoded as graphs. Nodes represent schema elements and are traversed in a combined bottom-up and top-down manner. The matching algorithm consists of three phases and operates only with tree structures to which non-tree cases are reduced.

The first phase (linguistic matching) computes linguistic similarity coefficients between schema element names (labels) based on morphological normalization, categorization, string-based techniques (common prefix, suffix tests) and a thesauri lookup. The second phase (structural matching) computes structural similarity coefficients weighted by leaves which measure the similarity between contexts in which elementary schema elements occur. The third phase (mapping elements generation) computes weighted similarity coefficients and generates final alignment by choosing pairs of schema elements with weighted similarity coefficients which are higher than a threshold.

COMA. COmbination ofMAtching algorithms (COMA) [225] is a composite schema-matching tool. It provides an extensible library of matching algorithms, a framework for combining obtained results, and a platform for the evaluation of the effectiveness of the different matchers. Matching library is extensible, and contains six elementary matchers, five hybrid matchers, and one reuse-oriented matcher. Most of the matchers implement string-based techniques (affix, n-gram, edit distance, etc.) as a background idea; others share techniques with Cupid (thesauri look-up, etc.); and the reuse-oriented matcher tries to reuse previously obtained results for entire new schemas or for its fragments. Schemas are internally encoded as DAGs, where the elements are the paths. This aims at capturing contexts in which the elements occur. Distinct features of the COMA tool with respect to Cupid are a more flexible architecture and a possibility of performing iterations in the matching process.

NOM. Naive Ontology Mapping (NOM) [227] adopts the idea of composite matching from COMA [225]. Some other innovations with respect to COMA are in the set of elementary matchers based on rules exploiting explicitly codified knowledge in ontologies, such as information about super- and sub-concepts and super- and sub-properties. At present the system supports 17 rules. For example, one rule states that if super-concepts are the same, the actual concepts are similar to each other. NOM also exploits a set of instance-based techniques.

QOM. Quick Ontology Mapping (QOM) [226] is a successor of the NOM system [227]. The approach is based on the idea that the loss of quality in matching algorithms is marginal (to a standard baseline); however, improvement in efficiency can be tremendous. This fact allows QOM to produce mapping elements fast, even for large-size ontologies. QOM is grounded in matching rules of NOM. However, for the purpose of efficiency the use of some rules has been restricted. QOM avoids the complete pairwise comparison of trees in favor of an incomplete top-down strategy. Experimental study has shown that QOM is on par with other state-of-the-art algorithms for the quality of the proposed alignment, while outperforming them with respect to efficiency. Also, QOM shows better results than approaches within the same complexity class.

OLA. OWL Lite Aligner (OLA) [228] is designed with the idea of balancing the contribution of each component that composes an ontology (these include classes, properties, names, constraints, taxonomy, and even instances). As such it takes advantage of all the elementary matching techniques that have been considered in the previous sections except the semantic ones. OLA is a family of distance-based algorithms which converts definitions of distances based on all the input structures into a set of equations. These distances are almost linearly aggregated (they are linearly aggregated modulo local matches of entities). The algorithm then looks for the matching between the ontologies that minimizes the overall distance between them. For that purpose it starts with base distance measures computed from labels and concrete datatypes. Then, it iterates a fix-point algorithm until no improvement is produced. From that solution, an alignment is generated which satisfies some additional criterion (on the alignment obtained and the distance between aligned entities). As a system, OLA considers the alignment as a solution to a clearly stated optimization problem.

Anchor-PROMPT. Anchor-PROMPT [234] (an extension of PROMPT) is an ontology-merging and alignment tool with a sophisticated prompt mechanism for possible matching terms. The anchor-PROMPT is a hybrid alignment algorithm which takes as input two ontologies (internally represented as graphs) and a set of anchor-pairs of related terms, which are identified with the help of string-based techniques (edit-distance test) or defined by a user, or another matcher computing linguistic similarity. Then the algorithm refines them by analyzing the paths of the input ontologies limited by the anchors in order to determine terms frequently appearing in similar positions on similar paths. Finally, based on the frequencies and user feedback, the algorithm determines matching candidates.

S-Match. S-Match [229] [230] [231] is a schema-based matching system. It takes two graph-like structures (e.g., XML schemas or ontologies) and returns semantic relations (e.g., equivalence, subsumption) between the nodes of the graphs that correspond semantically to each other. The relations are determined by analyzing the meaning (concepts, not labels) which is codified in the elements and the structures of schemas/ontologies. In particular, labels at nodes, written in natural language, are translated into propositional formulas which explicitly codify the label's intended meaning. This allows for a translation of the matching problem into a propositional unsatisfiability problem, which can then be efficiently resolved using (sound and complete) state-of-the-art propositional satisfiability deciders. S-Match was designed and developed as a platform for semantic matching, namely, as a highly modular system with a core of semantic relationship computations, where single components can be plugged, unplugged or suitably customized. It is a hybrid system with a composition at the element level. At present, S-Match libraries contains thirteen element-level matchers and three structure-level matchers.

Table 7.3. Summary of schema-matching approaches

	Element Level Matching		**Structure Level Matching**	
	Syntactic	**External**	**Syntactic**	**Semantic**
SF	string-based, datatypes, key properties		iterative fix-point computation	
Artemis	domain compatibility; language based	common thesaurus, broader term, related term	matching of neighbors via clustering	
Cupid	string-based, language-based, datatypes, key properties	auxiliary thesauri, synonyms, hypernyms, abbreviations	tree matching weighted by leaves	
COMA	string-based, language-based, datatypes	auxiliary thesauri, synonyms, hypernyms, abbreviations alignment reuse	DAG matching with bias toward children of leaf nodes; paths	
NOM/ QOM	string-based, domains and ranges	application specific vocabulary	neighbor matching, taxonomic structure	
Anchor-PROMPT	string-based, domain and ranges		bounded path matching	
OLA	string-based, language based, datatypes	WordNet	iterative fix-point computation, neighbor matching, taxonomic structure	
S-Match	string-based, language based	WordNet, sense-based, gloss-based		propositional SAT, DLs

7.3 Ontology Driven Information Integration

In order to achieve semantic interoperability in a heterogeneous information system, the meaning of the information that is interchanged has to be understood across the systems. Semantic conflicts occur whenever two contexts do not use the same interpretation of the information. The use of ontologies and metadata descriptions for the explication of implicit and hidden knowledge is a possible approach to overcome the problem of semantic heterogeneity. Uschold and Gruninger mention interoperability as a key application of ontologies, and many ontology-based approaches [239] to information integration in order to achieve interoperability have been developed.

In this section we discuss existing solutions for ontology-based information integration presented in [240]. Various approaches to intelligent information integration have been adopted in systems such as SIMS [241], TSIMMIS [242], OBSERVER [147], Carnot [21], InfoSleuth [243], KRAFT [244], PICSEL [245], DWQ [246], Ontobroker [247], SHOE [248], Crossvision Enterprise Information Integrator by Software AG [413] and others. Most of these systems use some notion of ontologies for integration across information resources. An evaluation of these approaches is presented based on the following criteria:

Use of Ontologies: The role and the architecture of ontologies heavily influence their representation formalism.

Ontology Representation: Depending on the use of the ontology, the representation capabilities differ from approach to approach.

Use of Mappings: In order to support the integration process the ontologies have to be linked to the underlying schemas used to store the data. If several ontologies are used in an integration system, inter-ontology mappings between classes in different ontoologies is also important.

Ontology Engineering: Before an integration of information sources can begin the appropriate ontologies have to be acquired or be selected for reuse. How does the integration approach support the acquisition or reuse of ontologies?

We begin with a discussion on various ontology-based architectures and the role of plau. This is followed by a discussion of the use of different representations, i.e., different ontology languages for information integration. Mappings used to connect ontologies to information sources, inter-ontology mappings, and associated methodologies for ontology engineering for information integration are also discussed.

7.3.1 The Role of Ontologies in Information Integration

Ontologies can be used in an integration task to explicitly describe the semantics of data an information stored in the underlying information sources. This can be achieved by identification of corresponding concepts from ontologies. The uses and roles played by ontologies in information integration is discussed next.

Explicit Semantic Descriptions

Different approaches for using ontologies for information integration can be characterized as: *single ontology approaches*, *multiple ontology approaches* and *hybrid ontology approaches*, and are illustrated in Figure 7.2. Some approaches provide a general framework where all three architectures can be implemented (e.g., DWQ [246]). A discussion of the three main architectures is as follows.

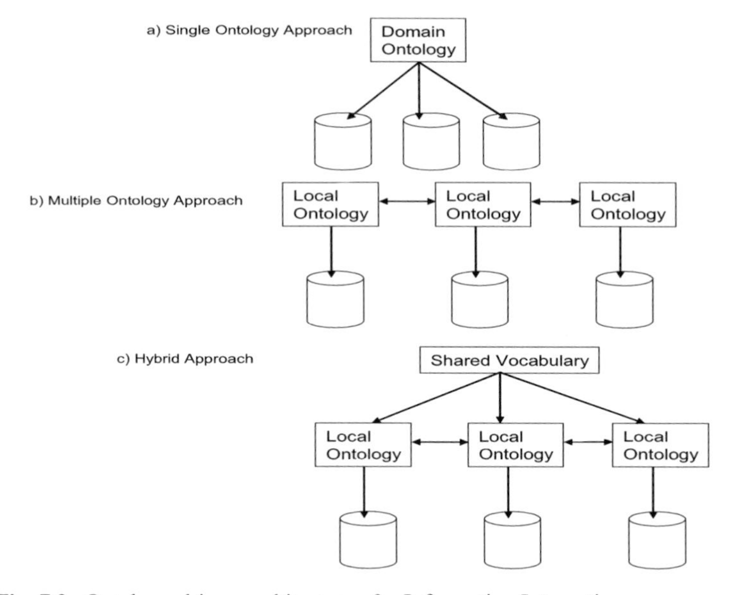

Fig. 7.2. Ontology-driven architectures for Information Integration

Single Ontology Approaches. Single ontology approaches use a domain ontology providing a shared vocabulary for the specification of the semantics. All information sources are related to one global ontology. We have adopted the single ontology approach in the solution design presented for the clinical use case and scenario. A prominent approach of this kind of ontology integration is SIMS [241]. The SIMS model of the application domain includes a hierarchical terminological knowledge base with nodes representing objects, actions, and states. An independent model of each information source is described for this system by relating the objects of each source to the global domain model. The relationships clarify the semantics of the source objects and help to find semantically corresponding objects.

Single ontology approaches can be applied to integration problems where all information sources to be integrated provide nearly the same view on a domain. But if one information source has a different view on a domain, e.g., by providing another level of granularity, finding the minimal ontology commitment [89] becomes a difficult task. For example, if two information sources provide product specifications but refer to absolute heterogeneous product catalogs which categorize the products, the development of a global ontology which combines the different product catalogs becomes very difficult. Information sources with reference to similar product catalogs are much easier to integrate. Also, single ontology approaches are susceptible to changes in the information sources which can affect the conceptualization of the domain represented in the ontology. Depending on the nature of the changes in one information source it can imply changes in the global ontology and in the mappings to the other information sources. These disadvantages led to the development of multiple ontology approaches.

The domain ontology can also be a combination of several specialized ontologies. A reason for the combination of several ontologies can be the modularization of a potentially large monolithic ontology. The combination is supported by ontology representation formalisms, i.e., by importing other ontology modules (e.g., Ontolingua [89]).

Multiple Ontology Approaches. In multiple ontology approaches, each information source is described by its own domain- or application-specific ontology. For example, in OBSERVER [147], the semantics of an information source is described by a domain-specific ontology. In principle, the domain ontology can be a combination of several other ontologies but it cannot be assumed that the different domain ontologies share the same vocabulary. The Crossvision Information Integrator supports a multiple ontology approach. It relies on information models which are organized using ontologies, which are managed in a metadata repository (CentraSite). A semantic inference engine (Semantic Server) then allows the raw data to be aggregated dynamically, perfectly tailored for the individual business user's needs.

At a first glance, the advantage of multiple ontology approaches seems to be that no common and minimal ontology commitment around an ontology is needed. Each ontology could be developed without respect to other information sources or domain ontologies — no common ontology with the agreement of all information sources/ontologies are needed. This ontology architecture can simplify the change, i.e., modifications in one information source/ontology or the adding and removing of information sources/ontologies. But in reality the lack of a common vocabulary makes it extremely difficult to compare different source ontologies. To overcome this problem, an additional representation formalism defining the inter-ontology mapping is required. The inter-ontology mapping identifies semantically corresponding terms of different ontologies, terms which are semantically equal or similar. But the mapping also has to consider different views on a domain, i.e., different granularities of the ontology concepts. Issues of semantic heterogeneity may also occur in defining inter-ontology mappings.

Hybrid Ontology Approaches. To overcome the drawbacks of the single or multiple ontology approaches, hybrid approaches were developed. Similar to multiple ontology approaches the semantics of each source is described by its appropriate domain ontology. But in order to make the source ontologies comparable to each other they are built upon one global shared vocabulary [249] [250]. The shared vocabulary contains basic terms (the primitives) of a domain. In order to build complex terms of ontologies the primitives are combined by some operators. Because each term of an ontology is based on the primitives, the terms can be easily mapped to each other, than in multiple ontology approaches. Sometimes the shared vocabulary is also an ontology.

In the COIN system [249], the local description of a piece of information, the so-called context, is simply an attribute value vector. The terms for the context stem from the common shared vocabulary and the data itself. In the MECOTA system [250] each piece of source information is annotated by a label which indicates the semantics of the information. The label combines primitive terms from the shared vocabulary. The combination operators are similar to the operators known from the description logics, but are extended for the special requirements resulting from integration of sources, e.g., by an operator for aggregation. In the BUSTER system [251], the shared vocabulary is a (general) ontology, which covers all possible refinements. For example, the general ontology defines the attribute value ranges of its concepts. A domain ontology is one (partial) refinement of the general ontology, e.g., restricting the value range of some attributes. Since domain ontologies only use the general ontology, they remain comparable.

The advantage of a hybrid approach is that new sources can easily be added without the need of modification in the mappings or in the shared vocabulary. It also supports the acquisition and evolution of ontologies. The use of a shared vocabulary makes the source ontologies comparable and avoids the disadvantages of multiple ontology approaches. The drawback of hybrid approaches, however, is that existing ontologies cannot be reused easily, but have to be redeveloped from scratch, because all domain ontologies have to refer to the shared vocabulary.

Ontologies as a Query Model

Integrated information sources normally provide an integrated view. Some integration approaches use the ontology as the query schema, e.g., the SIMS system [241]. The user formulates a query in terms of the ontology. The system reformulates the query into subqueries for each appropriate source, collects and combines the query results, and returns the results. Using an ontology as a query model has the advantage that the structure of the query model should be more intuitive for the user because it corresponds more to the user's appreciation of the domain. However, the user has to know the structure and the contents of the ontology.

Ontologies as Verification Mechanism

During the integration process several mappings must be specified from a domain ontology to the local source schema. The correctness of such mappings can be considerably improved if these can be verified automatically. A subquery is correct with respect to a query if the local subquery provides a part of the queried answers, i.e., the subqueries must be contained in the global query (query containment) [246][245]. Since an ontology contains a (complete) specification of the conceptualization, the mappings can be validated with respect to these ontologies. Query containment means that the ontology concepts corresponding to the local subqueries are contained in the ontology concepts related to the query.

In the DWQ system [246], each source is assumed to be a collection of relational tables. Each table is described in terms of its ontology with the help of conjunctive queries. A query and the decomposed subqueries can be unfolded to their ontology concepts. The subqueries are correct, i.e., are contained in the query, if their ontology concepts are subsumed by the ontology concepts. The PICSEL project [245] can also verify the mapping, but in contrast to DWQ it can also generate mapping hypotheses automatically which are validated with respect to a global ontology.

The quality of the verification strongly depends on the completeness of an ontology. If the ontology is incomplete, the verification result can erroneously imply a correct query subsumption. Since in general the completeness can not be measured, it is impossible to make any statements about the quality of the verification.

7.3.2 Ontology Representations Used in Information Integration

Various approaches to intelligent information integration based on ontologies have predominantly used variants of description logics in order to represent ontologies. The CLASSIC system [165] has been used in the OBSERVER system [148] and by database researchers investigating semantic heterogeneities and interoperability [252]. The SIMS system makes use of the LOOM description logic [253]. Other terminological languages used are GRAIL [254], used in the TAMBIS system [46], and OIL [256], which is used for terminology integration in the BUSTER system [255].

Besides the purely terminological languages mentioned above there are also approaches using extensions of description logics which include rule bases. Some examples are the use of CARIN [167], a description logic extended with function-free horn rules in the PICSEL system [245]. The DWQ project [246] uses AL-log [257], which combines simple description logics with Datalog and the logic DLR, a description logic with n-ary relations. The integration of description logics with rule-based reasoning makes it necessary to restrict the expressive power of the terminological part of the language in order to maintain decidability.

The second main group of languages used in ontology-based information integration systems are classical frame-based representation languages. Examples for such systems are COIN [249], KRAFT [244] and InfoSleuth [243]. There are also approaches that directly use F-Logic [82] with a self-defined syntax (Ontobroker [247] and COIN [249]).

7.3.3 The Role of Mapping in Information Integration

The task of integrating heterogeneous information sources provides a use case for ontologies. Ontologies may be viewed as the glue that puts together information of various kinds. Mappings refer to the connection of an ontology to other parts of the system. Mapping are a critical requirement for information integration for (a) connecting ontologies with the information source they describe; and (b) connecting different ontologies used in a system. In Section 7.2, we discussed a representative set of techniques to identify and discover mappings between two schema or ontology like artifacts. In this section, we discuss how these mappings, once generated can be used in the context of Information Integration.

Mapping Ontologies to Information Resources

Different approaches used to establish a connection between ontologies and information sources are as follows.

Structure Resemblance. A straightforward approach to connecting an ontology with the database schema is to simply produce a one-to-one copy of the structure of the database and encode it in a language that makes automated reasoning possible. The integration is then performed on the copy of the model and can easily be tracked back to the original data. This approach is implemented in the SIMS mediator [258] and also by the TSIMMIS system [242].

Definition of Terms. In order to make the semantics of terms in a database schema clear it is not sufficient to produce a copy of the schema. There are approaches such as those used in the BUSTER system [255] that use the ontology to further define terms from the database or the database schema. These definitions can consist of a set of rules defining the term and are, in most cases, described by concept definitions.

Structure Enrichment. This is the most common approach for relating ontologies to information sources, and combines the two previous approaches. A logical model is built that resembles the structure of the information source and contains additional definitions of concepts. A detailed discussion of structure is presented in [252], which is used in OBSERVER [147], KRAFT [244], PICSEL [245] and DWQ [246]. While OBSERVER uses description logics for both structure resemblance and additional definitions, PICSEL and DWQ define the structure of the information by (typed) horn rules. Additional definitions of concepts mentioned in these rules are given by a description logic model. KRAFT does not commit to a specific definition scheme.

Meta-annotation. An interesting approach is the use of meta-annotations that add semantic information to an information source. This approach is particularly relevant in the context of the integrating information on the Web, where annotation may be viewed as a natural way of adding semantics. Ontology-based integration approaches developed for the Web context are the Ontobroker [247] and SHOE [248] systems.

Inter-ontology Mapping

Some information integration systems such as [148] [244] use more than one ontology to describe the information. The problem of mapping different ontologies is a well-known problem in knowledge engineering. We now discuss approaches that are used in the context of information integration systems.

Defined Mappings. In the KRAFT System [244], translations between different ontologies are done by special mediator agents which can be customized to translate between different ontologies and even different languages. Different kinds of mappings are distinguished in this approach starting from simple one-to-one mappings between classes and values to mappings between compound expressions. This approach allows great flexibility, but it fails to ensure a preservation of semantics: the user is free to define arbitrary mappings even if they do not make sense or produce conflicts.

Lexical Relations. An attempt to provide at least intuitive semantics for mappings between concepts in different ontologies is made in the OBSERVER system [148]. The approaches extend a common description logic model by quantified inter-ontology relationships borrowed from linguistics. The relationships used are synonym, hypernym, hyponym, overlap, covering and disjoint. While these relations are similar to constructs used in description logics they do not have a formal semantics. Consequently, the query translation algorithm is probabilistic in nature.

Top-Level Grounding. In order to avoid a loss of semantics, one has to stay inside the formal representation language when defining mappings between different ontologies (e.g., DWQ [Calvanese et al., 2001]). A straightforward way to achieve this is to relate all ontologies used to a single top-level ontology. This can be done by inheriting concepts from a common top-level ontology. This approach can be used to resolve conflicts and ambiguities. While this approach enables establishment of connections between concepts from different ontologies in terms of common superclasses, it does not establish a direct correspondence. This might lead to problems when exact matches are required.

Semantic Correspondences. An approach that tries to overcome the ambiguity that arises the previous approach, is to identify well-founded semantic correspondences between concepts from different ontologies. In order to avoid arbitrary mappings between concepts, these approaches have to rely on a common vocabulary for defining concepts across different ontologies. One approach uses semantic labels in order to compute correspondences between database fields. Another approach is to represent concepts from different ontologies in a description logic model of terms and use subsumption reasoning to establish relations between different terminologies. Approaches using formal concept analysis also fall into this category, because they define concepts on the basis of a common vocabulary to compute a common concept lattice.

7.3.4 The Role of Ontology Engineering in Information Integration

Since ontologies play a crucial role semantic information integration, it is crucial to support the ontology engineering process, especially that part which is likely to have an impact on the information integration process.

Ontology Development Methodologies

Example information integration systems and their approaches for developing ontologies are discussed as follows.

InfoSleuth. Ontologies in InfoSleuth are defined primarily manually using Entity-Relationship (E-R) models. Approaches for semi-automatic construction of ontologies from textual databases have been proposed in [259]. The methodology is as follows: first, human experts provide a small number of seed words to represent high-level concepts. The system then processes the incoming documents, extracting phrases that involve seed words, generates corresponding concept terms, and classifies them into the ontology. During this process the system also collects seed word candidates for the next round of processing. This iteration can be completed for a predefined number of rounds. A human expert verifies the classification after each round. As more documents arrive, the ontology expands and the expert is confronted with the new concepts. This is a significant feature of this system, called the "discover and alert" feature.

KRAFT. Ontologies in KRAFT are built based on two methods: manual construction of shared ontologies and extraction of domain or information source ontologies. KRAFT offers two methods for building ontologies:

- The steps of the development of shared ontologies are (a) ontology scoping, (b) domain analysis, (c) ontology formalization and (d) top-level ontology. The minimal scope is a set of terms that is necessary to support the communication within the KRAFT network. The domain analysis is based on the idea that changes within ontologies are inevitable and the means to handle changes should be provided. The authors pursue a domain-led strategy, where the shared ontology fully characterizes the area of knowledge in which the problem is situated. Within the ontology formalization phase the fully characterized knowledge is defined formally in classes, relations and functions. The top-level ontology is needed to introduce predefined terms/primitives.
- A bottom-up approach to extract an ontology from existing shared ontologies was introduced in [260]. The first step is a syntactic translation from the KRAFT exportable view (in a native language) of the resource into the KRAFT schema. The second step is the ontological upgrade, a semi-automatic translation plus knowledge-based enhancement, where the local ontology adds knowledge and further relationships between the entities in the translated schema.

Ontobroker. There are three classes of Web information sources [261]: (a) Multiple-instance sources with the same structure but different contents, (b) single-instance sources with large amount of data in a structured format, and (c) loosely structured pages with little or no structure. Ontobroker uses two ways of formalizing knowledge. First, sources from (a) and (b) allow it to implement wrappers that automatically extract factual knowledge from these sources. Second, sources with little or no knowledge have to be formalized manually.

SIMS. An independent model of each information source is described in the SIMS system, along with a domain model that must be defined to describe objects and actions. The SIMS model of the application domain includes a hierarchical terminological knowledge base with nodes representing objects, actions, and states. In addition, it includes indications of all relationships between the nodes. Scalability and maintenance problems on addition of a new information source or change in domain knowledge are addressed. As every information source is independent and modeled separately, the addition of a new source is relatively straightforward. A graphical LOOM knowledge base builder (LOOM-KB) is used to support this process. The domain model is enlarged to accommodate new information sources or new knowledge.

Tools for the Annotation Process

Some of the systems discussed in this chapter provide support with the annotation process of information sources, leading to a semantic enrichment of the information. Some tools used in the process are OntoStudio (previously known as Ontoedit, discussed in Chapter 6.3), the SHOE Knowledge Annotator and the I-COM tool used in the DWQ project. With the help of the SHOE Knowledge Annotator tool, the user can describe the contents of a Web page [262]. The Knowledge Annotator has an interface which displays instances, ontologies, and claims (documents collected). The tool also provides integrity checks. With a second tool called Expose the annotated Web pages are parsed and the contents stored in a repository. The I-COM tool [111] was developed within the DWQ project. This tool uses an extended entity-relationship (EER) conceptual data model and enriches it with aggregations and inter-schema constraints.

Ontology Evolution

Support for ontology evolution is a critical piece of functionality in the context of an information integration system. An integration system and the ontologies must support adding and/or removing sources and must be robust to changes in the information source. The SHOE system is one system that takes these issues into account.

Once the SHOE-annotated Web pages are uploaded on the Web, the Expose tool has the task to update the repositories with the knowledge from these pages. This includes a list of pages to be visited and an identification of all hypertext links, category instances, and relation arguments within the page. The tool then stores the new information in the PARKA knowledge base. The problems associated with managing dynamic ontologies through the Web have been presented in [248]. By adding revision marks to the ontology, changes and revision become possible. The authors illustrated that revisions which add categories and relations will have no effect, and that revisions which modify rules may change the answers to queries. When categories and relations are removed, answers to queries may be eliminated.

7.4 Summary

In this chapter we presented applications of metadata and ontologies, such as semantic annotations, mappings and information integration. These applications are enabled by semantic descriptions of data and resources on the web-based and other repositories; and themselves enable new functionality on the web and on internal organizations' intranets. We presented tools and techniques for annotation of Web resources with semantic metadata annotations. Two types of metadata data annotations are considered: (a) structured and semi-structured metadata annotations of unstructured Web content; and (b) structured metadata annotations of structured Web content. It was noted that the latter corresponds to mapping the schemas underlying the structured content to domain-specific ontologies, and a discussion and taxonomy of schema-matching techniques was also presented. Finally, we presented various approaches adopted for ontology driven information integration, including a discussion on various types of architectures, the role played by ontologies in the creation of mappings, specifying queries and as a verification mechanisms.

Part III
Process Aspects of the Semantic Web

8 Communication

The Semantic Web would be impossible without the advent of simple and efficient communication networks that allow any user connected to the Web to access any public Semantic Web site without effort, very efficiently and extremely fast (most of the time). The basis for this ease of access is a very simple data format for specifying Semantic Web pages and a very simple communication protocol for their access. Both can be easily implemented on any computing platform. This ease of implementation ensures that everybody can participate independent of their particular computing equipment.

Complementing the Semantic Web, machine-to-machine communication (in contrast to serving up content for human consumption) is addressed by Semantic Web Services. Semantic Web Services are the mechanism for software-to-software communication and coordination (whereas the Semantic Web is for human users). Semantic Web Services are not a disruptive new paradigm, instead, they leverage existing communication knowledge, conventions and technologies and improve on them.

This Section builds the fundamental basis for Semantic Web Services. It discusses the concepts of communication from a principled perspective in Section 8.1. Based on these fundamental concepts, major communication paradigms are listed in Section 8.2. Long-running communication in the context of B2B integration and EAI integration is reviewed in Section 8.3. In Section 8.4 a particular type of communication, Web Services, is emphasized as the focus of the following Chapters. Section 8.6 summarizes this Chapter.

8.1 Communication Concepts

Communication in its basic form allows two or more parties to exchange data that for them has value (at least equivalent to the effort spent on communication). There are basic forms of communication like synchronous or asynchronous communication that provide the fundamental basis. In addition, for senders and receivers to understand each other, data formats have to be agreed upon as well as the possible range of content for those data formats so that the communication partners can understand each other: a date with the value of 01-02-07 can be misunderstood easily if its precise semantics is not captured (one possible interpretation is July 2nd, 2001). Finally, in order for senders and receivers to synchronize the sending and receiving of data, they have to follow specific communication protocols. This Section outlines the basics of communication and builds the foundation for the remaining Sections in this Chapter.

8.1.1 Fundamental Types

When parties are communicating they need to establish a communication channel over which the data is communicated between them. All communication channels can be classified into only a few basic classes that define the basic properties of communication. The three basic forms are as follows:

- **Synchronous Connection**. A synchronous communication channel requires all communicating parties to be part of the communication channel concurrently in order to communicate. The parties exchange data between each other. The sending party puts the data on the synchronous channel (or connection) and the receiving party or parties receive the data. In a synchronous connection it is possible that the receiving party starts receiving the initial data while the sending party still sends the remaining data. If one party leaves the synchronous connection, it cannot participate in the communication any more. If the leaving party is one of the last two parties on the connection, the communication finishes (or is disrupted) as the communication requires at least two concurrent communication partners. Elaborate synchronous connections allow parties to send and receive concurrently in both communication directions; less advanced connections can be used only for one direction of data transfer at any given point in time. Examples for synchronous connections are the ancient telephone for humans or the remote procedure call between software systems.
- **Asynchronous Connection**. Asynchronous connections are very different in nature from synchronous connections. An asynchronous connection does not require all communicating parties to be concurrently connected to the connection itself. At any point in time a sending party can put data on the asynchronous connection and at the same or different points in time a receiving party can take data from the connection (as long as data is present and as long as the connection itself is available). The asynchronous connection itself stores the data. In this sense it is stateful and through this mechanism allows the independent presence of sending and receiving parties. If a sending party puts several separate pieces of data on the asynchronous connection, it depends on the particular implementation of the asynchronous channel if the order of the data is preserved or not. If it is not preserved and the receiving party depends on the correct order, the data must contain some information about the order so that the receiving party can reorder the data appropriately independent of the asynchronous communication channel. In the general case data on the asynchronous connection is consumed by the receiving party once it takes the data off the connection. In this sense the reading is removing ("destroying") the data on the channel.
- **Shared Variable**. Communication over shared variables is the third type of connection. A shared variable is accessible by a sending as well as a receiving

party. A sending party can put data into a shared variable any number of times at any point in time. Each time the sending party puts data into the shared variable it overwrites the previous value. A receiving party can take data from a shared variable. When it does so, the data is not consumed; instead, the data remains and other receiving parties can access the data in the shared variable. The receiver can write data to the shared variable, too, of course. In this case, if the receiver does not want any other party to read the shared variable, it can put a "null" on it, i.e., overwriting its value. Putting data into a shared variable and reading a shared variable are asynchronous to each other. It is also not guaranteed that a receiving party sees all values of the shared variable. If the sending party writes very often, it might very well be the case that the receiving party does not read fast enough and misses intermediate values. Since the shared variable allows concurrent access it needs to ensure that the read or the write is atomic to avoid that senders and receivers are interfering while operating on the shared variable.

All specific implementations of communication technology can be reduced to one of the three fundamental types discussed above. For example, communicating through a database is a shared variable communication. Communicating through queues is an asynchronous communication. A remote procedure call is a synchronous connection. Some of the major technologies are introduced later in Section 8.2..

8.1.2 Formats and Protocols (FAP)

A communication channel of either type is minimally required in order for parties to communicate with each other. If the communication itself should be successful, meaning, the communicating parties understand each other and have a constructive communication with a defined outcome, more has to be agreed upon then just the communication channel. There are two major aspects of communication that need to be in place for a meaningful communication: formats and protocols.

Formats refers to the data structure and data content that is communicated. The sender as well as the receiver have to agree on the data structure and content in order to understand each other. Structure refers to the particular data elements and their relationship that is communicated whereas content refers to the values in the data elements. Only if sender and receiver agree on structure and content, can they "make sense" out of each other's data and have a meaningful conversation. This agreement that has to be in place and needs to cover all possible values and structures. As the communication has to work under all allowed combinations, the agreement is quite difficult to achieve in general, as it is practically impossible to enumerate all possible combinations to prove that the communicating parties understand each other for each combination.

Protocols refer to the exchange sequence of the data in their particular formats. A communication in the general sense requires that sender and receiver exchange several distinct sets of data by sending them to each other. This requires a specific order to ensure that both, sender and receiver understand where they are in the communication and what data has to be exchanged next in the sequence of exchanges. The involved parties only make constructive progress during their communicating if they follow the correct exchange sequence or one of several correct sequences (if several are permissible).

No matter which fundamental type of communication channel is used, the formats as well as protocols have to be agreed upon so that all involved parties can participate in the communication in a meaningful way. Furthermore, if formats are transmitted that cannot be understood, or if protocols are violated, then the communication must be able to recognize this error situation and try to get back to a meaningful state. For example, if formats are not understood, then the receiver must be able to send back a "not understood, please send again" response. Otherwise, if a communication error is not detected, no repair is possible and the communication has ended unsuccessfully.

If the protocol is violated, the violation must be detected, the overall communication must stop and synchronize on a state that all parties agree to as a consistent state from which to continue. This might happen if, for example, certain data messages are lost in the communication and the receiver is waiting for a specific exchange that the sender assumes happened already.

Both formats and protocols play an important role in Web Services as well as Semantic Web Services. Formats are described using Semantic Web languages whereas protocols are defined through various elements that the Semantic Web Services efforts provide.

8.1.3 Separation of Interface and Logic

Formats and protocols have to be implemented as software code in order to make communication over communication channels work. The data formats and data values that are sent over communication channels have to be independent of the software of the sender or the software of the receiver to achieve maximum independence. In any communication setup it is impossible to guarantee, ensure or enforce that both, sender and receiver use the same software from the same vendor for sending and receiving the data. Consequently there needs to be a distinction between the implementation of how to produce the data or consume the data (implementation or logic) and the definition of the data itself (including its possible contents). This distinction follows the well-established separation of interface and implementation in computer science. Later on when Web Services and Semantic Web Services are discussed this distinction becomes a very important aspect.

The same applies to the behavior of communication. The order of formats sent and received by the communicating parties must be described in such a way that all communicating parties can agree to it independently of the software used to implement and enforce the behavior. This means that the definition of behavior must be done in such a way that the behavior can be inferred from the language used to describe the behavior (instead of examining the code of the software that implements the behavior). In consequence, this allows both, the sender and receiver to agree on the behavior while implementing it in their preferred software technology or with technology of their preferred software vendor. Again, when talking about Web Services and Semantic Web Services this aspect becomes important in the formalisms and languages used.

8.1.4 Communicating Parties

Communication cannot take place without communication partners engaging in the communication by sending and receiving data from each other. In a given communication there is always a sending partner (sender) and at least one receiving partner (receiver). The sender sends out data that the receiver obtains by taking part in the communication. Several receivers are possible in a communication and all of them receive the data sent by the sender.

During a communication the role of sending and receiving can change if the communication is conversational. Once a sender has sent out data, and after the receivers have received the data, one of the receivers can assume the sender role and send out data. This is especially the case in the situation where the communicating parties have a dialog in the sense that formats are sent back and forth in order for both parties to accomplish the goal of the communication. In a multiparty communication it is possible that several parties start sending at the same time as they do not know about each other's state and intent. In such a situation it is important to ensure that either the protocol does not allow such a conflict to happen or that a dynamic mechanism is available at run time that enforces only one sender at a time.

Some communication channels allow only one sender at any given point in time. In this case there can only be a single sender for a given communication and no coordination has to be enforced through the protocol. However, some communication channels allow the concurrent sending of data by several senders. In this case several senders can send data, but for the communication to be meaningful, the participating receivers needs to be able to receive data from different senders concurrently. If a channel allows several concurrent senders it is not necessary to enforce the one-sender-at-a-time policy, of course.

Another dimension opens up when a single communication channel can "host" several independent communications. In this case it is necessary to distinguish the communication not only by communication channel, but also by identifier within one communication channel. In computer science this case can take place when asynchronous technology like queueing technology is used. In this technology it is possible to send messages across a queue that originate from different senders and are addressed to different receivers. In this case each message must either carry a communication identifier to identify the communication or each message carries a receiver identifier so that the respective receiver knows which messages to read.

If the approach is followed that messages carry the identifier of a communication then the notion of an "instance of communication" is important. This can be further formalized by associating senders and receivers (which are instances, too) to communication instances. Going forward we assume this notion. At any given point in time, when senders and receivers communicate, they do this in context of an instance of a communication. Consequently, it is possible that the same senders and receivers open up a separate instance of communication. In addition, the same senders and receivers can be participants in different instances of communications, either concurrently or sequentially.

This notion of communication is independent of the fundamental types of communication as outlined in Section 8.1.1. They have to agree on a fundamental type (by selecting a given communication technology). Of course, the separation of interface and implementation is important, as stated in Section 8.1.3. The relevance here is that no matter in how many communications a given participant (or party) takes part, it has to maintain the separation.

And, furthermore, the communicating parties have to agree on the FAP, as outlined in Section 8.1.2. Each communication in the general case follows a defined FAP. Different communications can follow different FAPs, as agreed upon by the participants. This ensures that every communication over any channel is meaningful for the participants.

A further generalization is possible, although not really used widely. If a FAP is defined (for example as a standard), then in many cases the definition might be sufficient for the communication requirements of several parties. However, sometimes a given FAP might not be sufficient. This could be if data formats are missing that are required for a particular case. The parties, in order to overcome this problem, can either extend or modify the FAP, or they can change the FAP during a communication and switch over to a different FAP. The switch over then enables the set of participants to use as many FAPs in one communication as required to make the communication work and meaningful. While this is a very interesting generalization, it is not usually done.

If two parties are involved in a communication it is called a binary communication. If more than two parties are involved, it is called a multi-party communication. A multi-party communication enables several parties to take place in a communication. So far it was assumed that there is one communication instance and all parties are related. Furthermore, it was assumed that all parties are aware of each other and all concurrently receive data from a sender. However, this is not necessarily always the case. A multi-party communication can actually be achieved by one party being the "communication coordinator" and all other parties engaging in a binary communication with the coordinator. So only the coordinator is aware of all the parties, but each party is only aware of the one coordinator. This requires that the coordinator is part of the communication for its whole length, can receive all data from all senders, and can relay sent data to all parties that are not sending at a given point in time. This also allows having the coordinator engage with different parties using different FAPs.

8.1.5 Mediation

Formats and protocols are agreed upon between the parties of a communication. Once they have agreed upon it, they will follow it precisely as otherwise the communication will most likely not be meaningful and therefore unsuccessful. The communicating parties have no interest in this situation. Therefore, they will do everything necessary to comply with the FAPs.

In the general case, the FAP is determined by the interface that the communicating party can support (see Section 8.1.3). This interface determines what the communicating party can support, and hence this interface allows the selection of one or more FAPs that comply with this interface.

As in the general case, the interface has to be implemented in order to support it at run time. Therefore, the interface is implemented by software. If the internal data processing environment, however, supports different data structures and data content, then there is a mismatch between the interface and the software used to achieve the implementation. Why would this ever be the case? Why would a party not define the interface in such a way that the interface can be implemented easily?

In the world of communication over world-wide networks across company boundaries there are established practices of FAPs (often referred to as B2B protocols). [263] discusses quite a number of those. In order for a given party to easily participate in a communication it is wise to support the FAPs of a given industry. Therefore, the party is probably inclined to solve the discrepancy between the interface it needs to support and its available implementation technology rather then supporting an interface that does not allow it to easily participate in given FAPs.

Bridging the data structures and data content on the interface and the implementation software is the data mediation problem. This problem is well-studied and many attempts are made to structurally overcome it [263].

The same is true for the protocol aspect. The behavior that a given FAP demands might be the same or might be different from the behavior the underlying software for implementing it exposes. In addition to data mediation as described above the concept of protocol mediation is required. [264] describes the protocol mediation problem in detail. Only if data and protocol mediation are both supported it is possible to map the interface to the implementation within a given party of a communication.

Not all parties can support at their interface all the FAPs that they need to in order to participate properly in the various communications. In this case it is necessary to move the mediation (data and protocol) outside the interface. So instead of mediating the difference between the interface and implementation within a party, the mediation is done outside the interface between parties. This makes the concept of a "middle man" necessary. The middle man is a party to the communication for specifically mediating between communicating parties. The middle man established a communication channel with all participating parties and uses different FAPs for that. The FAPs used are those that the parties can support. The middle man itself mediates between the FAPs as it passes along the data from the sender to the receivers.

The benefit of this approach is that parties can participate in communications that they could not support directly. Of course, the middle man has to be available and able to mediate appropriately.

The most flexible party to a communication is the one that has a declarative way of mediation within its boundaries between its interface and its implementation. If the mediation is declarative, existing mediations can be changed or new mediations can be added. If the change or addition of mediation is fast and flexible, the party can define additional interfaces as required by FAPs as it can build the mediation to its implementation easily. It is therefore assumed for simplicity that this is the approach going forward in the remaining Chapters about processes and the Semantic Web. If a given party cannot implement mediation this way, the way out is the middle man.

8.1.6 Non-functional Aspects

Ideally, communication is secure, reliable, recoverable, fast, and has many other "nice" properties that make it convenient for the communicating parties. Properties like security, reliability, recoverability, performance, and others are called non-functional communication properties (as they are related to the communication system behavior, not the semantics of a communication). Different implementations of communication channels have various support for non-functional communication properties. Depending on the particular needs of the communicating parties they have to select the most suitable mechanism.

A few properties are discussed in the following. The list is not complete and only highlights the most important aspects:

- **Security**. Security has many different facets. The most relevant are that data communicated should not be visible to any party not involved in the communication. This is usually achieved by either encrypting the data packets themselves that are communicated or encrypting the whole communication channel instead of the individual packets. Furthermore, no other attack should be possible like taking data packets off the communication channel, or introducing additional ones in order to cause disruption in the protocol. Another aspect is authentication and authorization of the parties that want to join a communication channel. Not all parties should be easily able to join a communication just like that. The originator of the communication should be able to restrict access as necessary and have parties authenticate themselves in order to allow the proper authorization.
- **Reliability**. Reliability is important in the presence of failures. In case of a failure it must be clear what status the communication is in and which of the last data transmissions succeeded successfully and which did not. This allows after a failure to continue the communication from a consistent state forward.

- **Transactionality**. Transactional behavior of communication is important in the presence of fatal errors. If a server goes down and has to be restarted, if a network fails or a communication software stops working, then it is important that the communication can be recovered to its last consistent state. This is important as it allows the communicating parties to continue the communication without having to execute any recovery strategy itself, let alone compensating actions that would modify already achieved states.
- **Throughput**. If the data sent is of high volume or if many communications are ongoing in parallel then the communication channel might become a bottleneck in the sense that it cannot support all communication as fast as in a low load situation. In this case the communication system degrades in terms of performance. Throughput is important and the ideal situation is that degradation happens only under very high load. Furthermore, it should be gradual, not sudden.
- **Performance**. Performance is related to the speed of data transmission. In many situations speed is of high importance, for example, when communication takes place over synchronous connections that require a fast response. In other situations performance is not as important as the communicating parties do not have to operate within the bounds of specific time lines.
- **Availability**. We are used to the immediate and constant availability of the phone system. Whenever we want to make a call we expect the phone be available and ready. Connections to the Internet are also expected to be "always-on". In this sense every party is expecting to be able to engage in a communication whenever they need to. High availability of the communication channel is important. Of course, this does not mean that all communicating parties are always available; a phone call might not be taken by the intended recipient.

This discussion of non-functional communication properties concludes this Section. The fundamental communication concepts have been introduced that form the conceptual basis of communication as related to Semantic Web Services. In the next Section specific communication paradigms that are based on the fundamental types are introduced.

8.2 Communication Paradigms

Based on the fundamental types of communication, namely shared variables, synchronous and asynchronous communication, different specific communication paradigms were developed over time. Leaving the postal mail approach of storing data on a storage medium like DVDs and sending them by postal mail aside (i.e. "physical communication"), the most important current communication paradigms (like client/server or queueing) that are en vogue are discussed throughout this Section.

For each paradigm, the FAP as well as the number of communicating parties are discussed as well as to which basic type it belongs.

8.2.1 Client/Server (C/S)

The client/server communication paradigm is one of the oldest paradigms and is part of the synchronous connection type. This approach distinguishes a provider of functionality, called server, from the consumer of functionality, called client. Clients and the server can be on the same computer or they can be on different computers. In the latter case, communication is established over a network (be it a local network or a wide-area network).

The FAPs for this paradigm are determined by the server. The server defines and specifies the possible invocations a client can make and their order, it defines the data structures as well as the data content. And it defines the behavior, too. The client has no ability to influence any of these definitions, it can only use whatever the server provides and allows at any given point in time.

A server can serve many clients. The exact number depends the server's capacity and the size of the computation its clients request. The clients do not know about each other, so it is not possible to have a multi-party communication; instead, all communication is binary between a client and the server.

8.2.2 Queueing

The queueing paradigm became popular in recent times with the advent of explicit queueing systems as a separate architecture and technology component or implicit database or application server functionality. Queueing is of the asynchronous connection type as it decouples the communicating parties.

Queueing is in principle a one-way communication mechanism where a sender submits messages to a particular queue using an enqueue operation. The messages will be stored in the queue, generally in the order of receipt. The receiver takes messages from the queue using a dequeue operation. In its simplest form, queues maintain the message order and operate under the first-in-first-out (FIFO) mode.

In this situation the determination of the FAP becomes an interesting topic as it depends on the viewpoint of who decides on the FAP. In the majority of cases the queueing paradigm takes the form of an asynchronous client/server model where the receivers determine the FAP and the senders have to comply. However, this is not necessarily the only possible viewpoint. An alternative viewpoint is that the sender is an information source and publishes its message to a queue and is not really interested in which receivers pick up the message content. In this viewpoint fundamentally the receiver is the one that is "interested" in the message and has to comply to the senders FAP accordingly.

Over a queue many senders can communicate with many receivers. In order to establish a two-way communication two queues can be put in place, each queue for one direction of communication. Alternatively one queue can be used and the senders and receivers both put messages and read message from the same queue. If it is important to know in this case what messages are response messages, they have to be marked accordingly either through typing the message or an attribute in the contents of the message. In the case of two queues, one can be marked as the request queue and one as the response queue. Still, in any case one sender and one receiver communicate with each other. The reason is that on dequeue the message gets removed from a queue, meaning, the only one receiver can receive a single message.

However, more advanced queueing systems allow more than one receiver to receive the same message. The queueing system in this case ensures that all receivers receive a copy of the same message. This is accomplished by receivers declaring interest in specific messages and the queueing system notifying all receivers about the advent of those. These systems are called publish/subscribe systems. Receivers (subscribers) then will receive those messages from the senders (publishers) and can proceed with whatever processing they need to do.

8.2.3 Peer-to-Peer (P2P)

The peer-to-peer paradigm is very similar to the client/server paradigm and is of the type synchronous connection. The major difference is that the server in the client/server paradigm is usually stationary and in a central location to which all clients connect to. In the peer-to-peer paradigm this is not the case. Two communicating parties (in this case called peers) communicate directly with each other and establish a connection directly, not going through some central location like a stationary server at all. Each party can become server whenever it wants to and can become client whenever necessary. Fundamentally, every peer is a server or a client at any point in time. In this case any two pairs of parties that know about each other can establish a direct connection at any time, establishing peer-to-peer links.

In the peer-to-peer paradigm, as each party can become a peer at any time, all have to agree on the FAP in order to be able to establish direct connections. More elaborate peer-to-peer protocols allow peers to communication with each other that do not have a direct communication link. In this case other peers act as relay station forwarding the communication. Two peers therefore communicate directly, whereby the actual data transport is over other peers as intermediaries (invisible to the communication channel itself).

It is possible that one peer communicates concurrently with several other peers. In addition, several peers can communicate with each other at the same time. So a true multi-party communication can be established where the peers know about each other.

8.2.4 Blackboard

The blackboard paradigm is an approach to further decouple senders and receivers, even more then a queueing system allows to do. A blackboard architecture provides a space where data can be posted, changed or removed by a sender. There is no specific guarantee about when data is made available and how long it will reside there. Receivers can read data at any point in time and as often as they want or need to. The basic protocol of how to post data and how to read data is determined by the blackboard. However, the FAP between senders and receivers is not determined by the blackboard, that remains within the control of senders and receivers. Still, the FAP that senders and receivers use between them and the FAP of the blackbaord has to match within the constraints of the blackboard. For example, if the FAP requires versions, but the blackboard does not support versions, the protocols are incompatible.

There can be any number of senders and receivers, and there can be many receivers participate in the same communication as the blackbord does not restrict data to be accessible only to specific senders.

8.2.5 Web Services

Web Services is a new communication paradigm that is centered around the public Internet as communication transport layer. Web Services have three aspects to it. First, the interface of the communicating parties is described using an interface definition language (called the Web Service Definition Language (WSDL)). This formal language allows the definition of the messages a communicating party sends as well as receives. Message sending and receiving is based on operations that have messages as input and output parameters. This approach differs from B2B protocols where messages are sent and received without the notion of operations.

Second, an explicit transmission protocol is defined, called SOAP (initially standing for the Simple Object Access Protocol, however this has since been dropped). This protocol is abstract in the sense that it defines how a message as defined in WSDL is structured when it is sent using a concrete transmission protocol. A separate binding is defined to bind the SOAP protocol to a real transport. Bindings exist (amongst others) for HTTP as well as MIME. This is interesting as HTTP is a synchronous protocol based on the synchronous communication type whereas MIME is an asynchronous protocol based on the asynchronous base type. The interesting aspect is that the interface definition is independent of the actual communication mechanism.

Third, a publication mechanism is defined that allows communicating parties, if they so wish, to publish their interface definitions in public or private directories for others to look up. This supports the detection of communicating parties based on their defined interfaces.

The FAP are in part predefined by the notion of operations with input and output parameters. However, the sequence or order of operations that has to be called is undefined and open for the communicating parties to agree upon. The data structure and content is free for the participating parties to decide, however, the specification language is XML Schema. With XML Schema the communicating parties can agree on structure and content to the extent XML Schema supports the definition.

Web Services are a bilateral communication mechanism that allows two parties to communicate with each other. A multi-party communication is not supported by the current Web Service standards or technology.

8.2.6 Representational State Transfer (REST)

REST (representational state transfer) [265] is a particular style of enabling communication based on the principle that all data as well as operations on data are enabled using strictly static URLs based on the HTTP protocol. The fundamental approach is to see data and operations as identifiable resources and the identification mechanism is URLs. As such, resources like data or operations are identified by URLs. Accessing a particular car (identified for example as 45671) from a car selling web site (for example, www.sellyourcar.com) could be www.sellyourcar.com/car/45671. The response to issuing the URL would be the data format and data content representing the car identified with 45671. Searching a car could be www.sellyourcar.com/findCarForm. This would provide the client to obtain the data format with the search criteria to be filled for searching a particular car.

REST implements a "classical" client/server model where the client requests action from a server through particular structured URLs. A server provides a response to clients if a well-formed URL is transmitted to it. The communication is synchronous and over the HTTP protocol. The static URLs ensure that the data or functionality can be accessed with the same URL at any point in time.

The form of communication is binary as only two parties can participate in one communication. However, a server can provide responses to several clients, of course. The FAPs are not explicit, but implicit (analogous to the client/server model). In this communication style there is no explicit definition of the data formats, permissible values or the protocols as interface as the interface is not explicitly defined. Instead, all aspects, including the URL structure, are defined by the server in implicit form (as opposed to a WSDL definition). This follows closely "normal" HTML page requests and responses.

8.2.7 Agents

Agents is a concept stemming from the area of Artificial Intelligence (AI). Agents are autonomous entities that form a perception of the world around them and that can communicate with other agents. The communication allows agents to achieve their task by asking other agents to contribute. From a communication viewpoint agents do not only communicate data, but explicitly ask other agents to perform specific tasks. The asked agent can execute the task, delegate the task to another agent or refuse to engage in executing the task (effectively rejecting it). In this sense there is an explicit notion of acceptance as well as refusal of tasks.

An agent can communicate with any number of other agents. However, all agent communication is bilateral in the sense that each agent communicates with one or more other agents directly, never in a multi-interaction way.

The formats are not predefined by an agent protocol. Agents can agree on the formats and data content they want to use. They even can engage in communication without having agreed upfront on the specific formats as an agent can always respond with the "I don't understand" message back to the message originator in order to indicate that the communication will not be possible due to data format or data content misunderstanding. However, in order for every agent to exchange messages with any other agent, a basic protocol for at least exchanging messages needs to be in place. Otherwise any communication is impossible. Also, minimally the "I don't understand" message needs to be agreed upon upfront, too, for agents to be able to tell each other that they could not understand. Otherwise the response message could not be interpreted either, making any communication impossible.

8.2.8 Tuple Spaces

Tuple spaces are like blackboards where communicating parties can add tuples into a space that can be read by other parties. The data structure is predefined as tuples and all communicating parties have to follow this structure. The tuple space itself is a space that exists on its own without communicating parties to be connected to it. Tuple spaces are therefore following the shared variable basic principle. Access to the tuple space is concurrent, however, each individual tuple is accessed atomically for consistency reasons.

The formats are open for the parties to determine or to define. The basic protocol of tuple management is determined by the tuple space. Any protocol beyond that, i.e., the number and order of specific tuples written is solely in the discretion of the communicating parties.

Any number of parties can communicate with each other at the same time using tuple spaces. In this approach bi-lateral as well as multi-party communication is supported. Interestingly enough, from the viewpoint of an individual party, it is not clear at all if there is a bilateral or multi-party communication. It can be the case that one party writes tuples that are never picked up by any other party. In this sense tuple spaces (like blackboards) allow a one-party communication; this is really an oxymoron, unless the storing and reading of tuples by the same party is considered communication with itself.

8.2.9 Co-location

Co-location is a communication paradigm that allows parties to communicate without crossing remote networks for the purpose of the communication. Co-location is based on the principle that the communicating parties share their communication code so that a party communicating with another one really does a local invocation instead of a remote invocation. This allows the existence of a structured communication without incurring the network overhead for sending data between the parties.

However, one must ask the question how data actually ever gets transferred between the two parties as in the end of the day they are in separate environments? The basic assumption in the co-location paradigm is that the communication code implements database updates as side effects. If this is the case, if one party calls another party's communication code, that updates the database of that party. So in reality the communication on a communication protocol level is local, however, the remote access part is "pushed down" to the database access layer.

This paradigm is relevant especially within organizations where remote database connectivity is possible. In such an environment all communication is limited to a one-hop database invocation without additional remote invocations across a network. This reduces the remote invocations while keeping the database connectivity constant.

Any number of parties can participate in such a co-location as each party uses the communication code of every other party it communicates with. In addition, the formats and protocols can be freely agreed upon as the shared communication code can be invoked as needed by the sender.

8.2.10 Summary

Many communication paradigms exist, each having its very own properties. In a given communication situation, some might be more appropriate then others. However, at the end of the day, all allow the transmission of data from a sender to a receiver.

8.3 Long-Running Communication

Communication between two parties is not restricted to only a single individual exchange of data. In many cases several exchanges take place, from sender to receiver and back. These exchanges usually take place one after another. If this communication is following a protocol for the whole duration of the communication and is about the same business process or about the same business objective (like for example clarifying the insurance coverage of a patient) then it is considered a long-running communication.

The term long-running comes from the fact that the individual communications are related to each other and not arbitrary. Furthermore, if one individual communication fails, then only the failed one needs to be repeated or corrected, not the whole communication from the beginning up to this point. In a long-running communication each individual successful communication is regarded as a consistent state. So if the last individual communication fails, and as the last consistent state is persisted, it can be retrieved and taken as the restart point for continuing the long-running communication.

Long-running communication is widely used, especially in context of inter-organization communication in form of B2B protocols as well as in intra-organization communication in form of Enterprise Application Integration (EAI), also called Application-to-Application (A2A) integration. In the following each is discussed separately in turn.

8.3.1 Business-to-Business (B2B) Protocols

B2B communication takes place when company boundaries are crossed while data is passed back and forth between the communicating parties. A typical situation in the supply chain industry is when a buyer sends a purchase order to a seller and the seller responds with an acknowledgment that the order will be fulfilled. Later on the seller would send an invoice for the goods shipped, expecting a payment from the buyer. Another situation in the healthcare domain would be the communication about clinical tests between different healthcare providers.

B2B communication is about sending and receiving meaningful business data that allow businesses to act upon or to react to. The formats have to be agreed upon so that the communicating parties can understand each other. The same is true for protocols as the communicating parties have to comply to the protocols in order to send or to wait for a message at the precisely correct time.

As outlined in [263] there are several standards organizations maintaining and further extending standards that define the formats and protocols. Examples are RosettaNet, HIPPAA, EDI, just to name a few.

The challenges in setting up proper B2B communication are manifold. The main challenge is the semantically correct interpretation of the various data elements in the messages that are exchanged. As the data sent across is in general coming from various back end application systems, it reflects many different data models. As the formats themselves represent a data model the correct interpretation depends not only on the structure of the messages but also their contents.

Another challenge is to design the long-running B2B interactions and to ensure run-time compliance to the agreed long-running protocol. Depending on the complexity of the long-running process it is possible that many correct executions of the processes exist. Plus, many different error situations can happen that require compensation in order to get back on track to a correct execution.

Since important business data are communicated other aspects are of importence. Security is a big and important item, and so is reliability. Reliability refers to the guarantee that a message was not only passed, but also received. Nirvana in this case would be an exactly-once transmission so that both, sender and receiver are guaranteed that each message was transmitted exactly once. This ensures that every message is accounted for and not lost. In addition, neither sender nor receiver have to worry about error detection, handling and recovery on a message passing level.

In addition to the reliability of a single data exchange, the consistency of the overall long-running process is very essential as both communicating parties rely on the consistency in order to do successful business with each other. If one transmission in a process fails (and keeps failing) then it might be the time when the overall process needs to be abandoned. As it is a long-running process many states were committed along the way clearly requiring compensation to undo already achieved work.

At run time, speed and throughput is of concern as well as security. However, these are non-functional properties not relevant to the discussion in context of semantics.

8.3.2 Application-to-Application (A2A) Protocols

Application-to-Application integration is also referred to as Enterprise Application Integration (EAI). A2A protocols are very similar to B2B protocols in that data is passed back and forth. In contrast to B2B protocols, the endpoints are applications within an enterprise, not systems across a network between companies.

The issues and problems are the very same as in B2B protocols, including security. This might be a surprising statement as in general applications are invoked over their interfaces. However, applications may still provide purchase orders or wait for acknowledgements at their interfaces. In this sense there may be a long-running process implemented inside that requires compliance. Transformation is necessary as the data model inside the application might be different from that of the B2B protocol or other applications that are integrated. As enterprises, especially larger ones have different physical locations communication between applications is possibly going over the Internet, so security becomes an important aspect. In this sense, A2A and B2B integration are very similar.

Some of the problems can be solved a lot easier due to the fact that the integration is within an enterprise. For example, the exactly-once semantics in data transmission can be achieved using transactional communication systems like transactional RPC or transactional queueing systems. Also, the endpoint of the integration are applications within the same enterprise, so supposedly the communication between the governing groups should be a lot easier then across company boundaries.

From a semantics viewpoint there is no difference between B2B and A2A protocols at all. This is the reason why not distinction is made in the remainder of the book.

8.4 Web Services

Web Services are the current silver bullet for (remote) communication in context of the Web as well as within enterprises or governmental organizations. Aside from the fact that Web Services is a relatively new development (only a few years old) the initial charm lay within its perceived simplicity. To keep things simple, there was the notion of an interface, a transport protocol and a mechanism to register interfaces. And that was all there was to it initially. Everybody liked this simplicity despite the fact that everybody knew from the very beginning that additional features and functionality are needed like security, transactions, policy, processes, and so on.

The initial run-time model was that of a client/server model. A server defines services by means of interfaces in a XML-based language call WSDL (Web Service Description Language). These interfaces are made available in a registry that clients can lookup. The initial effort was UDDI (Universal Description, Discovery and Integration). Clients wanting to invoke services, could lookup UDDI repositories, retrieve interfaces and (having this knowledge) invoke those interfaces dynamically. Of course, there was no security, no reliability, no policy or any additional functionality.

Over time, the realization set in that these features were necessary leading to the development of WS-* (spoken "Web Service Star" or half-jokingly "Web Service Death Star") which refers to a (relatively large) set of standards and proposals that together lead to an acceptable set of technologies for communication. This set of standards is in the meanwhile so complex that efforts are underway to simplify this (leading to a split and proliferation of multiple activities).

All these efforts have as common denominator a few technical concepts and principles that will be discussed in the next Chapter of the book in more detail. The next Chapter discusses also the most important standards from this set in more detail to outline the overall complexity that is to be mastered in order to implement useful Web Services.

One of the efforts that wants to enhance the state-of-the-art are Semantic Web Services. This is a very important activity as it strives to incorporate semantics at the level of description models, mechanisms and languages. A separate Chapter is devoted to those to introduce current efforts in this space and yet another Chapter of the book looks at standards activities for Semantic Web Services.

8.5 Clinical Use Case

The clinical use case (see Chapter 2 and Chapter 13) is fundamentally a distributed application system with many concurrent activities. Doctors retrieve and enter data, clerks schedule appointments, test results are forwarded or inferencing takes place to derive information.

All the various subsystems of the clinical use case implementation are independent in their data management and they communicate with each other in order to exchange data. Chapter 2 outlines the basic use case as well as its requirements. Chapter 13 outlines the complete use case in detail and shows how semantic technologies are used to define the functionality.

Here we discuss how communication technologies are used to support the clinical use case implementation. First of all, as the use case takes place in a clinical environment, security and reliability play very big roles. Security ensures that the patient specific data is only accessible to those authorized to see it. For the communication technology this means that only those communication technologies and their implementations can be used that support security.

The second big requirement in clinical environments is reliability. It is essential that any data that is collected manually is transmitted and stored in such a way that it cannot get lost. Transactional communication technologies ensure this requirement. Also, any derived information as well as information gathered through test results have to be reliably communicated so that it is ensured that data is not lost during communication, be it a one-step communication between two systems or a multi-step (long-running) communication between several systems.

For the latter case, when several systems have to cooperate, long-running communication is the best approach to ensure that the cooperation finishes and does not stop undiscovered at a partial state. Long-running communication maintains the state during the cooperation and can restart or continue after a failure from the last consistent state.

The next few chapters will introduce Semantic Web Service technology. Services will be discovered, put together into an orchestration as well as invoked. This technology is in principle independent of the underlying communication technologies as they can use any available one. For the clinical use case this means that all remote invocations of services is based on secure and reliable communication. Web Service orchestration will be executed on long-running communication so that state is not lost.

Some of the systems, like for example the system that discovers services, does not have to be reliable. In these special cases, when a discovery fails, it can be rerun in order to obtain a good result. In more general terms this means that idempotent functionality does not have to be reliable as it can be re-executed without loss of information.

When it comes to the selection of a sepecific technology implementation no general advice can be given as in every real life implementations the choice is determined mostly by the already existing infrastructure. For example, if a hospital has already a transactional queueing system from a specific vendor, than this is the one that needs to be deployed in Semantic Web Service implementations, too, for pragmatic and financial reasons. Also, if some of the remote connections are based on B2B protocols using healthcare document or message standards than this is the one to be chosen. However, independent of the specific technology that is available, the aspects of security and reliability are essential and must be achieved.

The use case in Chapter 13 focusses on the functional definition of the clinical use case and does not further go into the specific communication technologies any more.

8.6 Summary

In summary, communication is a highly utilized and very well researched area of computer science. Communication between computers is at an all time high as the networks around the globe get more tightly integrated every day. The latest development, Web Services, made communication a lot easier (initially) leading to a lot of development of remote services.

Various communication styles and mechanisms exist that address different functionality. However, common to all is the lack of semantic description languages and concepts, as pointed out in this Chapter. The following Chapters focus first on Web Services and later on Semantic Web Services that in the end strive to overcome the semantic description problem of dynamic behavior expressed as long-running processes.

9 State of the Art in Web Services

This Chapter introduces the current state-of-the-art in Web Services in the form of various standard proposals that are being worked on currently or that are already implemented in software products. It starts with a brief introduction into the history of Web Services in Section 9.1 and continues with the description of traditional Web Services in Section 9.2. Section 9.3 provides an overview of the standards activities that are ongoing in the area as well as already achieved standards. Web Service composition, as a specifically important area, is called out separately in Chapter 10.

Web Services are a set of technologies that have to be put in context of an architecture in order to be used properly in a given implementation. This new architecture paradigm is called Service-Oriented Architecture (SOA) and will be introduced after the Web Service discussion in Section 9.4. Web Services are syntactic in nature not actively trying to advance into the space of semantic technology at this point in time. However, semantics remains one of the biggest challenges in the proper design of software systems, especially when the software system is composed of interoperable services that are not originating from the same set of authors. Therefore, a brief overview of the semantics aspects are provided in Section 9.5 leading up to the following Chapter on Semantic Web Services. Section 9.7 summarizes.

9.1 History

With the emergence of the Internet infrastructure in conjunction with the simple HTTP protocol and XML as a data representation language the question arose if this light and simple (and world-wide available) infrastructure can be used instead of the more heavy-weight environments like CORBA for "long-distance" remote communication across the Internet and organizations' boundaries.

This idea resulted in the XML RPC work of Dave Winer together with Microsoft [352] in 1998. In its simple form it uses HTTP as the transport protocol and uses XML as the platform-independent encoding of the data passed between a sender and a receiver. The specification of the XML RPC protocol is very brief and herein lies its power.

Out of this XML RPC work arose the more sophisticated SOAP transport protocol afterwards [353]. Dave Winer was involved in this development, too, together with members of Microsoft. Originally SOAP was the acronym for Simple Object

Access Protocol. The acronym interpretation was dropped when SOAP became a W3C Recommendation in 2003 [354]. Today virtually all Web Service implementations support the SOAP protocol.

In parallel to the SOAP development the Web Service Description Language (WSDL) was introduced. While SOAP is concerned about the data representation and data protocol binding at run time ("on the wire"), WSDL is concerned with the functional definition of endpoints (in WSDL terminology called operations) that can be remotely invoked. As example, retrieving a patient's name based on an identifier like the SSN would be an operation. WSDL is also based on XML and is using the XML syntax for defining endpoints: their names, parameters, bindings, and so on. In 2001 a W3C Note [355] defined WSDL 1.1 and in 2006 WSDL 2.0 became a W3C candidate recommendation [356].

A third standard, Universal Description Discovery and Integration (UDDI) was in the works in the same time period [357]. UDDI is a standard that defines how to store web service definitions in a repository as well as how to discover them. It takes the role as a public Web Service directory that can be used to find or "discover" services. Service consumers are the customers of such directories. Corporation internal directories exist, too, for non-public use of Web Services that are for within the corporation only. In 2005 version 3.02 of UDDI was approved by the OASIS organization [358].

The three standards, SOAP, WSDL and UDDI form the initial basis or "traditional Web Services" foundation. These three were a very small and simple basis to start from. At the same time it became clear very rapidly that many necessary features remained out of scope, like security, reliability, and other important functional and non-functional aspects. This in turn caused a flurry of additional standards proposals in the Web Service area that started to address important additional areas, but at the same time introduced a vast complexity that software vendors, corporations, researchers and standards organizations are struggling with currently as it is not clear yet how all these various standards relate to each other, especially their implementations from a technology component point of view.

9.2 Traditional Web Services

The basic idea of Web Services is centered around the idea of separating interface, implementation and deployment. The interface of a Web Service defines the names of the operations of a particular service and for each operation the input and output data in the form of messages. In addition, the possible invocation locations are provided so that actual Web Services can be invoked at run time. The definition format is defined as Web Service Description Language (WSDL).

If all goes well in a general sense, there will be a huge number of Web Services available in the Web, as this will be the only way of communication in the future. In order to find Web Services it is necessary to devise a way for their discovery. UDDI is such a mechanism and is part of the "traditional" Web Services idea. Finally, at run time, actual invocation data and return data needs to be transported over the network using various forms of communication technology. SOAP defines the message layout for these data transfers.

There have been many articles and books written about Web Services, for example [359]. Because of this high availability of information the following descriptions are kept very short.

9.2.1 WSDL

WSDL has an abstract part that defines the operations as well as their input and output messages. Figure 9.1 is the example as shown in [356].

```
<?xml version="1.0" encoding="UTF-8"?>
<wsdl:description targetNamespace="http://example.org/TicketAgent.wsdl20"
xmlns:xsTicketAgent="http://example.org/TicketAgent.xsd"
xmlns:wsdl="http://www.w3.org/2006/01/wsdl"
xmlns:xs="http://www.w3.org/2001/XMLSchema"
xmlns:xsi="http://www.w3.org/2001/XMLSchema-instance"
xsi:schemaLocation="http://www.w3.org/2006/01/wsdl
                         http://www.w3.org/2006/01/wsdl/wsdl20.xsd">
<wsdl:types>
     <xs:import schemaLocation="TicketAgent.xsd"
                namespace="http://example.org/TicketAgent.xsd" />
</wsdl:types>

<wsdl:interface name="TicketAgent">
     <feature ref="http://example.com/secure-channel"
              required="true"/>

     <wsdl:operation name="listFlights"
                     pattern="http://www.w3.org/2006/01/wsdl/in-out">
          <wsdl:input element="xsTicketAgent:listFlightsRequest"/>
          <wsdl:output element="xsTicketAgent:listFlightsResponse"/>
     </wsdl:operation>

     <wsdl:operation name="reserveFlight"
                     pattern="http://www.w3.org/2006/01/wsdl/in-out">
          <wsdl:input element="xsTicketAgent:reserveFlightRequest"/>
          <wsdl:output element="xsTicketAgent:reserveFlightResponse"/>
     </wsdl:operation>
</wsdl:interface>
</wsdl:description>
```

Fig. 9.1. Abstract WSDL example

First, the namespaces are declared that are in use in the definition. The next section defines types in form of XML schemas. Then the abstract interface definition follows: two operations are specified, one called “listFlights” and the other is called “reserveFlight”. Each operation has an input message and an output message specified the defines the data required by the operation as well as sent back by the operation. A feature is specified, too, that indicates what particular non-functional feature the client of these operations has to support, in this case a secure channel.

[356] and [360] are the full specification of WSDL version 2.0 which is a Recommendation of W3C.

9.2.2 SOAP

The SOAP recommendation [354] states that "*SOAP Version 1.2 (SOAP) is a lightweight protocol intended for exchanging structured information in a decentralized, distributed environment. It uses XML technologies to define an extensible messaging framework providing a message construct that can be exchanged over a variety of underlying protocols. The framework has been designed to be independent of any particular programming model and other implementation specific semantics.*"

The SOAP specification defines the structure of a SOAP message. This structure defines different parts like a header and a body. Figure 9.2 shows an example from [354].

```
<env:Envelope xmlns:env="http://www.w3.org/2003/05/soap-envelope">
<env:Header>
     <n:alertcontrol xmlns:n="http://example.org/alertcontrol">
          <n:priority>1</n:priority>
          <n:expires>2001-06-22T14:00:00-05:00</n:expires>
     </n:alertcontrol>
</env:Header>
<env:Body>
     <m:alert xmlns:m="http://example.org/alert">
          <m:msg>Pick up Mary at school at 2pm</m:msg>
     </m:alert>
</env:Body>
</env:Envelope>
```

Fig. 9.2. Example SOAP message

The message shows in the header section information for the receiver in the "alertcontrol" tag. The body contains an "alert".

In addition to the message structure, SOAP also defines how to bind messages to network protocols like HTTP or SMTP. Furthermore, the specification allows for extensions that are specific so particular senders and receivers, specifically non-functional properties like security or reliability. Finally, the fourth part that the SOAP recommendation defines is a processing model of what it means to send and receive SOAP messages.

The relationship between SOAP and WSDL is defined in the WSDL specification adjunct [361]. This recommendation shows how a Web Service defined in WSDL is mapped to the SOAP message format at run time. This binding supports therefore a platform and processing language independent communication across networks. A Web Service provider publishing a WSDL specification of its services therefore will be interoperable with a Web Service consumer that is WSDL and SOAP compliant. The WSDL document will ensure that both, Web Service provider and Web Service consumer have a common understanding of the Web Services.

9.2.3 UDDI

UDDI is the third element of the "traditional Web Services" trio. The purpose of UDDI is to provide an environment where Web Service providers can register their services (i.e., "advertise" them) and where Web Service consumers can find Web Services according to their needs. UDDI can be seen as a market place where Web Service providers and consumers find each other. The technical underpinning is called a UDDI registry that contains the various Web Service descriptions that can be queried by Web Service consumers.

The UDDI specification [358] defines data structures and access interfaces to interact with a UDDI registry. For a UDDI registry to be useful Web Service providers have to store their Web Service definitions in the registry. Once those are stored they are available for search or "discovery". A Web Service provider can update its services as well as retract them.

Web Service consumers can access a UDDI registry to search for Web Services based on keywords and other information. Once they found a set they can then decide which particular Web Service to invoke. Of course, between invocations, they have to be aware of any changes so that subsequent calls to the same service are successful and not result in failure.

Services that have side effects (e.g. those updating a database) are treated specially in the sense that Web Service providers do not really provide the particular invocation endpoints in UDDI. This is not mandated by UDDI, however, Web Service providers do not open up all details because of fear of misuse and denial-of-service attacks. A Web Service provider in general only puts the functional definition of its Web Services into UDDI leaving the connectivity information unspecified. A Web Service consumer, once he found the services he wishes to use, will then have to contact the Web Service provider off-line in order to establish the connectivity. This usually is governed by a contract that contains provisions for service level agreements as well as course of action for misuse of the services.

9.2.4 Summary

In summary, SOAP and WSDL provide the infrastructure for Web Service communication independent of computing platforms and independent of those transport protocols that have bindings available for SOAP and WSDL. These two standards are sufficient for basic communication.

UDDI adds independence in the sense that it allows Web Service consumers and providers to find each other in a registry or broker setting so that neither consumers nor providers have to search each other through other means.

However, while these three standards are powerful and useful, they do not provide functionality that is usually required in enterprise computing. Aspects like security or reliability are relevant in context of business data, especially when communicated across organization's boundaries. In the following a brief overview of additional standards work is given that aims to address the missing functionality of SOAP, WSDL and UDDI.

9.3 Emerging Web Service Specifications (WS*-Stack)

Once the three fundamental specifications (WSDL, SOAP, UDDI) took shape and were available for use, it immediately became clear that these three specifications, although very simple and straight forward, will not be sufficient for serious communication of valuable and critical business data over the Web. "Industry-strength" properties like security, reliability, transactions, long-running communication, policy, and so on were unsupported by the traditional Web Service technology. A multitude of standards have been worked on in the space in general that address industry-strength requirements. They are introduced next.

9.3.1 Standards

More or less immediately after the realization that the traditional three Web Service standards are not sufficient, various companies and standards organizations started working on extending the traditional Web Services with additional functional and non-functional properties to make Web Services really useful in industrial settings.

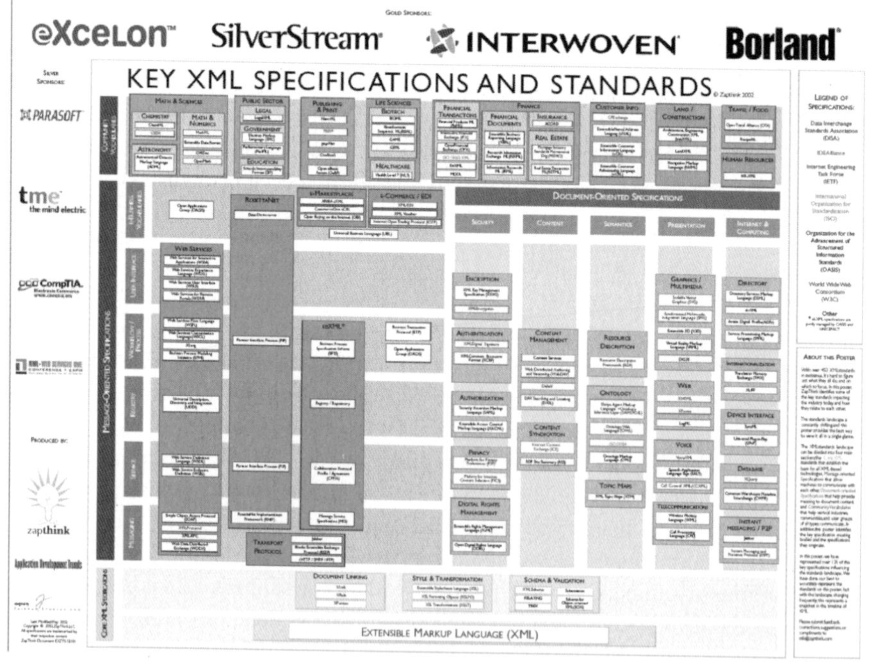

Fig. 9.3. Zapthing Poster of Standards [362]

Zapthink [362] made an attempt a few years back to arrange all efforts on one big poster which can be seen in Figure 9.3. This poster does not only show Web Service related standards at that time, but also efforts of other organizations clearly indicating that Web Services are only one of many efforts. At that time the poster gave a very good overview of the complexity and sheer number of standards being worked on across all industries and on all levels. Some of those went away, new standards appeared, but in general the complexity and variety remained. In context of the Semantic Web and Web Services only the Web Service specific standards are of detailed interest. These are introduced next in more detail.

9.3.2 Web Service Standards

While the Zapthink poster categorizes standards from all standards organizations, way beyond pure Web Service standards, Figure 9.4 shows the a subset focussing on Web Service standards only. This newer representation is taken from [364] and is one of many representations that are available. There is not (yet) a single one everybody agrees on as a reference representation at this point in time. Figure 9.4 also categorizes the various standards into various categories. This classification is again not a standard one as there is no agreement in this area, either.

Business Domain Specific extensions	Various	Business Domain
Distributed Management	WSDM, WS-Manageability	Management
Provisioning	WS-Provisioning	
Security	WS-Security	Security
Security Policy	WS-SecurityPolicy	
Secure Conversation	WS-SecureConversation	
Trusted Message	WS-Trust	
Federated Identity	WS-Federation	
Portal and Presentation	WSRP	Portal and Presentation
Asynchronous Services	ASAP	Transactions and Business Process
Transaction	WS-Transactions, WS-Coordination, WS-CAF	
Orchestration	BPEL4WS, WS-CDL	
Events and Notification	WS-Eventing, WS-Notification	Messaging
Multiple message Sessions	WS-Enumeration, WS-Transfer	
Routing/Addressing	WS-Addressing, WS-MessageDelivery	
Reliable Messaging	WS-ReliableMessaging, WS-Reliability	
Message Packaging	SOAP, MTOM	
Publication and Discovery	UDDI, WSIL	Metadata
Policy	WS-Policy, WS-PolicyAssertions	
Base Service and Message Description	WSDL	
Metadata Retrieval	WS-MetadataExchange	

Fig. 9.4. Web Service Technology Stack

Describing all standards here that are in scope of the Web Services "stack" would be too much as the sheer number is daunting. Details and specifications about the Web Service standards that are already published or are still in the works can be found with the respective standards organizations (which are described later in Chapter 12). An important resource that describes the ongoing developments in Web Service standards and others beyond those is Coverpages[4]. Not only does this web site collect information about standards, it also maintains a distribution list providing regular updates on a weekly basis.

9.3.3 Semantic-Web-Service-Related Standards

The interesting discussion is about those Web Service standards that are relevant for the Semantic Web Services work. This relevance is based on the focus to describe functionality semantically enabled so that interoperability is provided. Non-functional standards like transactions do not fall into this category.

- **Process Standards**. Figure 9.4 contains a few standards that are directly relevant to Semantic Web Services. These are the process related standards BPEL and WS-CDL (BPMN is omitted); these are discussed in Section 10.6. The reason for those being relevant is that processes express behavior and the behavior of different entities that need integration must match, otherwise the communication between those will fail.
- **Discovery Standards**. In context of Web Service discovery UDDI and WSIL are mentioned. UDDI has been discussed earlier in this Chapter.
- **Service Description Standards**. WSDL is important as WSDL supports the description of the interfaces of services and is hence very relevant. WSDL has been described earlier in this Chapter.
- **Transformation Standards**. Entities that communicate with each other follow different data models in general. In order for them to communicate transformation needs to happen from one format into another one. So far no standards organization took up the formation of a standard in the transformation space, although it is very relevant.
- **Business Domain Standards**. These standards focus on the vocabulary of specific business domains. For example, defining possible values of documents in the healthcare domain. These will be used by Semantic Web Services, but will not be the focus of improvement.

---

4. http://xml.coverpages.org (Accessed March 19, 2008).

9.4 Service-oriented Architecture (SOA)

Web Services, both individual and composite, are a methodology underpinned by technology. However, questions of granularity and design are out of the scope of the Web Service technology per se. The technology and standards introduced earlier allow the design and implementation of any type of service, fine or coarse grained, public or company-internal, reliable, transactional or just "plain". The technology components or the standards do not prescribe or impose a specific architecture or methodology, though. These are just technology components and "neutral" building blocks.

In real application environments, however, Web Services have to achieve a business goal, and so they have to fit into an existing computing and data infrastructure consisting of databases, application systems, security components, and so on. At the same time they have to be "enterprise-grade" in the sense that they are "always up" (meaning always available), reliable and transactional, semantically very clear, just to name a few non-functional requirements. While many of these attributes are a matter of proper software engineering, others can be established through a proper software architecture.

Using Web Service technology and standards properly demands an architecture approach around Web Services and this architecture approach is called Service-oriented Architecture (SOA). A brief description is provided next around SOA and the role of Web Services in this context.

9.4.1 Service Paradigm

An important aspect of software architecture is its independence from base technology on one side and from the functional business requirements on the other side. To draw the analogy, a relational database management system (RDBMS) implements its own architecture and functional model independent of the operating system it is running on. While a RDBMS has to adjust to the specifics of a given operating system (e.g. Unix vs. Linux vs. Microsoft Vista), it does not change its functional model. On the other "end" of its architecture a RDBMS implements through languages like SQL and APIs various ways of accessing its functional model (like relational tables and stored procedures). But it does not adjust or change this access layer when used in a specific domain like healthcare or supply chain.

The same is true for a Service-Oriented Architecture (SOA). A SOA needs to establish the model of what a "Service" is, how it relates to other services, which entities can provide services or use them, and how they relate to other elements like data or processes. Furthermore, depending on the scope of a concrete architecture, the effects of services can be part of a SOA if the SOA is concerned about the operational semantics.

The OASIS SOA reference model (RM) [363] provides such a SOA. It defines the notion of services and all its aspects. In addition, it does not assume a specific technology base at all therefore being agnostic toward a specific implementation. On the other hand, it does not restrict itself toward a particular application domain, either. In this sense it is a true software architecture.

This makes it very valuable since with the OASIS SOA RM it is possible to map the SOA architecture onto a technology base without "affecting" the business domain; and at the same time the particular services of a business domain can be specified independent of a particular implementation technology. This clearly indicates a sound conceptual base where the model of services is technology independent and at the same time expressive enough to represent the services of an application or business domain.

9.4.2 SOA and Web Services

Services are not necessarily the same as Web Services. Web Services commonly refer to services implemented based on Web technology and used in a Web context. For example, WSDL, implemented with XML as the encoding for data bound to HTTP/SOAP as the transport, is typically categorized as Web Service. At run time actual business data are transmitted as XML instances.

However, in many situations this is either not necessary or even counter productive. Within enterprises it is not necessary or useful to represent data in transit as XML instances or use HTTP as the transport protocol due to scalability and resource consumption requirements. It can very well be the case that a service is implemented using the Java programming language (and hence the data is encoded as Java objects) and the invocations are taking place over RMI or queueing systems. In such a situation services do not use Web technology, but the same concepts and principles of services as the more general concept apply.

Ideally, services can be defined using WSDL at design time, but at deployment time or at run time it may be decided to use Web technologies or enterprise engineering technologies for their invocation. First steps toward this approach can be found in efforts like WSIF that allow the specification of a service independent of its implementation. So it is possible to define the service interface using Web technologies while at run time it is possible to access the service using enterprise technologies like Java.

A good SOA therefore does not restrict the notion of "service" to Web Services at all, but makes a clear distinction between design time and run time and allows the "mix and match" depending on the particular requirements.

9.4.3 Open Issues and Technical Challenges

Despite the OASIS SOA RM much of the SOA space is unclear at this point and is to large extent technology driven. The technology focus is not a surprise since all software vendors have technology components that address the (Web) services space. In addition, specific perceptions exist that are worth mentioning here. The following issues have to be discussed and resolved in every single service implementation, often very implicitly. Only the most important and imminent ones are called out in the following.

- **Queues, queues, queues and queues**. With services comes the perception of a "plug-and-play" functionality in the sense that services can be developed and made available at any time. Once available, services can (or cannot) be used, as needed by possible Web Service requesters. This plug-and-play idea on a conceptual or design level must be mirrored on a run-time level. The supposedly simplest way to achieve this is to make services available as entry points through queueing systems. A service can be made available for use by providing a request queue for it. Using it then means to enqueue messages into the request queue and enclosing the name of a response queue so that the service knows where to place the result (or error message). At the same time queues introduce asynchronous communication, adding to the complexity of error recovery. Furthermore, there can be many queue communications between services if they call each other. It is not very clear how a service notion based on asynchronous communication plays with user interactions that are preferable synchronous in nature.
- **Transactions**. In corporate environments transaction controlled modifications of data is paramount to be recoverable and consistent at any time. Web Service technology does not make it easy at all to support this base functionality. Many technology components like HTTP connectivity or XML based implementations do not easily (if at all) participate in distributed (or even local) transactions.
- **Data and Services**. Services in their nature are "functional", meaning, they are optimized for the description and execution of functions. In many situations, however, data manipulation is a lot more important then function execution. If everything in an IT architecture is defined as a service, then data access and retrieval as well as modification is consequently implemented as a service, too. For example, issues arise if services have to invoke each other to derive to a data result set that could have been achieved by a single SQL statement. In this situation either the service calls are made (potentially being inefficient), the SQL statement is implemented and made accessible as a service (but then functionality is possibly duplicated) or the database access layer is presented as a service accepting any SQL statement (but that would hide the functionality in the invocation of the layer). The important point here is that the relationship between data and services as concepts is to be clarified in projects from an architectural viewpoint so that the particular way of accessing data is done consistently.
- **Fractal Nature**. In the sense of the Mandelbrot structure from mathematics, services are fractal. A service internally can call services that in turn can call services. Looking at the interface of a service it is not visible at all how deep the invocation chain or tree is going to be. If helper services like logging, monitoring and SLA management are services themselves, then there will be a lot more service invocations going on at run time then purely those required to bring

about the business functionality. If service invocations are expensive in this case (like remote queue invocations), service use can become quite costly and slow.

- **Design Time vs. Run Time**. As discussed earlier, when Web technology is used for defining services a decision has to be made if the same technologies are used at run time or if the run-time technologies should be selected with different criteria in mind. Convenience will lead down the path of using the same technologies; but that is not necessarily the best choice. In general, run-time requirements address performance, throughput and reliability whereas design time requirements address structure, consistency and correctness. In order to achieve those conscious decisions have to be made.
- **Management**. Services have to be managed. Each service being brought online might have to be made available at a specific and scheduled point in time. In other cases, a service might have several versions concurrently, the oldest one possibly scheduled for depreciation. In this case it is important to determine the service consumers and to be able to notify them about the pending depreciation. If errors happen at run time it is important to detect these and analyze those to determine the root cause quickly. Service management has to worry about all these topics and every project or enterprise has to not only be aware of those, but also address them.
- **SLA Supervision**. While service management focusses on the functioning of a service, service level agreement (SLA) supervision focusses on the proper functioning within specific performance and throughput boundaries. If a service degrades it has to be not only flagged, but also notifications might have to be sent out. In addition, additional computing power might have to be brought online in order to compensate for problems.

These topics are a few important ones that have to be clarified in service implementation projects. Ideally, a SOA addresses those in a specific project or enterprise so that all services follow the same approach, be it their design time specification, their run-time behavior or their management.

9.5 Semantics and Web Services

The whole Web Service community works currently on a syntactic technology stack, meaning, various technology components that together allow the definition and execution of services addressing the design as well as run-time requirements. All standards and specifications revolve around syntactic means to define interfaces, data formats as well as non-functional properties. Their correct use and interpretation is left to the engineers. For example, a service requester has to ensure that the data format required by a service is the same as that it uses when invoking this service. If such a data is syntactically correct, but semantically incorrect, the service infrastructure cannot detect it automatically and instead of throwing a "semantic error" it allows the service invocation to take place.

Ensuring that implementations of Web Services are interoperable is governed by the Web Service Interoperability Organization (WS-I)[5] determining if the software is interoperable on a syntactic level. Major software vendors are engaged as it is in their very interest that their software interoperates properly.

The reason why ideally interoperability has to be mechanically proven is that the syntactic standards do not contain enough semantics so that the different software technologies are interoperable by construction. This is the basic reason for trying to enhance Web Service technologies with semantic technology. The goal is to make sure that Web Services are semantically interoperable (in addition to being syntactically interoperable) so that any semantic mismatch can be detected by the software infrastructure. Of course, this means that semantic technology will also be used at design time to define the services properly from a semantics perspective.

9.5.1 Semantics, What Semantics?

On a critical note, it sounds like services and their interoperability will not work unless the semantic technology is applied properly to the areas of services. Many research papers state that without semantics service invocations will fail, data will be inconsistent, and so on.

At the same time, organizations have their software systems interact for a very long time already over networks using various standards in the B2B space and more recently using Web Services. Not only data is transmitted for informational purposes, but real transactional data like money transfers, payments, orders, and so on.

The situation looks like an obvious contradiction where one community expects big failures while the other community seem to use Web Service technology without any problems.

Looking at the situation more closely it is not a contradiction at all. Rather, it is the case that different communities solve the same problem in very different ways. Looking at industry first, it can be observed that the service descriptions are mostly of syntactic nature (aside from language elements like enumeration types, and these can be debated, too). Any semantic mismatch is left to be detected and sorted out at run time. Furthermore, the communication infrastructure is not responsible for the content and its correct semantics, but the business application systems like home grown applications or off-the-shelf ERP systems. Any mismatch will come up in the business context and will be corrected there. This in principle means that the semantic correctness is ensured during the "last mile" in service invocations and error are reported back to the "first mile".

The research community, however, is working on solving the semantic mismatch detection into design time or at most to the invocation initiation time. This means that if a service consumer and provider already mismatch on a definition, an

5. http://www.ws-i.org/ (Accessed March 19, 2007).

invocation will never take place. If the semantic match cannot be 100% determined based on the service definitions because run-time data values are needed, then additional matching checks take place at service invocation initiation when the run-time values are present. If a mismatch is detected, the service invocation will not take place.

From an industrial perspective this means that (ideally) the business applications will never have to deal with semantically incorrect or mismatching data as the service infrastructure has caught any issues long before. From an industrial perspective this is certainly desirable. At the same time it moves a lot of development effort upstream into the design time of services. It remains to be seen if this is practically possible in industrial settings.

A service invocation has many aspects and from a functional side data, processes and service selection can be called out as important areas for applying semantic technology. These three areas are discussed next in more detail.

9.5.2 Data Semantics

Although semantic interoperability is a very difficult property to achieve, even the simplest example highlights the severity of the problems. Let's use as example an international traveller who resides in the US and is on a trip to Japan. There he has to visit a hospital and that hospital sends a message to the traveller's healthcare provider in the US asking for the coverage. As part of this transmission the date of admission into the hospital is included. The data structure on a syntactic level has and attribute "admission date" with elements "day", "month" and "year". An example date would be "11" for the day, "June" for the month and "2007" for the year. However, the message contains "11", "June" and "20". Syntactically this is a correct date, semantically there is a mismatch as in the US "2008" is expected (Gregorian calendar), but the Japanese hospital put in the Imperial year "20" which is equivalent to "2008" semantically.

This mismatch cannot be detected at a service interface level as the mismatch is based on the data values. Consequently the invocation succeeds and the software implementation has to deal with this semantic mismatch.

This type of semantics is called data semantics as it is about the interpretation of data values. The goal is to ensure that the sender of data (or the service invoker) and the message receiver (or the service provider) have the same understanding of the data that is communicated back and forth between them. Instead of leaving it to the implementation of services to detect any semantic data mismatches the goal must be to detect them at design time as well as service invocation time. This reduces the risk of service invocation failure, and in addition relieves the service implementations from establishing the correct data semantics as the service run time system would ensure that service invocations only take place if the data is semantically correct. The goal of Semantic Web Services is to exactly achieve this.

A next step would be data consistency. Data consistency refers to correct data in the sense of its pragmatics. For example, an admission date of June 11th, 1898 would be a correct data, but it would not be in a pragmatic sense as this date is very old and cannot be a valid admission date for a patient living today. However, in a historical data exchange about a patient back then it would be consistent data. This example highlights that the correctness of data also depends on the context in which it is used, not only on its syntactical structure and correct data values.

Another important topic that plays a role in this context too is the trustworthiness of data. Trustworthiness depends a lot on where data came from and which parties were involved in the establishment, change and forwarding of the data. We do not discuss trust in more detail here, but the area of trust is getting more and more attention and is focussed on more and more in various communities.

9.5.3 Process Semantics

Process semantics is another fundamental aspect in context of Web Services. Process semantics is the semantic match of behavior of a service provider and its service consumers. For example, a service provider might require a three step invocation in order to determine the coverage of a patient. A first invocation is to establish the validity of the calling hospital, the second invocation is to establish the validity of the patient and the third invocation is to establish the coverage. The service provider only can provide the coverage if these three invocations are executed successfully and completely in precisely this order. A service consumer, like the hospital in Japan, has to conform to this sequence or "behavior" that is expected from it. If it cannot comply, the coverage of the patient cannot be established.

This match of the behavior exposed by a service provider and expected behavior of a service consumer is termed "process semantics". Like in the case of data semantics, it is possible to establish the match at run time when the service provider keeps track of the invocations and raises an error in case a service consumer initiates a wrong invocation. In this scenario the implementation of the services must be able to deal with all error situations.

Alternatively the match could be established at design time with a proper definition of the interface and at run time by the service framework that would only allow the invocation of correctly sequenced services. In this case the service implementations would not have to control the correct behavior; it would be done for them by the service infrastructure.

Semantic Web Services have process semantics as their second big goal so that services are not only properly defined, but also supervised at run time.

9.5.4 Selection Semantics

A third big topic in context of Semantic Web Services is the notion of service discovery. Service discovery is based on the assumption that there will be too many services available to know them all and in such completeness and detail that a service consumer can determine the perfect one. The assumption is that on a Web scale there will be so many services available that it is impossible to know them all. Furthermore, services will become available and unavailable, change their versions and service levels so that it is not the norm that once a service works it will do so forever. Based on these assumptions it is necessary to find or to discover services.

Research focusses on two approaches: one is to change the situation into a more stable one where services are made known and their availability be managed in repositories or directories. Instead of constantly being on the outlook on the Web for appropriate services, a service consumer looks up a structured repository and searches for what it needs. Service provider have to take great care in publishing their services in the relevant directories. The upside of this approach is the structured way of dealing with the service detection and the completeness assumption that if a service is not in a directory, it does not exist.

A second approach tries to acknowledge and to work with the nature of the Web and discovers services dynamically in "search" style. Services are detected on the Web (as web pages are) in form of Web searches. This is not necessarily reliable or complete, but follows the architecture of the Web. The idea here is that a reliable and useful service will be detected and will be reliable in its own interest.

There are many pro and con arguments for both these focus areas which are not discussed here except one. The proponents of the dynamic discover claim that the Web is by nature distributed and using a centralized directory contradicts this approach. The proponents of the directory style of service discovery claim that services have to be a lot more precisely defined and continuously searching the Web for new services is not practical. Despite the upsides and downsides of each approach it can be expected that both approaches will become relevant in different contexts. A possible scenario is that the directory approach will prevail within organizations where services are under the organization's control. The discovery approach might be used on a Web scale for finding services initially.

However, the important aspect is that no matter how a service and its provider was detected, it has to semantically match the expectations of the service consumer. The more semantically precise the web service definition provided by a service provider, the better the service consumers can search for services.

9.5.5 Other Types of Semantics

In addition to the major three areas for semantics described so far there are other areas, for example security, transactionality, trust, and more. Each of these additional areas require that a service consumer and a service provider understand each other perfectly well in their expectations and functionality. Therefore it is possible and necessary to ensure that semantically all parties involved in service invocation and execution understand and agree with each other perfectly well.

An observation is that areas that are commonplace today require less work or are already in a state where more work around semantics is not urgently required. Examples are transactions as they have been in mainstream research and industrial use for a long time. Other areas, like trust, are relatively new and require extensive work in order to be understood well and consistently.

9.6 Clinical Use Case

Unless it is possible to start building an application like the clinical use case from the very beginning, many infrastructure choices are made already by an organization and any new development has to be put into the existing context for organizational, financial and efficiency reasons. Currently it is safe to assume that many new developments in general are based on the traditional Web Service technology as the predominant infrastructure choice for some time to come. Traditional Web Service technology can be used in two roles. One role is being a distributed computing infrastructure and the other role is being interface technology to existing application system functionality. We discussed the former role first.

Semantic Web Services, as we will see later, are concerned about semantics on a higher level, describing interfaces, data and goals in context of the application functionality. They are not really concerned about the technical remote invocation infrastructure or data storage technologies themselves. We therefore assume that in the clinical use case the infrastructure is based on traditional Web Service technology. This is achieved by nicely abstracting away from it in a general system architecture by defining a clear layer that implements this abstraction. One example is the WSMX architecture (see Chapter 11). This architecture uses traditional Web Service technology as a foundation for executing Semantic Web Services. The Semantic Web Service designer does not have to be concerned about Web Service technology as a foundation as this is not visible on a design level. Only Semantic Web Services are visible and this is the only interface. The benefit of this approach is that proven technology is used while the Semantic Technologies are layered on top of those.

This is the reason that in the clinical use case traditional Web Service technology is invisible on an application definition level. Only Semantic Technology is visible and all requirements can be expressed through it. As Chapter 13 will show, all services as well as data structures are purely defined using Semantic Web technology. Any underlying non-semantic technology is hidden while being used.

The second role of traditional Web Service technology is interfacing technology to existing application systems. It can very well be the case that a clinic has one or more application systems that are already implemented and in use based on technology from a few generations ago. In this case it is possible to wrap the applications' interfaces through Web Service technology to make it accessible. Modern SOAs or even WSMX are able to connect to those bridging the existing systems to Semantic Web technology. In the clinical use case there is no system like this, however, it would be possible to include it.

In summary, for the clinical use case, we assume traditional Web Service technology as well as SOA as foundation layer for the execution of functionality. The specific clinical use case features are implemented using Semantic Web Technology that is transparently mapped to the foundation layer.

9.7 Summary

In summary, the notion of "Web Services" is around for a few years at this point in time and has entered mainstream development and use in commercial organizations. The underlying principles, of course, are known for a long time and the Web Service field can stand on those findings and results, take them up and refine them over time. This is currently done by building a whole set of Web Service technologies that allows remote communications including all necessary aspects like security, reliability, or trust.

Since Web Service technology is considered a collection of technology components, architectures for Web Services including methodology of how to use Web Services properly are needed and very much in the initial works at this point in time. This will make Web Services even more a basis for interoperable application development.

The semantic aspects of Web Services are currently not part of the core technology components. However, research efforts as well as standardization efforts exist already with the goal to influence and extend the Web Service technology components with Semantic Web technology. The following Chapter 11 outlines the current approaches in a lot more detail.

10 Web Service Composition

Web Services are often invoked in isolation, meaning, a Web Service operation provided by a Web Service provider is invoked by a Web Service requester. Based on the input message, the operation provides the result in form of an output message. However, in many cases an individual invocation is insufficient to achieve a more complicated communication or execution scenario that a Web Service requester needs to achieve, especially in the case of long-running communication as discussed in Section 8.3. In this case, several Web Service operations have to be invoked in a specified order to achieve the final outcome. For example, when an invoice for a patient is to be put together, several services are invoked to retrieve the patient's address, to obtain the line items that require payment, to compute the total amount the patient is liable for, and so on. The end state of the invoice is therefore the result of a series of service invocations. Web Service composition is the concept that provides a means for defining and executing explicitly long-running communication involving several Web Service invocations in an explicitly specified way.

Section 10.1 introduces the concept of composition in more detail. Section 10.2 outlines dynamic composition as an alternative approach to static composition. The Sections 10.3, 10.4 and 10.5 discuss how service composition is applied to B2B integration, EAI and complex business logic, respectively. Standards in the area of composition are discussed in Section 10.6 and Section 10.8 summarizes.

10.1 Composition

10.1.1 Motivation

Composition is not a new concept at all. On all levels of information systems engineering, be it on an architectural level or an implementation level, the approach of "divide and conquer" is predominantly used in order to structure a system at design time. In principle terms, the functionality is organized in well-defined and well-contained units that separate independent functionality from each other. Some are base functions like managing an address (the "divided" parts or components), others are tying the base functions together like constructing an invoice (the "conquer" part or composition).

The divide and conquer approach supports many design features concurrently, each of which is important and should be supported at the same time. One is functionality abstraction where the interface and the implementation are clearly separated. The interface defines the operations and (input/output) parameters in order to interact with a component, whereby the implementation is hidden behind the interface and implements the functionality of the component.

Modularization is the act of defining the components according to the containment principle, i. e., that a component should be responsible for one well-defined, contained and complete set of functionality. Depending on the particular information system in question the number of components might be low or high depending on the number of concepts and functions that have to be implemented for these concepts. Experience shows that the more the various components are independent of each other the easier the composition of these components will be and their future evolution.

Flexibility is another important feature for system design as business functionality changes usually require one or several system changes in many cases. Ideally, a specific change happens in a specific component, leaving other components intact so that changes are only local to a component. This provides a huge amount of flexibility as changes in many cases do not require revisiting a system as a whole, but only small parts of it, requiring only minimal effort with minimal overall system impact.

There are two main areas where composition is predominantly discussed in the context of Web Services, communication and complex business logic. Composition in the context of communication deals with the interaction of different independent entities like companies or application systems (the former one usually termed B2B integration, the latter EAI). This type of composition ties together existing systems as they are, independent of their location within an enterprise (the EAI case) or in different enterprises (the B2B integration case). The challenge with these systems is that they function independently and autonomously change their state (autonomous systems), so the composition has to be aware and cognisant of their autonomy.

Composition in the context of complex business logic does not coordinate independently acting systems, but invokes business functionality directly that is not part of another system, but only meant to be executed in the context of compositions. For example, retrieving a patient's address is used in many compositions, but is not executed independently outside those compositions.

Composition is usually assumed to be explicit composition, meaning, that there is a separate concept and implementation of "composition" that can be retrieved and managed independently of the components it invokes. Furthermore, it is assumed to be defined using a declarative language, for example WS-BPEL [266] or other languages like those discussed in Section 10.6.

However, explicit composition is not the only possible approach as implicit composition is possible, too. An implicit composition, for example using a programming language, invokes components, too, and in the specified order. However, it cannot be independently managed and is implemented by a programming language. While this is a perfectly valid approach, explicit and declarative compositions are preferred due to the clear separation of the composition from the components it invokes.

This motivation outlines the major corner stones that are relevant in the context of composition while the next few subsections provide a more detailed discussion of various aspects.

10.1.2 Definition of Composition

Independent of a composition being explicitly defined with a declarative language or implicitly implemented with a programming language, the same technical concepts have to be implemented and the same requirements fulfilled. Both cases are summarized in a concept that is referred to as composition (or composing object). The composition has an identity so it can be referred to it. At run time a composition can be instantiated many times so that it can be executed concurrently for different purposes or business cases. Following the example above, the invoices for many patients can be put together concurrently through as many concurrent executing compositions as necessary e.g. one for each patient.

This in turn means that the components that are composed by the composition can be invoked concurrently many times. Components that are invoked by a composition are referred to as composed objects or composed components. Since it is possible that a specific component invocation fails (retrieving an address for a non-existing patient), compositions and the composed components cannot only be in different states of execution, but also in different states of error or failure situations. It is assumed that the execution infrastructure is able to handle this concurrency without any problem.

A composition has to define the order in which it invokes components. This invocation order can be sequential, but can also include conditional branching or parallel branches as well as loops. If the order is not relevant, the order can be random, too, or unspecified (and it will be accomplished through other constraints like data availability or resource availability). Control flow constructs in a composition are very much like those that have been introduced by the workflow community for quite a while [269]. Other, more current efforts are for example WS-BPEL [266], OWL-S [272] or WSMO [276].

In order to make a composition useful the order of component invocations has to be established as well as the data that the components receive and return. And as different composed components work together, the output data of one might be the input data of another one. For example, retrieving the address of a patient requires a patient identifier. Once the address has been retrieved, it has to be passed on to the component that puts together the invoice in order for the correct address to show up for a given patient. So the output of the address retrieving component has to flow to the component that puts together the invoice.

Traditionally data flow is not modeled explicitly like shown in MOBILE [269], but implicit like in WS-BPEL [266]. However, explicit or implicit, all mandatory input parameters of the composed components have to be bound. Figure 10.1 shows the main concepts of composition.

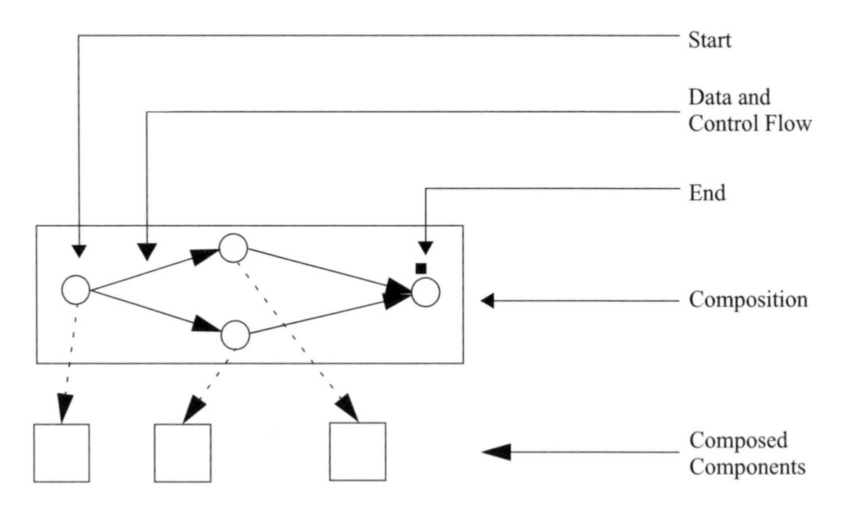

Fig. 10.1. Conceptual elements of Composition

A composition does not only retrieve data by invoking components, but components themselves also update data in data sources like databases or application systems. In enterprise settings this requires transactional control so that the database updates that a component executes take place completely or not at all. In case of failures data must stay consistent and not become corrupted. This means that components have to be transactionally controlled and executed in transaction contexts.

In addition, as discussed above, compositions themselves have state. A composition invokes components according to the defined control flow between the components. If one component's invocation is finished, the composition's state changed and it needs to advance to invoke the next composed components. This state must be stored, too, in order to be able to recover to this state in case of failures. For example, if the system crashes that executes the composition then the crash should not compromise the state the composition reached up to this point. At restart time, the composition needs to recover to the last consistent state it reached. Therefore, a composition's state change needs to be transactional, too, in order to establish a consistent state.

This transactional behavior, i. e., that a composition has internal states that are transactionally committed, is also called long-running transactions. Each step along its execution is recoverable while each state transition (invocation of a component) is transactional, too. In this sense, multi-step EAI or B2B transactions are long-running transactions. Complex business logic, however, does not necessarily have to follow this principle as in a complex business logic it might not be necessary to transact every state change, but all component invocations in one "big" transaction. The reason for this is that complex business logic might be invoking several components that all should be executed as a unit. For this to happen it is necessary that a composition can be configured to be a long-running transaction or to invoke all components inside a single, possibly distributed transaction (distributed transactions are needed in the case when several independent databases are involved in the composition).

As indicated above, each invocation of a component can fail and this error has to be recognized by the invoking composition. The composition has to decide how to respond to an error, by retrying or by passing on the error to its own invoker. If the error cannot be resolved, and if at the time of error several previous invocations were successful, it might be necessary to compensate for already achieved and committed invocations [268]. This compensation has to undo already committed states in order to ensure that after the compensation finishes it leaves a consistent data state. For example, if the insurance benefits structure of a patient changes at the time an invoice is put together it might be necessary to invalidate an already produced invoice (before it is sent out) to recreate it based on the new benefits structure. Since the original invoice was already stored and committed, it has to be marked invalid (the compensation) and superseded by a new version based on the new benefits.

The concept of composition is general and can be applied to Web Services, too. The next subsection will highlight the specifics in the context of Web Services.

10.1.3 Web Services and Composition

In short, a Web Service composition is a composition where the invoked components are Web Services. The composition therefore uses exclusively Web Services in order to achieve its own goals. All the general principles and definitions from the previous subsection apply.

Web Services following the WSDL model have several interaction patterns, like one-way or request-reply invocations. Compositions have to be able to use any of the invocation types, of course. From a composition viewpoint this means that it needs to be able to invoke Web Services as they are defined by their providers. For example, if the provider has a request-reply pattern, the composition must be able to issue a request message and be contacted by the Web Services for the reply message as this is an asynchronous pattern.

In addition it can be asked if the composition is a Web Service itself. In general, this can be the case, but does not have to be the case. A composition can be implemented as a Web Service itself. In this case the invoker assumes to be invoking a Web Service. However, as it is a composition, it in turn invokes other web services. But a composition does not have to be a Web Service itself, it can be of a different nature, like for example a java class invoking Web Services as invoked objects.

An interesting question is if a composition can mix and match components that are implemented as Web Services as well as implemented in other ways, like Java classes. The answer is a clear "yes" and "no". For example, the Oracle BPEL process manager [275] shows how this can work. It only allows Web Services to be integrated into the Web Service composition. However, through the Web Service Invocation Framework (WSIF) [284] it allows Java classes to appear as Web Services through different bindings. So from a composition viewpoint in this case only Web Services are invoked. However, from the viewpoint of the Java class it assumes it is invoked directly. Through the WSIF framework an indirection is provided that freely allows the use of different types of components, while the composition assumes that only Web Services are invoked.

10.1.4 Choreography and Orchestration

There is a lot of confusion and discussion around the terminology of choreography and orchestration [279]. Independent of this (ongoing) discussion, technically there are two sides of a composition that can be observed and are relevant and must be discussed independently from the discussion cited above.

A composition invokes invoked components one after another as specified by control flow and data flow. In this sense it makes components work together in a specific way (and the components are not aware of each other). The order of execution, the selection of which components to invoke and the particular handling of errors or exceptional situations are fully controlled by the composition itself and its internal knowledge. In order to use a composition successfully it is not necessary to know or even to understand its internals. For example, if invoices for patients needs to be put together, the requester is only interested in the invoices and in those cases where there was an error and why there was an error. The details behind the composition that puts together invoices are irrelevant for the requester of the invoices. So the outcome of the composition is important, not its internal implementation.

The other side of a composition is its behavior in relation to the requester of it itself. As mentioned in the example, the invoker of the invoice producing composition is interested in the possible outcomes, or in other words, the externally visible behavior. In this particular example the outcome can be a (potentially empty) list of invoices and a (potentially empty) list of patients for which it was not possible to put together an invoice together with reasons. In a more advanced case it could be that the composition asks back for details in order to put together an invoice. For example, if the insurance of a patient is not responsible any more the composition could ask back if an alternative insurance is to be used (instead of assuming that this situation represents an error).

In summary, a composition has an internal behavior that invokes components in a specific order and has an external behavior that defines how it interacts with its own invoker. One way to try to clear up the confusion would be to call the internal behavior "orchestration" and the external behavior "choreography", but that would be only one more contribution to the discussion. In this sense the jury is still out on the precise definition of the two terms.

10.2 Dynamic Composition

Compositions have to be designed and their internal behavior specified. The components or Web Services to be invoked have to be identified, their interfaces understood and then their invocation order defined using control flow and data flow directives. In general, this is a design task leading to a composition implementation. Independent of the explicit or implicit approach, in the end a software engineer or developer manually specifies the composition, tests it and releases it into production based on a release schedule. This specification can be created using a declarative or procedural programming language. Alternatively, graphical design environments are available for it, too.

When designing compositions manually it is the software engineer who has to ensure that the composition not only works syntactically, but also semantically in the sense that the execution of the composition at run time "makes sense" and leads to a desirable state of the business (i.e., its application systems as well as data and processes). In this case the graphical design environment or the run-time environment for compositions is not able to verify if the composition design is leading toward this goal. Semantic Web Services, as discussed in a later Chapter, will address specifically how Semantic Web technology is used in order to support not only syntactical but also semantically correct composition.

This manual specification of a composition leads to static compositions in the sense that their internal behavior is fixed at design time. At run time it cannot be changed and only instances of the fixed definition can be created and executed, as discussed already earlier. Any necessary change has to be applied to the static composition design and again released according to the release schedule into the production environment. Already running instances are not affected by the change.

Of course, through conditional branching the static compositions can adjust to conditions at run time so that specific components are only invoked when the conditions hold. However, in this case all conditions have to be defined at design time as well as all the components that might be invoked. While it is possible in many cases to enumerate all possible conditions, often it is not.

Manual composition design is not the only approach, however. Approaches from the planning community exist that allow a composition be dynamically constructed (and executed) during run time. In principle this works through an infrastructure that can be given a formal goal. The infrastructure, or planner, examines the goal and retrieves those components that either can fulfill this goal directly or are considered a partial solution to achieve the goal. In the former case the solution is given by executing the component. In the latter case the planner searches for another component that in tandem with the already found one can achieve the goal. This would be the beginning of a composition with two components. Of course, it can be that several additional components are necessary. Or, in the worst case, there are no components found, in this case the goal cannot be achieved automatically.

[267] puts the discussion of pure planning in the context of Web Services and outlines specific problems in context of this technology. One problem is the data heterogeneity. How does a planner know that e.g. a date of 2/20/2008 corresponds to a date of 20.2.20? To resolve this syntactic mismatch a data transformation engine is required that knows how to format a date in American syntax into a data in the Japanese Imperial date format and vice versa. In addition, Web Services allow side effects that are not known or visible at the interface. How can a planner take this into consideration? To resolve this issue a more elaborate interface definition is required that captures the side effect while the data changes are invisible at the interface level at run time.

Dynamic composition specification through planning is not available yet in industrial mainstream implementations of service composition. Currently it remains in the domain of research or very specific problem domains. Additional references in this space are [273] and [283]. [280] show additionally how a plan maps to composition languages like WS-BPEL (see Section 10.6 below). In the discussion that follows we assume a manual composition design, not an automatic one.

10.3 Business-to-Business Communication

The main characteristics of B2B communication is the conversational exchange of business data between trading partners. One trading partner starts the communication and the other trading partner responds. Many of these exchanges take place according to a predetermined protocol that in the end form a conversation with a defined end state that reach a positive (or negative) business agreement. Each interaction requires one trading partner to send a message containing the business data and the other trading partner to be ready to receive the message. In addition, depending on content or other conditions, different types of messages might be sent as well as a different number of messages. For example, a fulfillable order might be acknowledged, an unfulfillable one might be rejected, maybe containing data indicating what could be fulfilled instead.

Composition is an ideal approach for all involved trading partners to design their part of the overall communication protocol. Each trading partner has its own composition whereby each sending of a message of one trading partner corresponds to the receiving of a message at another trading partner. In this sense the compositions are complementary. If the composition of a trading partner does not accommodate a message that is sent by the composition of another trading partner, an error occurs and the conversation might not be successful.

If one trading partner can conditionally send different messages, the other trading partner needs to be able to receive any of those at this stage of the composition as it does not know the result of the condition evaluation. If each receiving activity is designed as a component invocation (for example, a blocking invocation that returns when a message is received), then several component invocations have to be concurrently take place in order to receive several messages.

In the context of Web Services, the sending as well as receiving of messages can be implemented as Web Service invocations that span across the Internet between the involved trading partners. A trading partner sending a document to another one does this through a Web Service invocation. For example, a hospital sending a claim to a health insurance company can do so by invoking a Web Service at the insurance and providing the invoice as input message to that Web Service.

Composition is therefore a good way to implement B2B protocols, especially in the Web Service context.

10.4 Application-to-Application Communication

B2B integration and Enterprise Application Integration (EAI) are very similar in nature to the extent where the exact same composition concepts can be applied. Like in B2B integration, different components are invoked that either retrieve from or update data into application systems. These invocations have to happen in a specific order, too, in order to arrive to a consistent business outcome. Application's internal processes or execution states have to be ready, too, to receive data, so that the order of communication between a composition and application cannot be random but must be well defined.

However, there are some interesting differences between B2B and EAI that are important to mention here. They do not impose specific requirements on the conceptual model of composition, however, they apply composition concepts in different ways.

EAI connects application systems and databases that are managed by enterprises internally. These are available on the local network and because of data consistency it is very important that any data access is done under transaction control. In an EAI context the invocation of application systems from compositions is performed under transactional control. Transactions across company boundaries are not ready yet for a wide deployment, so "access" to trading partners is non-transactional at present.

Due to missing transactions in B2B interactions the communication is based on the request/reply pattern. So every non-transactional message sent is answered by a non-transactional acknowledgement message as a general principle. Together with proper setup for time-out it is possible to at least guarantee that messages are received by the recipient. In an EAI context, the communication with application systems or databases is transactional. Therefore, the composition knows if an invocation was successful or not and, in consequence, one-way invocations are sufficient and no independent reply messages are necessary.

Another topic that is often seen as a differentiation between B2B and EAI is security. The perception is that security is really important when crossing companies' boundaries, but not when applications are accessed. However, in many cases, especially with larger organizations, accessing an application system is as secured as communicating with a trading partner. So in the general case, security is required and cannot be dropped or ignored at all.

In B2B integration every partner that is part of the B2B communication has its own composition that it is following, complementary to the one the trading partner it is communication with. So, for a binary B2B communication, two compositions are involved. EAI is assumed to be different in the sense that only one composition can connect all required application systems. There is no need for several compositions per se. While this is the common perception, in reality it does not reflect the nature of application systems. Those by themselves expose behavior that can be captured by a composition. The EAI case therefore is really tying together the compositions of the various application systems; [Bussler 2003] discusses this in detail.

Sometimes it is argued that application systems internally implement their own compositions. Looking at application systems in detail shows that application systems internally have business processes implemented as workflows. These are very similar to compositions in the sense that different application system internal components are composed so that the application system can implement complex internal logic. This internal workflow or composition, however, is not externally visible as such as it is completely executed internally within application systems. Only if application system logic needs data from outside the application system this need surfaces on the application system boundary. At that point the application systems in waiting for data from a source that is external to it. The same can happen if the application system makes data available to the outside world. In this case an external program or service has to access the application system to retrieve the data.

Only when the application system provides or requests data externally, can an external composition interact with it because only then does the application system logic wait for interaction. In all other cases the application system does not recognize outside activity and therefore the introduction of any required interaction with an external composition would require a change to the application system itself.

In summary, composition can be used to achieve application integration in the sense of EAI with the same concepts that can achieve the integration of businesses implementing B2B integration.

10.5 Complex Business Logic

In addition to B2B integration and EAI there is the third case where the required business logic is not implemented yet at all by trading partners or application systems. In this case a design has to be established that outlines how the business logic is implemented, as components, as composition, or both, possibly involving application systems or B2B partners.

Whenever different functionality has to be combined, like retrieving a patient's address and line items for an invoice the question arises if this should be done by a composition or hidden inside a component itself. On various levels of abstraction combination of business logic happens that needs to be implemented.

It would be very powerful to have one mechanism of composition for all levels of composition, be it on the highest level (like tracking and driving the process of how a patient is treated in a hospital) or the most detailed level (like retrieving the various elements of a patient's address). Having stated this wish it must be understood that the various levels have different characteristics.

On the higher levels compositions behave like long-running transactions where each component invocation is persisted recoverably. As the composition progresses through its component invocations it persists its own state after each component invocation so that the overall system state is consistent (and recoverable).

On lower levels this behavior might be the same, but there is also the other behavior required where the invocations of all components is tied together in one single transaction. Instead of long-running transaction behavior the behavior of a single transaction is required. If a composition is a single transaction then it appears itself as a component that does not have internal state that is persisted in individual steps.

Having a composition model that at the same time can implement long-running transaction behavior as well as single transaction behavior would be extremely helpful in system design. It would allow the system construction to be independent of the underlying technology and transaction behavior could be set according to the needs.

10.6 Standards and Technologies

There are several composition languages proposed by various groups for the explicit definition and execution of compositions. They can be roughly divided into syntactic and semantic languages. The semantic languages for composition are discussed in Chapter 11 (Semantic Web Services) and Chapter 12 (Semantic Web Standards).

The syntactic languages are briefly reviewed in the following as they do not address the main topic of the book. Also, many languages have been proposed over time (the last 20 years actually). However, most of them became obsolete in the sense that they never made it into software products (or if they made it into products, they are currently superseded by the standards discussed next). In the following we restrict ourselves to those that are currently very visible and in the process of being either implemented or already in use by enterprises.

10.6.1 Web Services Business Process Execution Language (WS-BPEL)

The Web Service Business Process Execution Language (WS-BPEL) [266] is currently the most prominent language for composition. Historically it has its roots in mainly two languages, XLANG [278] proposed by Microsoft and WSFL [271] proposed by IBM. Wikipedia [281] states:

"*IBM and Microsoft had each defined their own, fairly similar, 'programming in the large' languages, WSFL and XLANG, respectively. IBM and Microsoft decided to combine these languages into a new language, BPEL4WS. In April 2003, BEA Systems, IBM, Microsoft, SAP and Siebel Systems submitted BPEL4WS 1.1 to OASIS for standardization via the Web Services BPEL Technical Committee. Although BPEL4WS appeared as both a 1.0 and 1.1 version, the OASIS WS-BPEL technical committee voted on 14 September 2004 to name their spec WS-BPEL 2.0. This change in name was done to align BPEL with other Web Service standard naming conventions which start with WS- and accounts for the significant enhancements between* ***BPEL4WS*** *1.1 and* ***WS-BPEL*** *2.0. If you are not discussing a specific version,* ***BPEL*** *is sufficient.*"

In fundamental terms, BPEL is a XML-based language for the definition of long-running processes and compositions. BPEL is targeted to integrate services of one, two or more parties in order to achieve complex integration patterns. In order to differentiate between the observable behavior and the internal implementation of the composition BPEL supports the notion of abstract (or public) processes that define the observable behavior and executable (or private) processes that implement the processes. Abstract processes cannot be executed, while executable processes can be.

The services of different parties that are integrated through a BPEL process have to comply to the WSDL interface definition language. Using the WSIF [284] approach to binding of implementations ensures that not only Web Services can be integrated, but also other forms of implementations like programming language concepts (EJBs for example). Not only does BPEL utilize WSDL specifications of services, its internal data flow model is based on XML, too. This in turn makes BPEL use XPath as access path definition of data elements as well as XSLT transformation to transform data from one into another data model.

A process defined with BPEL is a reusable entity that can be used as subprocess within another process. This way it is possible to encapsulate specific process behavior in fragments of processes that can be combined later on to more complex and comprehensive processes.

Errors and faults can happen during process execution and BPEL provides for fault handlers that define the process behavior at execution time when an invoked service does not return or return with an error message.

BPEL has more detailed features as a complete process definition language that can be found in [266]. BPEL is currently "the" process execution language and it supported by all major software vendors.

10.6.2 Business Process Modeling Notation (BPMN)

The Business Process Modeling Notation (BPMN) [274] was developed with the clear direction and goal to be a modeling notation for modeling business processes so that several stakeholders have a common language to communicate requirements, solutions and insights. The scope clearly excluded related models that are models in their own right like organization structures, data models, and so forth. BPMN references those, but does not provide modeling language constructs for those.

Like BPEL it supports the notion of abstract (public) and private (internal) processes, but it also adds the notion of collaboration (global processes). Collaborations define how different parties relate to each other and how the message exchange patterns look like that are established between the parties.

BPMN provides a graphical notation with the expectation that this will make the understanding and communication a lot easier then a text-based language using XML or similar approaches. The graphical notation has four categories of diagrams: flow objects (like events and activities), connecting objects (like sequence and message flows), swimlanes (like pools and lanes) and artifacts (like data objects and annotations). These graphical modeling symbols are used to draw process diagrams. As these symbols are defined, the meaning of the process diagrams becomes clear. BPMN has a comprehensive set of modeling constructs that are introduced in the specification document [274].

Like in BPEL, it is possible to define compensating actions for activities in case they fail or result in an error. Having this specification at a process modeling level is as important as on an execution level as it is necessary to establish a consistent state in presence of failures.

The ultimate situation would be if a graphical notation like BPMN could be mapped to an execution-oriented language like BPEL. In this case it is possible to not only model the processes, but also to execute them according to the graphical model. A formal mapping from BPMN to BPEL exists in order to enable this mapping and several software vendors support the automatic generation of BPEL processes based on the BPMN notation.

10.6.3 Web Service Choreography Description Language (WS-CDL)

The Web Service Choreography Description Language (WS-CDL) [270] is a specialized language for defining the behavior between collaborating parties. It describes the common and observable behavior of the parties, independent of how this behavior is implemented internally within the involved parties. Once the common behavior is defined the description can be used as the basis for a contact between the involved parties.

WS-CDL is an XML based language that provides the language elements necessary to define behavior between parties. Amongst the elements are role type, relationship type, participant type, information types, activities, exceptions, and choreography. These various elements allow the definition of the precise interactions between parties.

WS-CDL is not (yet) implemented by the main software vendors. There is no official reason or justification for this lack of implementation. At the same time WS-CDL overlaps with BPMN as well as BPEL as the latter languages also claim to be able to model the behavior of interacting parties.

10.6.4 Java Business Integration (JBI)

The three above introduced languages, WS-BPEL, BPMN and WS-CDL, are languages that can implement compositions or processes. Once these languages are implemented and a run time environment is provided for their interpretation, compositions can be implemented.

Although these languages are generic with respect to a particular domain like supply chain, healthcare, and so on, it is not clear if they can implement all possible requirements from all domains. It might be very well the case that in a particular domain a requirement exists that cannot be implemented by these languages out-of-box. For example, [282] provides some control flow patterns that cannot directly be modeled by BPEL. In order to achieve those patterns a workaround would be necessary outside the language definition.

This situation is addressed by Java Business Integration (JBI) [277] in a very different way. Instead of defining a modeling language, JBI provides an integration programming language interface that allows a plug-and-play functionality of protocol implementations based on WSDL. This is based on the idea that basically all long-running protocols and interactions are based on fundamental one-way or two-way message exchanges. JBI provides the notion of an environment that supports basic message exchanges between components. The current and modern acronym is ESB (Enterprise Service Bus).

Any component can be plugged into this JBI environment as long as it conforms to the WSDL message exchange types. A component itself then can implement any long-running functionality that might be required. JBI in this way decouples the implementation of composition functionality from the communication with components.

For example, a JBI plug-in can be built that allows the execution of BPEL. At the same time another plug-in can be built that allows the execution of WS-CDL. Both of these plug-ins can communicate with a JBI implementation at the same time and messages can be exchanged as required. JBI therefore allows any number of composition languages and approaches be present at the same time.

From a customer perspective this provides the ability to implement compositions in the appropriate way, independent of the specific domains.

10.7 Clinical Use Case

Composition and all its aspects around dynamic discovery and mediation are a central requirement in the clinical use case. The detailed definition of the composition using Semantic Web Technology is precisely detailed out in Chapter 13.

The reason for using composition in the clinical use case is that functionally a process takes place starting with a patient encounter until the patient receives therapeutic guidance. All steps to derive to the therapeutic guidance are in a logical and causal sequence. Composition can express this nicely.

The composition defines the process flow while the services needed along the way are dynamically bound to the process. In contrast to workflow management systems, the flow is defined by goals so that dynamic binding of services becomes possible. In the end, while a composition is executed for a specific patient, the services needed are dynamically bound.

In addition, the clinical use case has several situations in which data has to be mediated between the composition's ontologies and those used by the services. This will be solved, too, with the composition definition. And, as in a real implementation, process mediation is necessary, too, as some services have a different external visible behavior then the composition definition assumed. Several process mediations are shown in the detailed definitions.

The composition as it is defined does not assume that only those services can be used that are locally available within the hospital. While this might be the first phase, it is possible to use external services with the same definition as the discovery of services is independent of their implementation location. This enables the hospital to enlist a variety of competing services over time without invalidating the composition needed.

In summary, the clinical use case requires the full power of composition and all its aspects. The detailed definitions in Chapter 13 show how all aspects are modeled and used for the use case.

10.8 Summary

Composition as a concept is a very powerful tool in order to structure complex business logic, be it B2B integration, EAI or complex computation. Long-running transactions are the norm in business information technology and implementations of composition through composition languages or frameworks are essential for the information technology infrastructure.

For the time being manual composition will be predominantly used in the general cases, dynamic or planning-based composition is not yet mainstream in software products. However, it might be very well the case that in the long run the sheer size of available services on the Web becomes useful and manageable.

The main problem in composition is the semantic consistency in terms of the data and services involved and most of the development time is spent to achieve semantic reliability. This area provides an enormous room for improvement that Semantic Web Services plan to achieve.

11 Semantic Web Services

Semantic Web Services focus on extending traditional Web Services such that their meaning is embedded in the syntactical description. A lot of work, especially in academia, is devoted to this space and the current status and achievements will be highlighted in this chapter. In Section 11.1 the reasons for Semantic Web Services and the main extensions to traditional Web Services are introduced. While some efforts are based on the development of Semantic Web languages, other efforts that are introduced in Section 11.2 use alternative approaches. Section 11.3 discusses in detail the current Semantic Web Service approaches that are based on the technologies developed in the Semantic Web community. Two very "hot topics" in the space are discovery and composition, both of which are discussed in Section 11.4. Section 11.6 provides a summary.

11.1 Semantics of Web Services

Semantic Web Services attempt to increase the usefulness of Web Services by extending them with semantic descriptions. The main areas for extension are interface descriptions incorporating semantic annotations, and the modeling of precise state information of Web Services (especially for long-running interactions).

11.1.1 Why Semantic Web Services?

Web Services, as a technology for application-to-application (A2A) integration over the Web, achieved a big step forward by using XML as its fundamental language. XML has revolutionized the way data is exchanged and represented across the Web. It provides a standard language for describing document types in any arbitrary domain, facilitating the sharing of data across different systems and, most notably, the Web. XML is flexible and extensible, allowing users to create their own tags to match their own specific requirements. As a result XML-based languages have been designed for use in many different fields. In the context of Web Services, WSDL is used to describe both interface and implementation details; SOAP is used to define messages sent to and from services while the datatypes, used in the content of the SOAP messages, are defined using XML Schema. Despite its universality, there remain serious deficiencies in XML as the language for Web Service interactions. XML is a language for defining the structure and syntax of data. It says nothing about the meaning, or semantics, that is associated with the data. This hinders a potential service client from using a particular Web Service because the need remains for a human to be involved to interpret both the XML descriptions of the Web Services (WSDL) as well as the XML descriptions of the data that the Web Services can exchange (XML Schema). The human has to decide if the service matches his or her needs and whether or not he or she can understand the data that should be sent to and received from the service.

As a consequence of the semantic ambiguity inherent in XML descriptions, a number of further problems arise, particularly when Web Services are to be considered as the basis for automated A2A integration. One problem is the location of Web Services for them to cater to a specific capability required by a potential client. The UDDI specification, amongst others, provides for a registry of service descriptions. However, the descriptions are not formalized and are only useful when interpreted by a human reader.

Automated data transformation is another issue. If the data definitions used by the Web Service do not match those used by the potential client, a transformation is required and this is typically encoded using the eXtensible Stylesheet Language (XSLT). For each pairwise transformation between a client and service two XSLT style sheets are required, one for each direction. Both must be hand-coded as there is no automated means for interpreting the data semantics. This has to be repeated for each client having heterogeneous data definitions.

The WSDL specification provides a means for all the publicly available operations offered by a Web Service to be described. In business processes, services are typically required to offer a complex behavioral pattern. RosettaNet is a B2B standard defining inter-company processes, including structure and semantics for business messages and secure transportation of messages over the Internet. RosettaNet defines various Partner Interface Protocols (PIPs). One example is PIP 3A4 for the exchange of purchase order (PO) request and confirmation messages between trading partners. The PIP defines four messages, PO Request, PO Confirmation and two signal messages to acknowledge the receipt of the request and the confirmation respectively. A Web Service conforming to PIP 3A4 needs to be able to define not only the messages and operations but also the control and data flow between the messages. This is not possible using WSDL alone as the definition of control and data flow between WSDL operations is not possible.

Moving on from the previous point, if one Web Service is defined to support the RosettaNet PIP 3A4 B2B protocol and a potential client supports a different B2B standard such as Electronic Business XML (ebXML), then mediation is required between the behaviors defined in terms of the two B2B protocols used by the client and the server respectively. As the public behavior offered by a Web Service is only informally described, there is very limited possibility for automating the mediation task.

11.1.2 Interface vs. Implementation

An enduring problem in computer science is how to model and design software solutions to problems in terms of the problems themselves rather than in terms of the specific computer machinery required. The notion of abstracting away from the underlying computer machinery is the basis for the evolution from programming at the assembly code level, to languages like C and FORTRAN, and further to declarative languages such as LISP, or object-oriented (OO) languages such as Smalltalk, C++ and Java. With OO languages elements in the domain of the problem are modeled as objects that can cooperate together to solve the problem at hand. Objects are typed, have states, and can send and receive messages to and from each other. Another underlying concept in OO programming is the separation of *what* an object can do from *how* the object achieves this functionality. This is provided for by the separation of interface and implementation. The interface is the public face of an object describing the data and behavior that the object makes available to other objects so that they can invoke its functionality.

Web Services also offer the separation of interface from implementation through the WSDL descriptions. The definition part of a WSDL document defines the datatypes, messages and operations that together define how to access the functionality offered by the service. WSDL also has a section for binding descriptions. This defines the Web location at which the service can be accessed in addition to the communication and message protocols that should be used for message exchange. Although, in this respect, Web Services seem similar to objects with separated interface and implementation, a major difference is that Web Services generally do not maintain state. In other words, rather than building on the OO-influenced remote method invocation (RMI), they build rather more on the earlier remote procedure call (RPC) technology.

As the WSDL definitions are fundamental in allowing potential users of a Web Service to determine how to interact with that service, it is imperative that this information be unambiguously defined. There are two aspects to consider, data and behavior. In WSDL, the recommended technology for the definition of datatypes is XML Schema. The WSDL schema itself provides the means for describing behavior. The drawback for Web Services is that as neither aspect of interface description results in unambiguous meaning, computer systems are blocked from being able to interpret and reason over the description. This, in turn, restricts the opportunities for automated Web Service discovery, composition and invocation, leading to a strong motivation for semantic annotations.

11.1.3 Modeling of State

Long-running business processes are quite common in the real world. Airline reservation systems usually allow holds to be placed on seats for a 48-hour period before they must be confirmed. In supply chain management software, an order for goods may be made from customer to supplier. Fulfilling the order might require parts to be ordered from a third party. The initial purchase order request remains open until a confirmation message can be sent back from the supplier. It is possible that this may take a number of hours or days. Additionally, the conversation between both partners may require several (possibly asynchronous) messages back and forth. In such cases, it makes no sense for the business partner making the request to block its software systems while waiting for a request to be completed. Instead a token is often shared between the trading partners that can be used to identify the state of the process at the provider's end.

As a result, it seems natural for Web Services that may be involved in long-running interactions to support the notion of state. This is sometimes considered as an attempt to map the distributed computing architecture of a specification like CORBA onto the Web, where a "factory" resource can be used to provide identifiers to new or existing sessions on request. However, there is no specific mechanism defined in WSDL for this purpose. In fact there is a strong argument from the Web community that Web Services, as resources available over the Web, remain stateless following the REST architectural style [299]. However the issue of how to deal with long-running Web Services remains. One approach, from the Grid community, is the Web Services Resource Framework (WSRF) [300], a family of specifications that combine a stateful resource to a Web Service through interfaces defined as part of a WSDL service description. WSRF relies on the WS-Addressing [301] specification as a means for providing the target endpoint for a message sent to a Web Service implementing the WSRF interfaces.

For businesses providing a service that, for example, handles purchase order requests, a hybrid approach involving document-style Web Services and business standard specifications is sometimes adopted. For example, if both partners to a business interaction agree to use the RosettaNet[6] specification for handling purchase orders (POs), they may agree to use a common order identification system. The WSDL definition for the service would then define that an incoming PO document should contain a typed XML element representing the PO identifier. The Web Service interface would remain stateless and simply pass received messages to the back-end systems designed to handle large volumes of concurrent PO requests.

Taking the example of the previous paragraph as typical usage of Web Services in stateful interactions, there is a strong motivation for the formalization of industry B2B specifications through the use of ontologies. The Semantic Web Services machinery could then be employed to automatically tackle interoperability issues between such ontologies at the conceptual level. Mappings would still be required but only to be established between concepts in related ontologies rather than on a one-to-one basis between data instances: the main drawback of using XSLT.

---

6. http://www.rosettanet.org/ (Accessed September 10, 2007).

11.2 Alternatives for Capturing Semantics of Web Services

Using the technology of ontologies to extend traditional Web Services is usually mainly targeted at the definition of the data contained in the messages exchanged by Web Services. One of the important characteristics of Semantic Web Services is their potential to improve on the existing Web Service model when it comes to describing behavioral semantics. Web Services are touted as the basis for a new wave of distributed computing over the Web, taking off from where CORBA [302] and DCOM [303] left off for intranet- and extranet-based distributed systems. Two aspects of behavior requiring a formal operational semantics for Web Services are:

- The external interface the Web Service offers to its potential clients
- The internal definition of the composition of other Web Services a particular service might use to achieve its objectives

Especially in the area of long-running Web Services and composition, a set of alternative technologies and formalisms is used to define precisely the meaning of execution. This is discussed in the following.

11.2.1 Finite State Machines

Finite State Machines (FSMs) are one of the oldest techniques [293][294] in computer science for modeling sequential behavior that depends not only on inputs but also on the state a system is in when an input is received. FSMs consist of five elements: states, state transitions, conditions, input events, and output events. A state provides information about something that has already happened. Transitions indicate a change of state and are described by a condition that, when fulfilled, results in the transition. Actions are activities that are performed at a given moment. For example, different types of activities are when a state is entered (entry), when a state is exited (exit), when a transition occurs (transition) and when an action takes place (action). FSMs can be drawn using statecharts or by state transition tables. Both illustrate the state that an FSM can move to given a current state and set of inputs.

Finite state machines are rule-based and thus are suitable for problem-solving algorithms. They are of two types. Deterministic FSMs are those for which, given a state and set of inputs, the next state can be predicted. Non-deterministic FSMs are those where the next state transition cannot be predicted given an initial state and set of inputs, and an unpredictable external event may affect the FSM. Additionally, there are two models defined for state machines, Moore and Mealy. Moore machines are those where outputs are a function of the state only. Mealy machines are those where outputs are a function of both the state and the inputs. In [295], Hendler recognizes FSMs as a potential useful means to model the process model for Web Services. Additionally [296] and [297] propose FSMs as a useful mechanism to model the internal and external behavior of services as a sequence of transitions between states.

11.2.2 Statechart Diagrams

In his paper [298], Harel notes that state machines are a natural medium for describing the dynamic behavior of complex systems where events may occur at run time, affecting the system's execution. However, he draws attention to the drawbacks of using FSMs for complex systems where the number of states may grow exponentially, resulting in unmanageable complexity and illegibility in the FSM diagrams. To counter these problems, he proposes statecharts as an extension of the notion of FSMs to include the concepts of hierarchy, state clustering, modularity and concurrency.

11.2.3 Petri Nets

Petri nets were invented in 1962 by C. A. Petri [304] as a mathematical model for concurrent, asynchronous, parallel behavior in distributed systems. Graphically, Petri nets are represented as bipartite graphs with place nodes, transition nodes and directed arcs (also called edges) that link them. Bipartite graphs contain a set of vertices that can be divided into two distinct disjoint sets such that no edge can have both endpoints connected to the same set. In the case of Petri nets, no edge can have both ends connected to places or both ends connected to transitions. Places can have one or more tokens. A place that is connected to an ingoing edge of a transition is considered an input for the transition. Similarly, a place that is connected to an outgoing edge of a transition is considered an output for the transition. Transitions may fire as long as sufficient tokens are available at the input places. When this happens, the transition is said to be enabled. Firing results in tokens being removed from input places and added to output places.

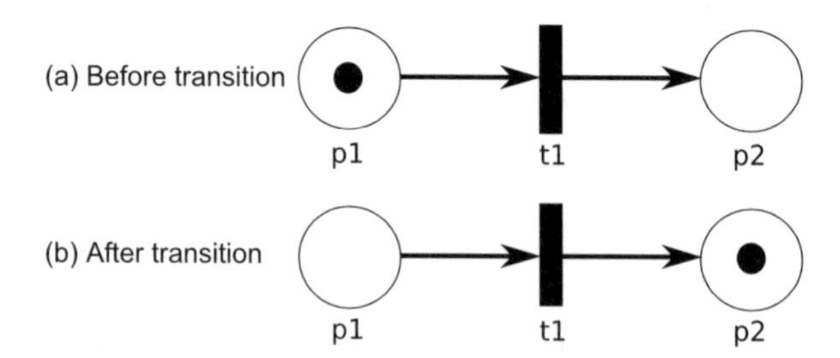

Fig. 11.1. Simple Petri net

Figure 11.1 shows two *markings* (a) and (b) for a simple Petri net with places, p1 and p2, and a transition, t1. A marking defines a possible state of a Petri net by defining what tokens are available at each of the net's places. In (a), there is a token at place p1 which means the transition, t1, is enabled and may fire. In (b), the transition, t1, has fired and a token has been removed from place p1 and added to place p2. Original nets allowed only one token to be added or removed from a place whenever a transition fired. Weighted Petri [305] nets are a generalization which allow multiple tokens to be added or removed from places.

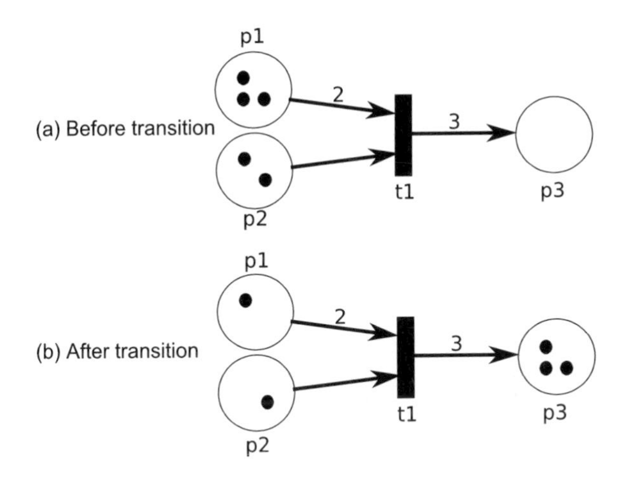

Fig. 11.2. Weighted Petri nets

Figure 11.2 shows a Weighted Petri net with the weights represented as positive integers labelling the edges. The edge from place p1 to transition t1 is given a weight of 2, the edge from place p2 to transition t1 is not labelled, implying a weight of 1, and the edge from t1 to p3 is given a weight of 3. When the transition t1 fires, two tokens are removed from p1, one token is removed from p2 and three tokens are added to p3.

There are other well-known extensions to the original Petri net model. These include colored Petri nets [306], timed Petri nets [307] and hierarchical Petri nets [308]. In traditional nets, the tokens have no types associated with them. The precondition for a transition to fire is that there be sufficient tokens available at the input places. The postconditions for a transition are that tokens be removed from the input places and added to the output places. With colored Petri nets, tokens are typed or *colored*. This is useful as the tokens in a Petri net model are usually modeling real-world objects that have associated attributes. The transitions in a colored Petri net can use the type and values of the consumed tokens to determine the type and values of the produced tokens.

When modeling real systems, it may be important to model temporal aspects of the system. In other words, there may be a need to model durations and delays. Timed Petri nets make this possible by associating time with tokens, places or transitions. For example, the model may be set up so that transitions take a certain amount of time to complete. When a transition is enabled, the tokens are removed from the input places. After a certain time duration, tokens are added to the output places. A consequence is that the state of the system is not always clearly represented.

Petri nets for large systems can easily become very complex and difficult to analyze. This difficulty can be addressed using hierarchical Petri nets which allow a hierarchy of subnets to be constructed, each of which can be used to analyze one particular area of the system. Each subnet can be considered as a black box that may accept inputs from, and provide outputs to, other parts of the system being considered. It can be mathematically proven that the combination of subnets for a hierarchical Petri net have the same behavioral semantics as if the entire system were modeled as one very large single net. The main benefit they offer is the ease of use of Petri nets when modeling large and complex systems.

Modeling asynchronous distributed systems using Petri nets allows the model to be checked for a number of potentially undesirable properties. According to [309], these include:

- **Termination.** Does the Petri net terminate?
- **State reachability.** Are all possible states for the Petri net reachable?
- **Immediate reachability.** Is a particular state reachable when a specific transition fires?
- **Partial deadlock.** Is there a state where there is at least one transition that can never fire?
- **Deadlock.** Is there a state where no transition can fire?
- **Livelock.** Is there a set of states where the only transitions that can fire move between the states so that the Petri net never terminates?

11.2.4 Process Algebras

Process algebras (or process calculi) are algebraic languages that provide a formal foundation for modeling programs which can run concurrently in parallel, and which can interact with each other. In the case of such paralell systems, it's insufficient to say that each program can be simply modeled as an input/output function because the interaction between them affects their respective behaviours. Baeten [387] provides a pragmatic description by focussing on the individual definitions of the words "process" and "algebra". He points out that "process" refers to the behaviour of a system, or the total events or actions that the system can perform, the order in which they are executed and various aspects of this execution. In the context of modeling systems it is useful to keep the focus on certain essential aspects of the behaviour possibly ignoring other real-world considerations so that process models describe an observation of the behaviour of interest. The word "algebra" indicates using a generalized axiomatic approach in describing the process model. With a process modeled using algebraic equations, it becomes possible to apply algebraic laws to allow descriptions to be manipulated and analysed, and also provide a basis for formal reasoning about the process.

Petri nets preceded the conception of Process algebras by about a decade. The first Process algebra was devloped by Milner in the early 1970s and published as A Calculus for Concurrent Systems (CCS) [390]. Pi-calculus [391], which has become a popular Process algebra, has CCS as its theoretical starting point. In the examples later in this section, we use Pi-calculus as a representative process algebra. However, there are many others and a good starting point for further reading is in Baeten's work at [387].

Although Petri was the first person to develop models of interacting sequential processes, the focus of Process algebras is slightly different. A high-level difference is that Petri nets are bipartite graphs, while CCS (as a representative Process algebra) is a more textual, linear-like set of equations using an algebra that includes operators for concurrency, parallelism, conditions and functions (or data buffers). Van der Aalst [388] points out many notions for Petri nets have been translated into process algebra and vice versa. He argues that an important difference is that the notion of invariants devloped for Petri nets do not exist for Process algebra. In [389] the authors highlight that, although both approaches model concurrent systems, they tend to be used by different communities. Petri nets are popular with system and control engineers interested in issues around liveness and dynamic invariants of system design. Process algebras, such as CCS, are more popular with computer scientists who have some interest in liveness and invariance but are more interested in comparing the behaviours of systems. Another difference is that the Process algebras define systems as a collection of independent agents communicating with each other. Petri nets allow systems to be defined whose actions depend on internal and external inputs but it is not always easy to identify individual agents within the net.

As explained by Milner in his tutorial at [392], fundamental to the Process algebra of CCS, and more recently of Pi-calculus, is the notions of *naming*. In the basic version of Pi-calculus, there are only two types entities: *names* and *processes*. Names have no structure and there can be infinitely many of them. Processes, in Pi-calculus, are built from names using the syntax:

$$P \ ::= \ \Sigma_{i \in I} \ \pi_i \cdot P_i \ \Big| \ P|Q \ | \ !P \ | \ (vx)P$$

There are four parts to the right hand side of this definition. Each part is separated by a large vertical bars representing the logical *OR* operator. These parts are described briefly below in Table 11.1 while a full description is available in [392].

Table 11.1. Parts of Calculus Definition

$\Sigma_{i \in I}\ \pi_i\ .\ P_i$	The symbol pi is a prefix that represents an atomic action starting a process. *I* is an infinite prefixing set. There are two kinds of prefix. *x(y)* which means input a name *y* along a link (or channel) *x*. This binds *y* in the prefixed process. $\bar{x}y$ which means output the name *y* along the link *x*. In this case, *y* is not bound to the process. *x* and $\bar{x}$ are both names referring to links. *x* is used for input while its co-name, $\bar{x}$, is used for output. The summation in the expression represents a process that is able to take part in exactly one of several alternatives for communication. The choice itself is not made by the process.
$P\|Q$	The processes *P* and *Q* are concurrently active, can act independently and can communicate with eac other
$!P$	! is the replicator operator. This expression is shorthand for multiple copies of *P* running concurrently. (The calculus does not restrict this number.) Milner calls this "bang P"
$(\nu x)P$	Restricts the use of the name *x* exclusively to the process *P*. Milner calls this "new *x* in *P*".

The "**.**" (dot) operator indicates sequential actions. The final action for a process may be represented as a null action or **"O"** e.g. *x(y)***.O**, however this is usually omitted in favour of *x(y)*. We now explain a simple example of a processs described by Milner using Pi-calculus.

$$\bar{x}y \mid x(u).\bar{u}v \mid \bar{x}z$$

This process is equivalent to three concurrent processes, *P* | *Q* | *R,* where *P* represents *y* available for output on channel *x*, *Q* represents *u* expected as input on channel *x*, and *R* represents *z* available for output on channel *x*. One of two communications can happen on channel *x* but not both. Consequently there are two possible outcomes for the result:

$$\mathbf{O} \mid \bar{y}v \mid \bar{x}z \quad \text{or} \quad \bar{x}y \mid \bar{z}v \mid \mathbf{O}$$

To see how these outcomes are derived, we look in more detail at the first of them. In *P*, *y* is output on the channel *x* (i.e $\bar{x}y$), and is accepted by *Q* as input along the channel *x* (i.e. $x(u)$). The subsequent action uses the name *y* as the channel to output the name *v* (i.e. $\bar{y}v$). If this first set of actions takes place then *R* ($\bar{x}z$) remains unchanged.

A relatively recent application of Process algebras is to Web Service composition. In Section 10.1.4, we described how there are two sides to service composition. The first is where components, represented as services, are put together one after another, specified by control and data flow, to achive a particular task. In order to use such a composition, its not necessary to know its internals. The second aspect is the behaviour of the composition with respect to its requester. We pointed out that, in this way, service compositions can be seen to have both internal and external behaviour. Languages such as WSPBPEL provide a means of describing this behaviour structurally in tersm of XML. Process algebras (as do Petri nets) provide a formal language for describing the behaviours. One example of this application is in [393] where the authors focus on the use of Process algebra for simulation, property verification, and correctness of composition of Web Services. Another example in [394] describes a Process algebra called Finite State Process (FSP) for which the authors have developed a tool called LTSA-WS for the analysis of Web Service compositions described using WSBPEL.

11.3 Semantic Web Service Approaches

In this section, we look at the four leading ontology-based approaches for representing Semantic Web Services. These are OWL-S, SWSF, SAWSDL, WSDL-S and WSMO. In each case, the conceptual model is described and the languages used to express that model are explained.

11.3.1 OWL-S

Conceptual Model

OWL-S [310] is a Web Ontology Language (OWL) ontology, structured into three sub-ontologies, for describing different aspects of Semantic Web Services. The first aspect is the functionality a Web Service offers, including the constraints and non-functional properties that influence it. This is described using the ServiceProfile. Web Services enact their functionality through a behavioral model. Describing this is the aim of the ServiceModel. Finally, OWL-S seeks to build on top of WSDL and SOAP by mapping elements in the ServiceModel to elements in the WSDL description. This part of the OWL-S ontology is called the ServiceGrounding. We look at each of the three parts in the next paragraphs.

In OWL-S, the ServiceProfile describes what a Web Service does and provides the means by which the service can be advertised. As there is no distinction in the conceptual model of OWL-S between service requests and service provisions, the ServiceProfile is aimed equally at advertising services offered by providers and services sought by requesters. Owing to its genesis in the research area of artificial intelligence (AI), OWL-S defines the capability a service offers in terms of a state transition. It is possible to specify the inputs and outputs expected to be sent to and received from a service along with preconditions that must hold before the service can execute and the effects of the service executing. The intent is that along with arbitrary non-functional properties, this should be sufficient information for a discovery agent to be able to decide if a desired ServiceProfile matches any of the ServiceProfiles in the set of candidate OWL-S Web Service descriptions available to it.

The ServiceModel is used to define the behavioral aspect of the Web Service. This part of the service is modeled as a process in the sense that a service requester can view the process description and understand how to interact with the service in order to access its functionality. In some ways, this process model can be considered as a partial workflow where the service requester provides the missing parts. The ServiceModel allows for the description of different types of services, atomic, abstract and composite. Atomic processes correspond to a single interaction with the service, e.g., a single operation in a WSDL document. Composite processes have multiple steps, each of which is an atomic process, connected by control and data flow. Simple processes are abstractions to allow multiple views on the same process. These can be used for the purposes of planning or reasoning. Simple processes are not invocable but are described as being *conceived* as representing single-step interactions. A simple process can be *realized by* an atomic process or *expanded to* a composite process.

The final part of the conceptual model is the ServiceGrounding, providing a link between the ServiceModel and the description of the concrete realization for a Web Service provided by WSDL. Atomic processes are mapped to WSDL operations, where the process inputs and outputs, described using OWL, are mapped to the operation inputs and outputs, described using XML Schema. It is possible that a single OWL-S Atomic Process can be mapped to many WSDL operations (although this is not usually the case). Composite processes, being composed of atomic processes, are grounded in the same way with the additional requirement of an OWL-S process engine to interpret the defined control and data flow.

In many ways OWL-S was the first consensus-based ontology for describing Semantic Web Services. It is the product of merging earlier research from two separate languages, DAML [291] and OIL [292], resulting in an ontology initially called DAML-S but later renamed to OWL-S to emphasize the perceived layering of the ontology on OWL (a W3C Recommendation). The actual use of the description logics variant, OWL-DL, as the ontology language for OWL-S has some unwanted side effects noted in detail in [311]. In particular, OWL-S does not comply with the OWL-DL specification, which places constraints on how OWL-S ontologies can be reasoned over. A second problem is that variables are not supported within OWL-DL but are necessary when combining data from multiple cooperating processes in OWL-S.

Language

Although primarily the OWL-S ontology is defined using the Web Ontology Language (OWL), OWL-S is actually a mixture of a number of languages. This breaks to some extent the claim for OWL-S that it is layered on top of OWL (and so a natural candidate for standardization). The reason for the language mixture is that

Web Services are inherently associated with distributed computing on the Web through process definition and execution. OWL is simply not designed for this purpose. Rather, it provides an upper ontology for defining conceptual models. In particular, to take advantage of the most commonly available implemented logical reasoners, OWL-DL is used to define the domain models used in the Semantic Web Service descriptions.

When describing logical expressions for the preconditions and results of ServiceProfiles or ServiceModels, the modeler has a choice. The Semantic Web Rules Language (SWRL) [312] and Resource Description Framework (RDF) [42] treat expressions as XML literals while the Knowledge Interchange Format (KIF) [313] or the Planning Domain Description Language (PDDL) [314] can be used for treating expressions as string literals.

11.3.2 SWSF

The establishment of the Semantic Web Services Framework (SWSF) [315] was motivated by the recognition of some shortcomings of OWL-S as a conceptual model for Semantic Web Services. At the time OWL-S was developed, attention was focussed on how an ontology for Web Services could be described using OWL. OWL itself is layered on top of the Resource Description Framework (RDF), and it was considered an elegant solution to add OWL-S as a further layer. A significant problem, as indicated in Section 11.3.1, is that OWL (or more precisely OWL-DL) is not well suited to describing processes. This situation is unsatisfactory as the functionality offered by Web Services can be considered as a *partial* process involving the operations that the Web Service makes available to a client application. The process description is partial as the client itself provides the complimentary activities when it interacts with the service.

SWSF was devised to provide a full conceptual model and language expressive enough to describe the process model of Web Services. There are two parts to the SWSF. The first is a conceptual model called the Semantic Web Services Ontology (SWSO) axiomatized using first order logic, and the second is a language called the Semantic Web Services Language (SWSL).

Conceptual Model

SWSO defines a conceptual model for Semantic Web Services with a deliberate focus on extending the work of OWL-S to interoperate with and provide semantics for industry process modeling formalisms like the Business Process Execution Language (BPEL). The first-order logic axiomatisation of SWSO is called FLOWS (First-Order Logic Ontology for Web Services) and is based on the Process Specification Language (PSL) [316], an international standard ontology for describing processes in domains of business, engineering and manufacturing. One of the intentions of PSL was to provide a common interlingua for the many existing process languages, allowing interoperability to be established between them. As the number of conceptual models and languages for Semantic Web Services grows, there is a perceived need for such an umbrella formalism to facilitate interoperability in this area.

As mentioned, FLOWS is axiomatized in first-order logic and is expressed in a language called SWSL-FOL (Semantic Web Services Language for First-Order Logic). To enable logic-programming-based implementations and reasoning for SWSO, there is a second ontology available called ROWS (Rules Ontology for Web Services) and this is expressed in SWSL-Rules. ROWS is derived from FLOWS by a partial translation. The intent of the axiomatisation of ROWS is the same as that of FLOWS but in some cases it is weakened because of the lower expressiveness of the SWSL-Rules language compared to SWSL-FOL.

Service is the primary concept in SWSO with three top-level elements, derived from the three parts of the OWL-S ontology. These are Service Descriptors, Process Model and Grounding.

Service Descriptors. They provide a set of non-functional properties that a service may have. The FLOWS specification includes examples of simple properties such as the name, author, and textual description. The set is freely extensible to include the properties identified in other conceptual models such as WSMO non-functional properties or OWL-S service profile elements. Metadata specifications for online documents including Dublin Core[7] are also easily incorporated. Each property is modeled as a relation linking the property to the service. For example, Figure 11.3 shows FLOWS relations for service_name, version and reliability. Note that Web Service reliability is a subjective notion in the context of the quality-of-service (QoS) attributes a service may have. For it to be effective, a formal description of the meaning of reliability in Web Services is required. Some ongoing work in modeling this type of attribute using WSMO ontologies is described in [351].

```
name(service, service_name)
version(service, service_version)
reliability(service, service_reliability
```

Fig. 11.3. FLOWS Service Descriptor Properties

Process Model. The underlying objective of PSL is to provide a language and ontology that is expressive enough that all other process languages can be represented in it. If this is achieved then the integration of independent processes described with heterogeneous models becomes possible. FLOWS extends the PSL generic ontology for processes with two fundamental elements, especially to cater to Web Services:

- The structured notion of atomic processes as found in OWL-S

7. http://dublincore.org/ (Accessed September 10, 2007).

- Infrastructure for allowing various forms of data flow

The Process Model of FLOWS is organized as a layered extension of the PSL-OuterCore ontology. The primary layer is called FLOWS-Core and contains the two extensions just mentioned for Web Services. On top of this, five additional ontology modules are defined that are used to express different constraints on the occurrences of services and their subactivities. A simplified diagram of this layering is provided in Figure 11.4.

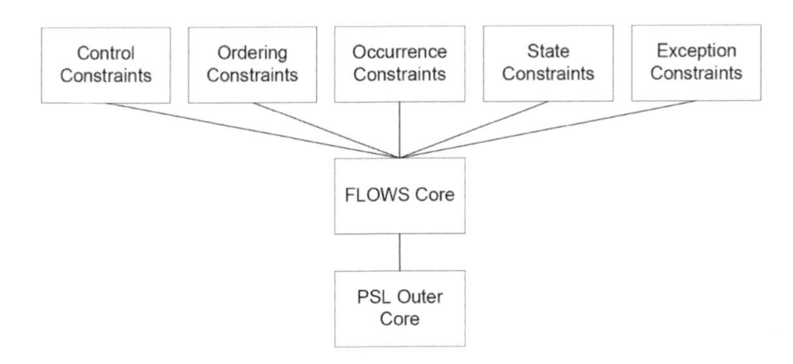

Fig. 11.4. FLOWS layered process model

As defined in the SWSF submission to the W3C submission, the layer has five additional ontologies:

- **Control Constraints** axiomatize the basic constructs common to workflow-style process models. In particular, the control constraints in FLOWS include the concepts from the process model of OWL-S.
- **Ordering Constraints** allow the specification of activities defined by sequencing properties of atomic processes.
- **Occurrence Constraints** support the specification of non-deterministic activities within services.
- **State Constraints** support the specification of activities triggered by states (of an overall system) that satisfy a given condition.
- **Exception Constraints** provide some basic infrastructure for modeling exceptions.

Four key terms defined by the FLOWS ontology are listed below:

- **Service.** A service is an object that can have an associated number of service descriptors as described above, and an activity that specifies the process model of the service.
- **Atomic Process.** An atomic process is generally a subactivity of the activity associated with a service. It is directly invocable, has no subprocesses and can be executed in a single step.
- **Message.** Messages have an associated message type and payload.

- **Channel.** A channel is an abstraction for an object that holds messages that have been sent but may not yet have been received. There is no restriction that all messages sent be associated with channels, but where this is the case there are additional axioms that must hold for the message.

Before leaving this brief description of the FLOWS Process Model, we draw attention to the fact that FLOWS allows the modeling of predicates or terms whose values may change in the course of an activity. The modeling elements are called *fluents* and can be imagined as providing a behavior similar to that of variables in a programming language, in that they allow processes to be chained together where a value from one process may be required by another. The absence of this was one of the observed drawbacks of the OWL-S process model.

Grounding. The SWSO approach to grounding follows very closely the grounding of OWL-S v1.1 to WSDL. The SWSO specification defines how the grounding must provide four things. These are:

- Mappings between the SWSO and WSDL messages patterns
- Mappings between message types as defined in SWSO and WSDL respectively
- Serialization from SWSO message types to the concrete message types defined by WSDL
- Deserialization from the concrete WSDL message types to the SWSO messages types

Language

We have already described that the Semantic Web Services Language (SWSL) comes in two variants: SWSL-FOL and SWSL-Rules. The starting point is SWSL-FOL which acts as a foundational ontology language with PSL as its foundation. SWS-Rules is derived as a partial translation to facilitate implementation and reasoning based on logic programming techniques.

Both variants share syntax but not the semantics of that syntax. In fact, neither language is a subset of the other, which means the two language variants are mutually incompatible (cannot be used together), which may somewhat complicate the understanding of how to use of SWSO/L. The modeler must decide which language best suits the purpose at hand. The decision is made simpler as each of the variants has a differing focus. SWSL-FOL is most useful for process-related descriptions while SWSL-Rules is geared toward the description of programming-like tasks such as discovery and contracting. Both variants comply with Web principles such as the use of URIs, integration with XML types and XML-compatible namespaces. Additionally both are layered languages where new features are incorporated at each layer.

A concise review of SWL-Rules is provided by the authors of [317]. As described in this report, SWSL-Rules is a logic programming language including features from Courteous logic programs [318], HiLog [319] and FLogic [320] and can be seen as both a specification and an implementation language. The SWSL-

Rules language provides support for service-related tasks such as discovery, contacting, and policy specification. It is a layered language as illustrated in Figure 11.5. The core of the SWSL Rules language is represented by a pure Horn subset. This subset is extended by adding features such as disjunction in the body and conjunction and implication in the head [321], or negation in the rule body interpreted as negation as failure (called NAF). Other extensions are (1) Courteous rules (Courteous), (2) HiLog, and (3) Frames.

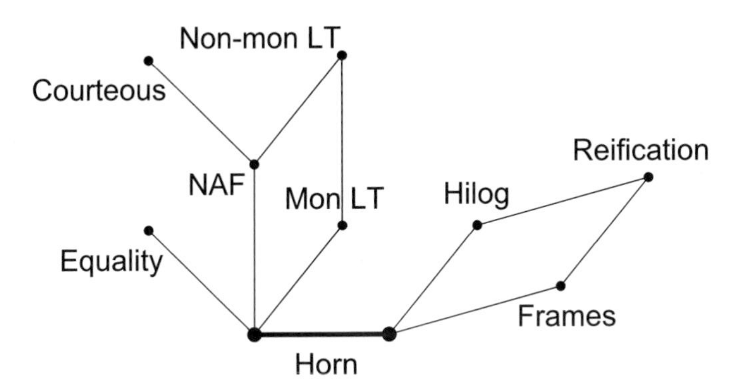

Fig. 11.5. SWSL-Rules Layers

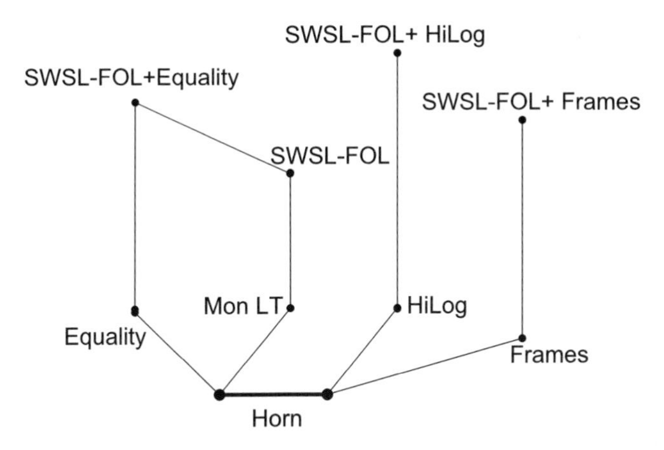

Fig. 11.6. Layers of SWSL-FOL and relationship to SWSL-Rules

On the other hand, SWSL-FOL, intended to describe the dynamic (process) aspect of services, is also layered. The bottom layer of Figure 11.6 shows the layers of SWSL-Rules that have monotonic semantics and therefore can be extended to full first-order logic. The most basic extension is SWSL-FOL but Figure 11.6 also shows three other possible layered variants that can be achieved by the relevant extension. Theses are SWSL-FOL+Equality, SWSL-FOL+HiLog and SWSL-FOL+Frame.

11.3.3 WSDL-S

WSDL-S [322] is a lightweight approach for augmenting WSDL descriptions of Web Services with semantic annotations. It is a refinement of the work carried on by the METEOR-S group at the LSDIS Lab, Athens, Georgia,[8] to enable semantic descriptions of inputs, outputs, preconditions and effects of Web Service operations by taking advantage of the extension mechanism of WSDL. WSDL-S is agnostic to the ontology language and model used for the annotations of WSDL.

In the following paragraphs we take a look at the approach of WSDL-S, the conceptual model representing the approach and the extensions to the WSDL language that realize the semantic annotations.

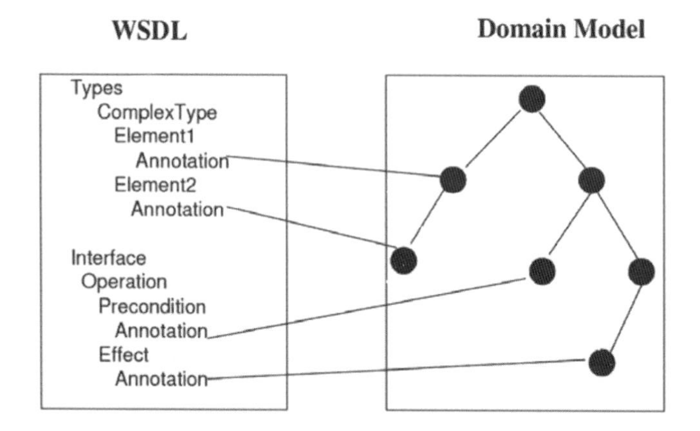

Fig. 11.7. Associating semantics with WSDL [322]

Approach. In contrast to the OWL-S, SWSO and WSMO, WSDL-S does not specify an ontology for the definition of Semantic Web Services. Rather, it takes a bottom-up approach with the appeal that potentially only a little additional effort on the part of service producers will provide a service description where the description of the data and operations of the service are bound to ontological concepts. WSDL-S intentionally builds directly on the existing Web Service technology stack.

8. http://lsdis.cs.uga.edu/ (Accessed September 10, 2007).

WSDL v1.1 allows for the definition of extension to its language. This is taken advantage of to provide an in-document link of certain WSDL elements to concepts in one or more ontologies (assuming that the concepts can be identified uniquely and that the links can be specified in legal XML). Figure 11.7 provides a high-level overview.

Embedding annotations into WSDL through legal language extensions does not affect the usage by the service provider of any other WS-* specifications or the usage of WSDL in the context of process description languages such as the Business Process Execution Language for Web Services (WSBPEL) [323]. Another feature is that where XML Schema is used as the data definition language for WSDL, it can be enhanced by linking XML Schema types to domain concepts either by a one-to-one mapping or through a transformation defined in a domain ontology.

Conceptual Model. WSDL-S defines its conceptual model using a simple XML Schema introducing five elements that extend WSDL. These are:

- **modelReference.** This is used for annotating both simpleTypes and complexTypes in XML Schema where there is a one-to-one mapping between the schema type and the ontological concept. For simpleTypes, it is a direct mapping. For complexTypes, it can be used in two ways, bottom-up and top-down. Bottom-level annotation involves describing every leaf element of the complexType with the modelReference attribute. Top-level annotation means that the complexType element itself is associated with a concept in the ontology. The assumption is that the subelements of the complexType will map directly to the sub-concepts and attributes of the domain concept.
- **schemaMapping.** Where there is no one-to-one mapping this attribute points to a transformation that links the XML Schema element to the ontology concept. For example, the value of the schemaMapping attribute might be a URI that identifies an XSLT transformation.
- **precondition.** At the level of a WSDL operation it is possible to point to a definition of the precondition that must hold before that operation can be executed. For simplicity only one precondition may be included and this may point to a set of logical expressions in the ontology language of choice.
- **effect.** Similar to preconditions, effects point to logical expressions that should hold after the execution of the service. In contrast to preconditions, WSDL-S allows for the definition of multiple-effect subelements of operations.
- **category.** This is adopted from OWL-S and is an extension to the WSDL Interface element of WSDL 2.0 (portType in WSDL 1.1). The intent is that category information can be included here that may be picked up by a Web Service registry implementation such as the one for UDDI.

Language. WSDL-S is defined using the XML-Schema listed in Figure 11.8

```
<?xml version="1.0" encoding="UTF-8"?>
<schema xmlns="http://www.w3.org/2001/XMLSchema"
 targetNamespace="http://www.ibm.com/xmlns/stdwip/Web-services/WS-Semantics"
    xmlns:wssem="http://www.ibm.com/xmlns/stdwip/Web-services/WSSemantics"
    xmlns:wsdl="http://schemas.xmlsoap.org/wsdl/">

    <attribute name="modelReference" type="anyURI" use="optional"/>
    <attribute name="schemaMapping" type="anyURI" use="optional"/>

    <element name="category" maxOccurs="unbounded">
        <complexType>
            <complexContent>
              <extension base="wsdl:documented">
               <attribute name="categoryname" type="NCName" use="required"/>
              <attribute name="taxonomyURI" type="anyURI" use="required"/>
              <attribute name="taxonomyValue" type="String" use="optional"/>
              <attribute name="taxonomyCode" type="integer" use="optional"/>
               </extension>
            </complexContent>
        </complexType>
    </element>
    <element name = "precondition">
        <complexType>
            <complexContent
                <restriction base="anyType">
                    <xsd:attribute name="name" type="string" />
                    <attribute name="modelReference" type="anyURI" />
                    <attribute name="expression" type="string" />
                </restriction>
            </complexContent>
        </complexType>
    </element>
    <element name="effect">
        <complexType>
            <complexContent
                <restriction base="anyType">
                    <xsd:attribute name="name" type="string" />
                    <attribute name="modelReference" type="anyURI" />
                    <attribute name="expression" type="string" />
                </restriction>
           </complexContent>
        </complexType>
    </element>
</schema>
```

Fig. 11.8. XML Schema for WSDL-S

11.3.4 SAWSDL

The Semantic Annotations for WSDL (SAWSDL) working group, formed recently by the W3C, provides a W3C Candidate Recommendation for Semantic Web Services based on a simplified form of WSDL-S. This is in the form of an incremental bottom-up approach to Semantic Web Services where elements in WSDL documents are provided with semantic annotations through attributes provided using standard valid extensions to WSDL. The approach is agnostic to the ontological model used to define the semantics of annotated WSDL elements. From SAWSDL's perspective, the annotations are valued by URIs. SAWSDL, like WSDL-S is targeted at WSDL v2.0 but it is also possible to use with WSDL v1.1 with an additional non-standard extension.

While WSDL-S specifies the attributes for modelReference, schemaMapping, precondition, effect and category, SAWSDL confines itself to attributes of modelReference and two specializations of schemaMapping, namely, liftingSchemaMapping and loweringSchemaMapping.

The modelReference attribute can be used to annotate XSD complex type definitions, simple type definitions, element declarations, and attribute declarations as well as WSDL interfaces, operations, and faults. The liftingSchemaMapping can be applied to XML Schema element declaration, complexType definitions and simpleType definitions. All attributes defined by SAWSDL are defined by the XML Schema, reproduced in Figure 11.9, to take a list of URIs as value.

```
<xs:schema
    targetNamespace="http://www.w3.org/2002/ws/sawsdl/spec/sawsdl#"
    xmlns="http://www.w3.org/2002/ws/sawsdl/spec/sawsdl#"
    xmlns:xs="http://www.w3.org/2001/XMLSchema"
    xmlns:wsdl="http://www.w3.org/2006/01/wsdl">

    <xs:simpleType name="listOfAnyURI">
        <xs:list itemType="xs:anyURI"/>
    </xs:simpleType>

    <xs:attribute name="modelReference" type="listOfAnyURI" />
    <xs:attribute name="liftingSchemaMapping" type="listOfAnyURI" />
    <xs:attribute name="loweringSchemaMapping" type="listOfAnyURI" />

</xs:schema>
```

Fig. 11.9. SAWSDL XML Schema

11.3.5 WSMO, WSML and WSMX

Both WSMO and OWL-S address the same problem space. After identifying perceived fundamental drawbacks of the OWL-S approach, the WSMO working group was formed to devise a more complete conceptual model for describing Web Services. Conceptually WSMO, unlike OWL-S, explicitly models separate concepts for goals and Web Services. Additionally WSMO models mediators explicitly as first-class elements capable of bridging heterogeneity issues. In contrast, OWL-S does not explicitly model mediators. Rather, they are considered as specific types of services. A detailed discussion of this rationale is provided in [310].

Conceptual Model

Of the models for semantically annotating Web Services described so far, WSMO and OWL-S are the most closely related. Both aim at the provision of a comprehensive conceptual model for Semantic Web Services. The authors of WSMO describe how an important foundation point of the work on WSMO was the model provided by OWL-S but maintain that OWL-S has a number of serious fundamental flaws that give rise to problems when attempting to use the ontology in practice. These are described in detail in [324] but we will describe a subset of them in this description of the WSMO conceptual model and in the following section describing the languages for WSMO that provide the formal definition of semantics for the conceptual model. WSMO is predicated on a number of underlying principles as defined in [325]. These are:

- **Web compliance.** Every element of WSMO is identified using Unique Reference Identifiers (URIs). Namespaces are supported. WSMO can be serialized to XML, the language of the Web, and WSMO service descriptions ground to the Web Service Description Language (WSDL).
- **Ontology-based.** Ontologies are used to model every element of WSMO.
- **Strict decoupling.** WSMO resources are defined in isolation of each other. There is no assumption that every resource must be defined using the same ontologies.
- **Strong mediation.** Strict decoupling is made possible by the attention paid to mediation in the WSMO model. Mediators are top level modeling elements that are used to bridge interoperability issues between independent, heterogeneous WSMO resources.
- **Ontological separation of roles.** In WSMO the viewpoint of service requester and service provider are distinctly represented by the complementary concepts of goals and Web Services. This separation is adopted from the research in the problem-solving domain and is a clear point of distinction between the OWL-S and WSMO models.
- **Description versus implementation.** The ontological information model defined by WSMO and the additional associated functionality layers on top of existing Web Service implementation technology.
- **Service versus Web Service.** The definition of service within WSMO is a superset of Web Services described by the WSDL language. WSMO is designed to cover all types of service that may be available on the Web.

There are four top-level elements defined by WSMO as necessary for a comprehensive semantic description of Web Services. These are Ontologies, Web Services, goals and mediators. Each of these elements is represented as a class with a number of attributes. Attributes have their multiplicity set to multi-valued by default. If an attribute is single-valued, this is explicitly stated. All WSMO elements have the attribute *hasNonFunctionalProperty*. This allows for the assignment of any non-functional properties (e.g., related to quality-of-service or price, or metadata regarding the owner of the element) to any element. The following paragraphs provide a brief description of each of the top level WSMO elements.

Ontologies. They are used to define the information model for all aspects of WSMO. Compared to structural languages used to define taxonomies such as XML, ontologies allow for the formal definition of concepts and attributes in addition to restrictions and rules constraining them as well as functions and relations that range over them. Two key distinguishing features of ontologies are the principles of a *shared conceptualization* and a *formal semantics*. Ontologies are only useful if the meaning they express corresponds to a shared understanding by its users. Likewise, the strength of an ontology is that the semantics of its elements are machine-understandable, made possible through the provision of a mathematical base for the language used to express the ontology. Ontologies defined in WSMO are part of the MOF model layer.

Web Services. From a simplified perspective, WSMO Web Services are defined by the functional *capability* they offer and one or more *interfaces* that enable a client of the service to access that capability. Capability is one example of an attribute of a WSMO class (i.e., Web Service) that is single-valued. In WSMO a Web Service is defined as offering exactly one capability. The Web Service class also has attributes for mediators (used to bridge heterogeneity problems), non-functional properties (as described above) and ontologies that are imported (providing domain models for some part of the description). We will focus on the capability and interface descriptions as these constitute one area where the similarities and differences between WSMO and OWL-S are apparent.

The capability of a Web Service in the WSMO model defines the functionality that the service can provide when invoked by a service requester. It is defined using a state transition model (similarly to OWL-S but with more intuitive semantics). Prior to a Web Service invocation, *preconditions* define the required state of the information space available to the Web Service and *assumptions* define the state of the world outside that information space. An example of a precondition when using a Web Service to purchase goods is that a creditcard number be valid or that a postal code be valid for the delivery scope of the service. An example of an assumption is that the address provided actually exist. Preconditions and assumptions are defined using sentences in a logical language known as axioms. Depending on the language used, the axioms can be more or less expressive.

Correspondingly, when a service executes successfully, *postconditions* are used to define the state of the information space, and *effects* describe the state of the world outside the information space. For example, a postcondition might be that a shipment confirmation message be sent to the service requester and an effect might be that the goods be physically put in a container and shipped.

All four types of condition are optional in the capability description. The service can be considered as one or more state transitions that move from the state defined by the preconditions and assumptions to the state defined by the postconditions and effects. An application wishing to locate a service for a specific task uses the capability description of a WSMO service to determine if it offers the requisite functionality. Universally quantified shared variables are used to allow information to be shared between the four conditions allowed in capability descriptions.

For example, the listing below shows a Web Service capability from our running translational medicine example for a service providing Therapeutic Guidance. The capability states that on provision of patient information and a set of results corresponding to that patient, a collection of proposed therapies will be returned by the service. The scenario is fully described in context in Section 13.1.4.

```
capability _"http://TherapeuticGuidelines/capability"

precondition
   definedBy
      ?patient memberOf Patient and
      ?listTestResults memberOf ListResultsTests and
      ?listTestResults[patient hasValue ?patient].

postcondition
   definedBy
      ?listTherapies memberOf ListTherapies and
      ?listTherapies[patient hasValue ?patient].
...
```

While the capability defines what a service offers, the WSMO Web Service *interface* elements describe views of external parties on how they can interact with the service. These are is subdivided into two further elements, choreography and orchestration. The interface choreography element describes how a service requester can interact with the service to achieve its goal, including message exchange patterns, the process model supported and the definition of the information types exchanged at the interface. The interface orchestration element allows for the definition of a Web Service as an orchestration of other cooperating services (or goals, which we describe later). The idea is not that all (or indeed any) of the details of how a service achieves its capability be made public, but rather an explicitly described orchestration, including control and data flow and data definitions, allow the separation of the description of how the Web Service achieves its aims from its implementation. Both choreography and orchestration elements of WSMO Web Services are modeled using ontologized Abstract State Machines (ASMs) [327]. ASMs were chosen as a general model as they provide a minimal set of modeling primitives (no adhoc elements), are sufficiently expressive, and provide a rigid mathematical model for expressing dynamics.

The listing below shows a WSML interface description for a Web Service for getting guidance on tests to order for a patient in our translational medicine example. This WSML snippet is broken down and explained in detail in the context of the full example described in Section 13.1.6

```
wsmlVariant _"http://www.wsmo.org/wsml/wsml-syntax/wsml-rule"
...
interface GetTestOrderingGuidanceInterface
orchestration TestOrderingGuidanceOrchestration
  stateSignature GetTestOrderingGuidanceSignature

  /* Concepts used as input and output to the orchestration */
  in Patient withGrounding {
     _"http://.../RulesEngineService.wsdl#wsdl.interfaceMessageReference
           (GetTestOrderingGuidance/GetTests/In)"}
  in Patient withGrounding {
     _"http://.../DataIntegration.wsdl#wsdl.interfaceMessageReference
           (PatientHistory/GetCardiacHistory/In)"}
  shared PatientCardiacHistory withGrounding {
     _"http://.../RulesEngineService.wsdl#wsdl.interfaceMessageReference
           (GetTestOrderingGuidance/GetTests/In)"}
```

```
shared PatientCardiacHistory withGrounding {
    _"http://.../DataIntegration.wsdl#wsdl.interfaceMessageReference
          (PatientHistory/GetPatientCardiacHistory/Out)"}
out ListTests withGrounding {
    _"http://.../RulesEngineService.wsdl#wsdl.interfaceMessageReference
          (GetTestOrderingGuidance/GetTests/Out)"}
/* Concept used to define sequential control */
controlled ControlState

/* transition rules define the state changes of the orchestration */
transitionRules

if (
   ?patient memberOf Patient and
   ?patientCardiacHistory memberOf PatientCardiacHistory and
then
   add(_# memberOf ListTests)
   update(?cs[value hasValue RulesEngineServiceCalled])
endif

if (
   ?patient memberOf Patient and
   ?cs[value hasValue initialState] memberOf ControlState)
then
   add(_# memberOf PatientCardiacHistory)
   update(?cs[value hasValue PatientCardiacHistoryAvailable])
endif
```

Goals. WSMO goals are used to describe, from their own perspective, the aims service requesters have when they wish to interact with Web Services. The separation of goal and Web Service descriptions in WSMO is the realization of the objective to separate concerns. Service requesters are free to specify the services that they require in their own terms. This is one of the distinctions between OWL-S and WSMO. In OWL-S the service concept is used to describe both services and requests for services. Although from a modeling viewpoint WSMO goals and Web Services contain the same structure, they represent different perspectives in the conceptual model and for this reason are kept separate. Like Web Services, goals are defined with attributes for non-functional properties, imported ontologies, mediators, capabilities and interfaces. All of these attributes are defined from the perspective of what a service requester would like to get from a Web Service. The matching of goal and Web Service descriptions (usually referred to as service discovery) may require logical reasoning if syntactically different, but semantically similar, terms are used by the two parties. Semantic mismatches may be resolved using one or more of the mediator types defined by WSMO to cater to interoperability issues.

Mediators. The last of the four top-level elements of the WSMO conceptual model are mediators. They are used to bridge interoperability between any two WSMO elements. A number of distinctions are drawn in the WSMO mediator model. The first is between the description of a mediator and its implementation. While WSMO Web Service descriptions say nothing about how the services are implemented (they ground to WSDL for this), the same holds true for mediators (they can be optionally grounded in a goal, Web Service or another mediator). They describe the bridge that is required between any two elements. A second distinction is between the *kind of* mediation that is necessary for Semantic Web Services and the *types* of mediator that are defined by the WSMO model. The former breaks down to three varieties of mediation:

- **Data mediation.** Handle mismatches at the data definition level.
- **Protocol mediation.** Handle mismatches between message exchange protocols. This relates to the choreography descriptions of Web Services.
- **Process mediation.** Handle mismatches between heterogeneous business processes such as those defined by the RosettaNet[9] or ebXML[10] standards.

Other varieties of mediation may also become necessary over time. The list above is not considered exhaustive. The latter distinction is represented by the four types of mediator defined by WSMO:

- **OOMediators.** Cater to differences in the descriptions of data models defined by ontologies.
- **WGMediators.** Handle mismatches between the definition of a service request as expressed in a goal and the definition of an offered service as expressed in a Web Service
- **GGMediators.** While a repository of goals is already available, GGMediators allow goals to be linked together where there are differences in their descriptions. For example, say a goal is already known to match to a given Web Service; a match of a weaker goal to the same Web Service may be facilitated through a GGMediator.
- **WWMediator.** Analogous to the GGMediator. While a given Web Service already is known to match a specific goal, a weaker or stronger Web Service could also be matched to the same goal through the use of a bridging WWMediator.

Language

Earlier Section 5.2.4 included a subsection providing a description and detailed references for the WSML fanily of languages that provide formal semantics for the conceptual model of WSMO. The languages are layered to provide different levels of expressiveness for the semantics of WSMO depending on the reasoning requirements.

Execution Environment

The evaluation of the conceptual model and formal languages provided by WSMO and WSML respectively, is made easier by the availability of a reference imple-

---

9. http://www.rosettanet.org/ (Accessed May 20, 2008).
10. http://www.ebxml.org/ (Accessed May 20, 2008).

mentation. The Web Service Modeling Execution Environment (WSMX) [330] [331] provides middleware functionality designed to take advantage of the semantic annotations of Web Services using the WSMO model. The implemented WSMX architecture provides an approach to the automated discovery, composition, mediation and invocation of Semantic Web Services. Other tools exist based on the conceptual models described earlier in this chapter but none provide a single coordinated platform capable of tackling all aspects of Semantic Web Service execution. Figure 11.10 shows a high-level overview of the WSMX architecture.

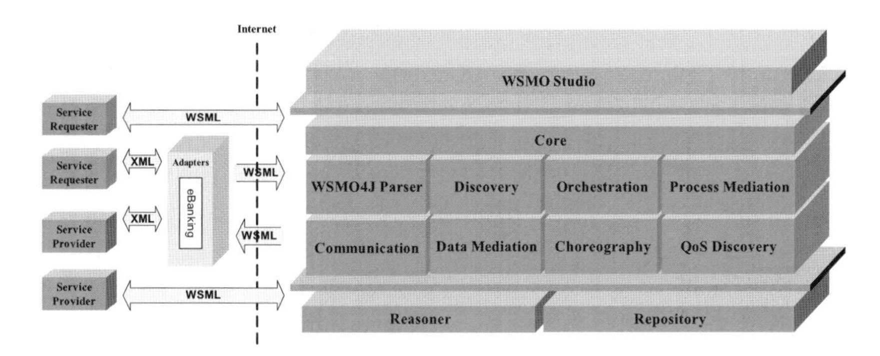

Fig. 11.10. WSMX Architecture

A detailed description of the WSMX architecture is available in [333]. Case-study-driven descriptions of its usage are available at [332]. In this section, we provide a brief description of the functionality of the various boxes in Figure 11.10 coupled with a description of some of the design decisions to create the platform to support this functionality.

The first point is that WSMX is intended as a middleware software layer *at the endpoints* of inter-service communications. This is an intent rather than a restriction. In other words, WSMX is not conceived as a third-party product that is independent of either a service requester or a provider but rather as a lightweight software layer that is positioned at the requester alone or at both the requester and the provider.

All information passed in and out of the WSMX boundary is represented in WSML. An adapter mechanism is provided to transform between non-WSML and WSML messages. All messages entering and leaving WSMX pass through the CommunicationManager which is responsible for handling any protocols relating to transport and communication. The WSMO4JParser is used to parse WSML descriptions to corresponding Java object models used as the internal data representation. Discovery takes care of matching goals to Web Services. The data and process mediation components take care of data, process and protocol heterogeneities where an appropriate mediator is available. The choreography and orchestration components are used to interpret and execute the abstract state machine models corresponding to the interface choreography and orchestration descriptions. Quality of service discovery (QoSDiscovery) acts as a further match-making mechanism between goals and Web Services based on ontologically defined QoS attributes. On the bottom layer of the diagram, the WSML Reasoner acts at the heart of the platform, being necessary for logical reasoning of WSML descriptions for the discovery, mediation, choreography and orchestration functions. At the top of the diagram, WSMO Studio[11] and the Web Services Modeling Toolkit (WSMT)[12] are two alternatives for a WSMO modeling environment.

WSMX is implemented using an event-based messaging mechanism based on Java Management Extensions (JMX)[13] and JavaSpaces[14]. These provide a light-weight community standards mechanism that allows all of the WSMX components to be decoupled from each other. WSMX components are implemented in Java and interact with each other through an event-based publish-and-subscribe messaging system. More information on this and the open source implementation code of each component is available at the WSMX SourceForge project Web site.[15]

11.4 Reasoning with Web Service Semantics

Semantic annotation makes it possible for computers to understand the meaning of data and make more accurate decisions on how that data should be processed. When we talk about computers being able to understand data, we mean that the data is expressed in a language based on some type of formal logic that a computer can reason over. The computational device that carries out this task is usually referred to as a reasoner. In the last section, we have reviewed the state-of-the-art efforts for Semantic Web Service ontologies and identified the logical formalisms on which they are based. In this section, we discuss three particular areas where reasoning with Web Service semantics provides significant value. These are discovery, composition and mediation.

11.4.1 Discovery

Both Preist [334] and Baida et al. [335] distinguish between the concept of a service and a Web Service. They define a service as something of value in a particular domain of interest. Web Services are considered as the agents that provide the

11. http://www.wsmostudio.org/ (Accessed September 10, 2007).
12. http://sourceforge.net/projects/wsmt (Accessed September 10, 2007).
13. http://java.sun.com/javase/technologies/core/mntr-mgmt/javamanagement/ (Accessed September 10, 2007).
14. http://java.sun.com/developer/technicalArticles/tools/JavaSpaces/ (Accessed September 10, 2007).
15. http://sourceforge.net/projects/wsmx (Accessed September 10, 2007).

actual service, while the details of how to interact with the Web Service are described using WSDL and the messages exchanged with the Web Service are formed using SOAP. In a broad sense, Web Service discovery means finding a provider agent (Web Service) that can offer something of value (service) in a particular domain that is of interest to the requester. In [336], the authors point out that WSDL Web Service descriptions provide the technical details for invoking a set of possible *concrete* services. For example, the Amazon Web Service allows for the purchasing of books, DVDs and CDs (amongst other things). The WSDL does not include any details of available titles. A requester looking for the concrete service "sell me a book with the title "The Lord of the Rings" would not find a direct match based on the WSDL Web Service description. Rather, he (or an agent operating on his behalf) would abstract his request to a search for WSDL Web Services that sells books. Once located, a set of such Web Services may be interrogated to check if they offer that particular title (or offer some concrete service). This leads to two stages of discovery, pointed out in [337], each of which may be strengthened through the use of semantic annotations. The first involves abstracting specific client requests, e.g., *from buy The Lord of the Rings* to *buy a Book*. The second is refining the results of the first stage so that a match with the specific request can be made. This second stage will usually involve interaction with the Web Service via the described interface.

In the rest of this section we look at existing efforts for Semantic Web Service discovery, paying attention to the underlying requirements for reasoning. We first look at keyword-based discovery using UDDI and then, in turn, look at subsumption-based matching using Description Logics (DL), request rewriting with algorithms for *best profile covering*, process querying and object-based discovery.

Keyword-Based Discovery

Keyword-based discovery is the basis of the first wave of efforts involving Web Services and the UDDI registry specification. Initially, UDDI was used much like a white-pages listing of available Web Services. Loosely structured information regarding the provenance of the Web Service providers is provided through six specific UDDI concepts:

- **businessEntity**: information about the business
- **businessService**: more detail on the service being offered
- **bindingTemplates**: each one describes a technical entry point for the service
- **tModels**: information regarding particular standards or specifications used by the service
- **publisherAssertion**: declare relationships between business entities
- **operationInfo**: metadata regarding the information in the other five categories, e.g., the time and date they were created.

Keyword-based Web Service discovery usually associated with the use of UDDI relies on string-matching techniques, and very often, with visual human inspection of the information returned either through a graphical user interface on the UDDI registry or with the use of a UDDI API. In either case, logical reasoning is not used to match syntactically different but semantically similar terms. There have been some efforts to build on the UDDI specification through the use of semantic annotation of the information contained in registry entries, e.g., categories linked to ontological concepts. The ontologies used for annotations may, for example, be referenced in the tModels for the entries. This was discussed in [338].

As is the case for all efforts using semantic annotation to aid Web Service discovery, the type of reasoning that may be used is dependent on the choice of ontological language. In particular, we described in Section 11.3.2 on SWSF and in Section 11.3.5 on WSMO and WSML how the various types of underlying logical formalisms reflect the reasoning that may be applied.

Subsumption-Based Discovery

The conceptual model for the OWL-S profile includes concepts for input and outputs of a Web Service. The formal logical language used for profile descriptions is OWL-DL (Description Logics). This is designed for the representation of complex hierarchies of information. In the subsumption-reasoning approach of [367], an advertisement matches a request when all the outputs of the request are satisfied by the advertisement and all the inputs required by the advertisement are provided by the request. The reasoner can infer from the subsumption hierarchy of concepts if particular concepts match even where there are syntactic differences. The underlying concepts of the inputs and outputs are used by the reasoner when computing potential matches. The assumption is that all concepts used in the description of the profiles of both requests and advertisements are defined in a specified registry of OWL-DL ontologies. If concepts are included from unknown ontologies, the reasoner will not recognize them or be able to reason over them.

Similar to the query-rewriting approach to Semantic Web Service discovery described in the next section, subsumption-based reasoning allows for degrees of matching, i.e., matching that recognizes the degree of similarity between advertisements and requests. Examples of degrees of matching are: exact match, plug-in match, subsumption match. Additionally, other algorithms take into account of the distance between concepts in a taxonomy tree. The amount of flexibility built into this kind of discovery is at the discretion of the designers of the matching algorithm.

Request Rewriting (with Best Profile Covering)

This approach builds on subsumption-based reasoning over the inputs and outputs of OWL-S service advertisements and requests. It is described in detail in [339] The algorithm extracts the inputs and outputs of the request, looking for *a combi-*

nation of Web Services that satisfies as much as possible the required outputs of the query, and that requires as little as possible of any inputs not provided by the query. The previous approach looked only for matches between one service request and one advertisement. The request is essentially rewritten into a description of the conjunction of Web Services from known OWL-S ontologies. Best profile covering means a much greater degree of flexibility is allowed in the matching algorithm. Two concepts are defined, *Profile rest* (Pres) and *Profile miss* (Pmiss). Pres is defined as the difference between the outputs defined in the query service profile description and the outputs defined in the advertisement service profile description. Pmiss is defined as the inputs required by the rewritten query (in terms of available Web Services) and the inputs provided by the service request.

Roughly speaking, the difference between two descriptions A and B (written A — B) means all the information that is part of A but not a part of B. In Description Logics, A — B may be a set of descriptions that are not semantically equivalent. In [339], the assumption is made that semantic equivalence holds and further references to how this can be achieved are provided. Best profile cover is defined as the situation where the size of Pres and Pmiss are minimized.

The inputs and outputs of the service requests and advertisements are normalized into clauses (where each clause is of a known concept). The best profile covering problem is then reduced to an interpretation of hypergraphs by defining the difference between two semantic descriptions as a set difference operation between the sets of atomic clauses of two semantic descriptions. Hypergraph theory is used so that the problem of discovering which Web Services best cover the query may be resolved by finding the minimum transversal of a hypergraph with the minimum cost. A hypergraph is constructed where each vertex represents a Web Service and each edge represents a clause (A) of the normal description of the output of the query. The edge is populated by services that have a clause A' in their output that is semantically equivalent to A.

To determine semantic equivalence, reasoning is essentially based on subsumption and consistency checking but the matching algorithm additionally provides a global reasoning mechanism, a flexible matching that goes beyond subsumption tests, and effective computation of missed information.

State-based Discovery

In Section 11.3.5 on WSMO, we described how Web Service and goal capabilities are modeled using preconditions, postconditions, assumptions and effects. As with inputs, outputs, preconditions and effects of OWL-S (Section 11.3.1), this represents a model for describing the state of the world before and after the execution of a Web Service. State-based discovery, as described in [337] for WSMO, seeks to takes advantage of these descriptions to check if the states described in the service request and advertisement, before and after the Web Service execution, match each other.

A state determines the properties of the real world and the available information at some point in time. An abstract service is considered as a set of state transformations. As described earlier, a Web Service description may be considered as abstract as it usually does not describe a single concrete service (e.g., sell books vs. sell the book with title "The Lord of the Rings"). A concrete service can be modeled as a transformation from one particular state to another. In [337], the authors describe a formal model for WSMO Web Services and goals, and based on this present a conceptual model for service location with four stages:

- **Goal discovery:** Locate a predefined goal that fits the requester's desire. The predefined goal is an abstraction of the requester's desire in a more generic and reusable from.
- **Goal refinement:** The goal is refined taking account of the specific information provided in the service request.
- **Abstract service discovery:** Using the capability descriptions of the goal and available Web Service descriptions (capabilities contain the conditions that define the states for before and after execution), Web Services that may be able to fulfill the service request are located. At this point there is no guarantee that the abstract capability of matching services will be sufficient for the request.
- **Service contracting:** The located services will be checked for their ability to satisfy the request. This will usually involve invocation of the services.

Additionally, the paper describes how abstract services and Goals can be represented as sets of objects during the discovery phase. Objects are both the outputs and the effects that can be observed by a requester as a consequence of delivery of a service. This is the key part of the discovery algorithm where the other parts of the capability description are used during the service contracting phase.

The layered family of WSML languages can be employed when defining capabilities, such that a greater degree of logical inference is available to implementations of the service discovery algorithm. This was discussed earlier in Section 11.3.5.

Process-Based Querying

Another approach to Web Service discovery uses the process ontology segment of Semantic Web Service descriptions. This is an important aspect of OWL-S, WSMO and, in particular, SWSF. The process models are queried using a process query language to determine if specific service advertisements match service requests. Such an approach is described in detail in [340]. For this purpose, process models are decomposed into the following concepts, against each of which a query can be made:

- **Attributes**: textual characteristics of the process
- **Decomposition**: a process may be composed of other subprocesses

- **Resource flows**: all process steps have input and output ports through which resources, used by the process, can flow
- **Mechanisms**: resources that are used by the process as distinct from resources that are consumed or produced
- **Exceptions**: characteristics of process failures

11.4.2 Semantic Web Service Composition

There is a significant relationship between Semantic Web Service discovery and composition. In general, the algorithms for composition depend on the availability of a set of Web Services that, when composed, provide functionality that matches that required by the request. Further, in the course of composition, one or more of the matching techniques, described in the last section, will be necessary to determine if a specific service matches the requirements of a particular stage in a service composition. That said, there is a substantial body of research into composition including and predating Web Service technology.

In this section, we look at a sample of the state-of-the-art approaches to Semantic Web Service composition using inference engines to assist in the composition by reasoning over semantic annotations. Specifically, we look at composition planning, constrained object models [341], process-based composition and workflow approaches.

AI Planning

Planning is a research topic adopted from artificial intelligence (AI) concerned with the realization of strategies by intelligent agents where the solution to the strategy is determined at run time based on information represented using some formal language. This is valuable as changes to the set of available services and additional information can be taken into account by the inference engine at each step of the planning. Broadly speaking, an initial and a final state are provided along with information (and constraints on that information) of actions that are available to the agent. Two common, broad approaches are adopted, forward chaining and backward chaining. In forward chaining, the agent starts with the initial state, looking for an action that can move the solution closer to the final state based on the available information, e.g., what actions can be executed where the inputs to that action are available in the information space. The process *chains* forward until the final state can be reached. Backward chaining starts with the desired final state and works backward to the initial state. As the models for Semantic Web Services presented in this chapter pay special attention to the formalization of the data consumed and produced by Web Services, as well as the constraints on that data, planning techniques, based on logical inference engines, are seen as a strong proposal to the problem of Web Service composition. A comprehensive review of AI planning is beyond the scope of this chapter. To give an indication of the variety of approaches, we provide brief descriptions and references to additional material.

In the work of McIlraith and Son [342], the authors propose the modeling of service requests and advertisements in terms of first-order situation calculus. Requests are represented as generic procedures while services are represented as actions that either change the state of the world or the information space. The logic programming language Golog is adapted and extended as a natural formalism for representing and reasoning about service composition in this context.

An interesting link between the Semantic Web Service and AI communities is through the relationship of PDDL and DAML-S (the precursor of OWL-S). DAML-S was strongly influenced by PDDL, resulting in a straightforward mapping between the languages (with restrictions). Consequently, an approach to Web Service composition proposed in [314] is based on the translation of DAML-S descriptions to PDDL and reuse of the PDDL planners.

In [343], the authors describe how Hierarchical Task Planning (HTN) is especially suited to composition of Web Services described by OWL-S, as HTN places particular focus on task decomposition and precondition evaluation, concepts that tailor well to the OWL-S process descriptions. In [344], the meta-model for automated planning from AI and the meta-model for process-based service enactment are merged in an effort to overcome the predominantly static nature of process descriptions favored by industry, such as those defined using BPEL. Overcoming the challenges involved in merging the meta-models allows for more dynamic compositions that can be flexibly enacted. Enactment means that the composed process itself is determined at run time based on the semantic description of the input and output data and relevant constraints. Once the composed process has been established, services are located for each activity and it is verified that the overall process is executable.

Workflow and Business Processes Technology

A popular approach to the composition of Web Services, from an industrial point of view, is through the use of business process modeling (BPM) where each step of a process can be performed by the execution of a Web Service. BPM itself shares a lot of its underlying theory with workflow modeling. Van der Aalst [345] provides a critical comparison Web Service composition language using a set of workflow patterns as the evaluation criteria. For a process or workflow to be established, the stages have to be identified and suitable activities selected. A control flow needs to be defined to ensure the correct sequence of invocation of each activity. Data flow also needs to be defined so that the correct datatypes are used to transfer data from one activity to another. Van der Aalst notes that Web Service composition languages adopt most of the functionality of workflow systems but show increased expressiveness and in particular put additional focus on communication patterns. He also points out the desirability of providing formal semantics for composition languages through mappings to established process modeling formalisms.

The Web Services Business Process Execution Language (WSBPEL)[16] provides an XML-based language for defining business processes in terms of operations provided by Web Services with WSDL descriptions. Although popular and maturing, BPEL essentially is a static means of describing processes made up of Web Service compositions. However, there is significant research activity to merge the theoretical aspects of workflow (and by extension BPM) with the rich expressiveness of Semantic Web Service descriptions in languages like OWL-S and WSMO.

For example, in [346], a BPEL process is defined manually as a *skeleton*. All candidate services with semantic annotations that may be used by the process (there may be multiple candidates for each step) are verified and then registered in a service container. The skeleton process can then be configured to use different combinations of services for different scenarios. A programmatic interface is used to carry out the configuration. For example, in a process that involves booking flights online, one airline's service may be replaced by another's without the need to modify the skeleton business process. As the process models including the input and output messages of each service are semantically described, inference reasoning comes into play where there are differences in the required inputs and outputs of messages for the various services. The reasoning engine can check for semantic compatibility and adjust the configuration of the process accordingly. A similar approach is described for the eFlow platform in [347] where the composite service is modeled as a graph. The graph consists of nodes for services, events and decisions. Arcs joining the nodes denote execution dependencies. The service nodes can be configured to resolve to a concrete service implementation either at design time or run time.

A related approach to process-based Web Service composition is goal-based orchestration [348] using the WSMO conceptual model. The key idea is that each stage in a process can be represented by a WSMO goal rather than a specific service identifier. The goals are resolved to concrete services at run time by a suitable execution environment such as WSMX. A three-tier model is proposed that allows the design of processes through a visual tool that can be mapped to a formal workflow language. The workflow language has then a direct mapping to the Abstract State Machine (ASM) formalism used to describe service behavior in WSMO orchestrations.

11.4.3 Mediation

A frequent, unstated assumption when tackling Web Service discovery and composition is that all artifacts (service requests and advertisements) use a common conceptual model for defining data, processes and protocols. In real-world conditions, it is highly unlikely that business partners can agree on this level of uniformity in

---

16. http://www.oasis-open.org/committees/wsbpel (Accessed September 10, 2007).

advance. Even within a single organization, where there are multiple operational units, each unit may use independent, heterogeneous conceptual models for legacy applications. In such a situation, both discovery and composition of services is very difficult without a defined means to bridge interoperability issues. Mediation is the activity of mitigating the problems of interoperability through ontology alignment. It has its origins in the significant history of research in the database community into schema mapping.

The formal description of data and process as promoted by Semantic Web Service technology provides the basis for mediation. Subsumption-based reasoning is used in the case of languages based on description logics such as OWL-DL while logic programming is used by WSML-Flight and rule-based reasoning is used for languages such as SWSL-Rule and WSML-Rule (which extends WSML-Flight with function symbols). Of the Semantic Web Service conceptual models discussed in this chapter, only WSMO defines mediators as a top level element. The other Semantic Web Service ontologies also recognize the necessity of mediation but do not model it explicitly within their scope. The four categories of WSMO mediators were identified in Section 11.3.5. In particular, current WSMO research efforts focus on design for data mediation and process mediation.

As defined in [349], data mediation is based on the definition of a formal model for ontology mappings. Mappings are created and stored using a formal language. The mappings are applied as needed when an issue of heterogeneity occurs. For example, in the WSMX execution environment, a goal may be defined in terms of one ontology while a candidate Web Service may use another. During the matching phase, the data mediation component checks for mappings between the two ontologies and applies mappings only as necessary. This means that usually only a subset of the mappings that correspond to the concepts used is required, helping the efficiency of the operation. The assumption is that the mappings between the ontologies have already been created.

In WSMO, process mediation deals with solving mismatches between the choreographies of interacting partners [331]. In other words, it is required where the requester's choreography (goal choreography) and the service's choreography do not match. Mismatches can appear not only when the requester and the provider use different conceptualizations of a domain (in which case data mediation is required), but also if they have different requirements for the message exchange pattern they wish to follow [350]. Essentially this means that one of them expects to receive/send messages in a particular order while the other has different messages or a different message order. The role of the process mediator is to retain, postpone and rebuild messages that would allow the communication process to continue.

11.5 Clinical Use Case

Using Semantic Web technologies to share the formal definition of the meaning of data models across the Web, so that they can be flexibly and powerfully queried, only tackles part of the problem with integrating independent heterogeneous applications. Such applications (and systems) interact with each other on the basis of the behaviour that they expose at their interfaces. This is well-recognized across various domains of interest, including medicine, where there are several specifications defining datamodels and behaviour within each respective domain.

In the running example, threading through this book from the use-case introduced in Chapter 2 to the detailed description in Chapter 13, we fous on a translational medicine scenario involving two of these sets of specifications: CEN 13606 and HL7-CDA. We look at how modeling both the data and the behaviour, defined for each specification, semantically can enable services to be located and combined more flexibly.

Applying semantics to Web Services means being able to speicfy the meaning of both the data and the behaviour that Web Services expose at their interfaces. The novelty of Semantic Web Services is not that it is the first technology seeking to define such semantics but that it applies existing semantic modeling mechanisms to Web Services rather than requiring that a new technology stack be built from the ground up.

In the last eight years multiple languages have emerged for the description of different aspects of the Semantics required by Web Services. The fundamental aims of these languages are very similar but each one either targets different specific aspects of Web Services or seeks to correct perceived inconsistencies in earlier efforts. There is also a varying degree of available tool support.

We choose WSMO as the conceptual model for the detailed example in Chapter 13 because of its support for mediation, the clear separation of modeling and ontological constructs for service requesters and providers, and its rule-based approach for the definition of behavioural semantics. Additionally, there is an open source execution engine available called WSMX[17] available for WSMO against which the model can be tested. WSMO has its own native language called WSML which we use in the example as it uses a frame-based syntax that is reasonably reader-friendly. It's important to note that WSML also can be expressed in RDF and can use the RDF examples included in other parts of this book.

The combination of Goals and rule-based process definitions for the sample translational medicine workflow means that the process designers can focus on what they want the process to do without having to worry at that point about the design implementation. Each step in the process is modeled as a Goal to be resolved to the most suitable service at run-time. For example, one Goal models

17. WSMX source and binaries are available at http://sourceforge.net/projects/wsmx, (accessed May 26, 2008).

the need to get guidance on the ordering of tests based on the symtoms and medical hsitory of a patient. Another Goal models the need to get therapeutic guidance depending on results returned from clinical and laboratory tests.

Each step may involve one or more independent services with possibly independent data and behaviour models. Mediation based on formally defined mappings act as bridges. These mappings do not come for free and require a design effort from domain experts. However, as they are defined between industry standards at the conceptual (rather than at the data-instance) level, they provide an extensible, flexible basis for reuse across multiple scenarios.

11.6 Summary

In this chapter we have taken a look at the state-of-the-art approaches for providing semantic annotations of Web Service descriptions. We started by looking at the motivation for applying semantics to Web Services, discussing the drawbacks of XML as a description language, in terms of providing machine-understandable and unambiguous semantics. As Web Service descriptions focus on process models for interacting with the software applications made available on the Web, we examined various approaches to capturing behavioral semantics. These included Finite State Machines, Statecharts and Petri-Nets. In the section on Semantic Web Service approaches we described in detail the four current prominent efforts, OWL-S, WSMO, SWSF and WSDL-S. Finally, as a strong motivation for the use of Semantic Web Services is the possibility to use logical inference engines to reason over semantic descriptions, we looked at three particular aspects of Semantic Web Service usage that may require reasoning support. These are discovery, composition and mediation. In each case, we discussed the prevalent underlying theories, most of which predate the introduction of the Semantic Web Services terminology.

Part IV
Standards

12 Semantic Web Standards

As in many areas it is important to work on standardization to allow widespread development of interoperable software. In the end, this is what counts as success. Consequently, the Semantic Web communities engaged in this process through various means are presented in this chapter.

12.1 Relevant Standards Organization

In the following subsections, we provide brief descriptions of major organizations involved in standardization activities. The list is not exhaustive and reflects the fact that consensus on the various aspects of Semantic Web Services, a relatively young technology, is limited and thus there are a limited number of relevant standards. For each organization, we describe the intent and scope and the kind of membership supported and summarize the standardization process.

12.1.1 International Organization for Standardization (ISO)

ISO, founded in 1947, is an international body constituted as a federation of national standards bodies. At the end of 2007, there were 157 members, with many of ISO's standards having legal status in participating member countries. The original intent was to promote common standards across national boundaries to facilitate the exchange of goods and services. The scope of ISO is broad, covering an ever-increasing set of goods and services that may be traded. For standards relating to electrical/electronic equipment, ISO adopts the work of the IEC (see Section 12.1.2). For standards relating to information technology, a joint committee established between ISO and IEC, called JTC1, takes responsibility.

Voting members are recognized national standards bodies that have paid their subscription to the ISO. Other categories of membership are possible with reduced rights. Standardization work occurs in three phases. First the technical scope of a future standard is designed. Second the voting members negotiate the technical details. Third, the standard is approved by the ISO itself.

12.1.2 International Electotechnical Commission (IEC)

The IEC was founded in 1906 as an international standards organization for all things related to electrical and electronic technologies. The genesis for IEC came from the British IEE and American IEEE institutions. Similarly to ISO, the voting members of the IEC comprise of national electrotechnical committees of subscribed countries. By the end of 2006, there were 67 voting members. A further 69 countries are affiliate members without voting rights. To create broad acceptance of their standards, IEC and ISO cooperate in a joint technical committee to promote standards that overlap between the two organizations.

The highest organizational unit in the IEC is the council made up of a general assembly of National Committees, who are members of the commission. The Council Board implements council policy and takes input from several Management Advisory Committees. In particular the Standards Management Board takes responsibility for the management of IEC's standardization work including the creation, management and dissolution of individual Technical Committees.

12.1.3 Organization for the Advancement of Structured Information Standards (OASIS)

OASIS is a non-profit consortium established for the promotion of eBusiness standards to facilitate processing of the increasing volume of business transactions carried out over the Web. Operationally, OASIS has elected directors and a full-time administrative staff funded by the consortium. Unlike ISO and IEC it is not made up of national bodies. Rather, any organization is free to join and, depending on the subscription level, members of an organization may obtain voting rights on the different stages of the technical specifications. Work is carried out by technical committees which, in turn, provide specification recommendations on which OASIS members can vote. Although OASIS does not produce international standards in the sense of ISO or IEC, its recommendations in the domain of eBusiness area are regarded important by industry. In particular, with its industry focus on the dynamic area of eBusiness, it strives to be as nimble and responsive as possible. In 2002 OASIS signed a memorandum of understanding with other standards organizations to coordinate the various standards efforts with the intent of avoiding overlap.

12.1.4 World Wide Web Consortium (W3C)

The World Wide Web Consortium was founded in 1994 by Tim Berners-Lee, the primary author of the URI, HTTP and HTML specifications on which the Web is founded. It has major hosts in the Europe, Asia and the US, in addition to several offices worldwide. According to the W3C home page, the W3C "is an international consortium where Member organizations, a full-time staff, and the public work together to develop standards". Its mission is to "to lead the World Wide Web to its full potential by developing protocols and guidelines that ensure long-term growth for the Web". To make this a reality, the W3C has already published over ninety standards, called W3C Recommendations. It is committed to keeping Web technologies non-proprietary to promote interoperability to the maximum extent, and to prevent market fragmentation.

As a member organization, W3C organizes itself in various entities. The Team is a group of about sixty researchers who lead the technical activities from a management perspective. The advisory committee consists of one member from each member organization and reviews proposals for new activities and proposed recommendations. The Technical Architecture Group was put in place to provide architectural guidance and to resolve architectural conflicts. Working Groups are established for technical developments, interest groups for general work and coordination groups for ensuring cross-group communications.

12.1.5 International Engineering Task Force (IETF)

The IETF is an all-volunteer organization for the development of Internet standards. Founded in 1986, its underlying motivation was provided by engineers of member organizations getting together to solve interoperability problems between their respective products. There are a number of activity areas including telecoms, security and transport. For each activity area, working groups are established for particular topics. Each working group has one or more chairs and closes down once work on that particular topic is complete.

IETF standards take the form of RFCs (Requests For Comments). The establishment of an RFC is through rough consensus, carried out over email following an agreed-upon decision-making procedure. The IETF meets three times a year and cooperates closely with ISO, IEC and W3C.

12.1.6 National Institute of Standards and Technology (NIST)

The NIST is a non-regulatory federal agency in the US. It has a broad scope in the promotion of standards and technology for the benefit of the US economy. Although NIST is not a standards body per se, it monitors technology standards and promotes their use in the US. This is particularly the case for US standards for which the NIST Standards Service Division (SSD) promotes recognition both domestically and globally to encourage US-driven trade.

12.1.7 The Object Modeling Group (OMG)

OMG is an international non-profit computer industry consortium founded in 1989 by Hewlett Packard, Sun Microsystems, Apple Computer, American Airlines, IBM and Data General. It was originally established to create standards for distributed computing but has evolved to include standards for enterprise integration and modeling. Some prominent examples are the CORBA specification for distributed computing and the UML specification for modeling object-oriented systems.

OMG has an open membership with members from both large and small computer industries. Each member organization has voting rights. OMG has a board of directors and a technology committee consisting of three sub-committees: the architecture board, the platform technology committee and the domain technology committee. Task forces are established under a responsible technology committee to create standards in particular areas. The process usually starts with an RFP (Request for Proposal) and goes through various phases before possibly being recommended as an OMG standard.

OMG maintains close liaisons with several other standards bodies. In particular, it is an ISO-PAS submitter, which enables OMG standards to be fast-tracked through the ISO standardization process. Several OMG standards have been accepted as ISO standards, including UML (ISO/IEC 19501), MOF (ISO/IEC 19502) and IDL (ISO/IEC 14750).

12.1.8 Semantic Web Services Initiative (SWSI)

SWSI is an ad hoc initiative of researchers in the area of Semantic Web and Semantic Web Services. SWSI is relevant in this section as the outcome of their work led to submission to the W3C of the Semantic Web Services Framework (SWSF) including the SWSL language (see Section 11.3.2).

12.1.9 United States National Library of Medicine (NLM)

NLM is the world's largest medical library, located on the campus of the US National Institute of Health, Bethesda, Maryland. It collects and maintains guidelines and standards relating to all aspects of healthcare in the United States. In this context, it has been the US coordinating center for standard clinical vocabularies since 2004.

The SNOMED CT (Systematized Nomenclature of Medicine - Clinical Terms) terminology was originally developed in the US by the College of American Pathologists and was licensed by the NLM for US use. Since April 2007, SNOMED CT has been owned, maintained and distributed by the International Health Terminology Standards Development Organization (IHTSDO) in Denmark. However, the NLM is the US member of IHTSDO and distributes SNOMED CT at no cost in the US under the terms of an IHTSDO international license.

Additionally, NLM funds ongoing development of LOINC (database facilitating the exchange of medical information) and maintains the RxNorm clinical drug vocabulary. SNOMED CT, LOINC and RxNorm have been designated as US-government-wide standards. In addition, NLM is working to align these three efforts with the international HL7 clinical messaging standard.

12.2 Semantic Web Content Standardization Efforts

In this section, we enumerate a set of content standards that seek to standardize representation of semantics to varying degrees. Some of these standards are markup languages that predate the Internet era, but form the basis from which the various Semantic Web markup languages have been derived. The intent is to identify some of the most prominent languages for describing information on the Web and to indicate earlier efforts from which these languages have derived. We also enumerate specifications for transforming and querying Semantic Web content, and a list of standardized vocabularies and ontologies that are currently in use.

12.2.1 Standard Generalized Markup Language (SGML)

The Standard Generalized Markup Language (SGML) [14] is a meta-language in which one can define markup languages for documents. SGML is a descendant of IBM's Generalized Markup Language (GML), developed in the 1960s. SGML was originally designed to enable the sharing of machine-readable documents in large projects in government, law and industry, which have to remain readable for several decades. Both XML and HTML originated as derivatives of SGML.

12.2.2 eXtensible Markup Language (XML)

The Extensible Markup Language (XML) [43] is a general-purpose markup language. It is classified as an extensible language because it allows its users to define their own tags. Its primary purpose is to facilitate the sharing of structured data across different information systems, particularly via the Internet. It is used both to encode documents and to serialize data. It started as a simplified subset of the Standard Generalized Markup Language (SGML), and is designed to be relatively human-legible. By adding semantic constraints, application languages can be implemented in XML. These include XHTML, RSS, MathML, GraphML, Scalable Vector Graphics, MusicXML, and thousands of others. Moreover, XML is sometimes used as the specification language for such application languages. XML is recommended by the World Wide Web Consortium. It is a fee-free open standard. The W3C recommendation specifies both the lexical grammar and the requirements for parsing.

12.2.3 eXtensible Stylesheet Transformation Language (XSLT)

XSLT [65] is a language for transforming XML documents into other XML documents. XSLT v1.0 is one of a family of three languages that collectively make up XSL. The other two are XSL Formatting Objects v1.1 and XPath v1.0. All three are W3C recommendations. XSLT v2.0 became a W3C Recommendation in January 2007. It was developed in parallel with XPath v2.0 and both share the same data model. While initially XSLT was developed to transform XML documents into other XML documents formatted for presentation, it has become much more of a general-purpose transformation language.

12.2.4 XPath

XPath [67] is a language that allows individual nodes in an XML document tree to be addressed. It is mainly used within a host language such as XSLT 2.0 or XQuery 1.0, both of which became W3C Recommendations in January 2007. It also provides basic facilities for manipulation of strings, numbers and booleans. XPath uses a compact, non-XML syntax to facilitate use of XPath within URIs and XML attribute values. XPath operates on the abstract, logical structure of an XML document, rather than its surface syntax. In addition to its use for addressing, XPath is also designed so that it has a natural subset that can be used for matching (testing whether or not a node matches a pattern); this use of XPath is described in XSLT.

12.2.5 XQuery

XQuery [64] is designed to be a language in which queries against XML data sources are concise and easily understood. It can be considered analogous to SQL for relational database systems and is flexible enough to query a broad spectrum of XML information sources, including both databases and documents. The use of expressions in XPath are augmented with the keywords FOR, LET, WHERE, ORDER BY, and RETURN, which resemble keywords in SQL. They enable queries to be built with embedded XML tags for the direct construction of valid XML documents based on the query's outcome.

12.2.6 XML Schema

XML Schema [87] can be used to express a schema: a set of rules to which an XML document must conform in order to be considered 'valid'. However, unlike most other schema languages, XML Schema was also designed for validation, resulting in a collection of information adhering to specific datatypes. An XML Schema instance is an XML Schema Definition (XSD) and includes the vocabulary (element and attribute names), the content model (relationships and structure) and the datatypes.

12.2.7 Resource Description Framework (RDF)

The Resource Description Framework (RDF) [96] is a language for representing information about resources on the World Wide Web. It is provided as a family of W3C Recommendations (current version from February 2004) and is particularly intended for representing metadata about Web resources, such as the title, author, and modification date of a Web page, copyright and licensing information about a Web document, or the availability schedule for some shared resource. However, by generalizing the concept of a "Web resource", RDF can also be used to represent information about things that can be identified on the Web (using Uniform Resource Identifiers or URIs), even when they cannot be directly retrieved on the Web. RDF is intended for situations in which this information needs to be processed by applications, rather than being only displayed to people. RDF provides a common framework for expressing this information, so it can be exchanged between applications without loss of meaning.

12.2.8 SPARQL

The SPARQL specification [83], a W3C Candidate Recommendation as of June 2007, defines the syntax and semantics of a query language for RDF. SPARQL can be used to express queries across diverse data sources, regardless of whether the data is stored natively as RDF or viewed as RDF via middleware. SPARQL contains capabilities for querying required and optional graph patterns along with their conjunctions and disjunctions. SPARQL also supports extensible value testing and constraining queries by source RDF graph. The results of SPARQL queries can be result sets or RDF graphs.

12.2.9 RDF Schema

RDF(S) or RDF Schema [88] is an extensible knowledge representation language, providing basic elements for the description of ontologies defined using RDF. Vocabularies declared using RDFS are described in terms of classes and properties using extensions to RDF provided by RDFS. The semantics are rather informal, aimed at providing users with a simple means to organize their RDF information. RDFS itself does not impose constraints on information as do the type systems, for example, object-oriented languages such as Java. Although it is a W3C Recommendation since February 2004, other standards such as OWL (Section 12.2.10) provide a means for describing ontologies whose semantics are well-defined and formalized in comparison with RDFS.

12.2.10 Web Ontology Language (OWL)

The Web Ontology Language (OWL) [59] is a language for defining and instantiating Web ontologies. An OWL ontology may include descriptions of classes, along with their related properties and instances. OWL is designed for use by applications that need to process the content of information instead of just presenting information to humans. It facilitates greater machine interpretability of Web content than that supported by XML, RDF, and RDF Schema (RDFS) by providing additional vocabulary along with a formal semantics. OWL is based on the earlier languages OIL and DAML+OIL, and has been a W3C Recommendation since February 2004.

12.2.11 Rule-ML

The Rule Markup Language (RuleML) [395] is a markup language developed to express both forward (bottom-up) and backward (top-down) rules in XML for deduction, rewriting, and further inferential-transformational tasks. It is defined by the Rule Markup Initiative, an active open network of individuals and groups from both industry and academia that was formed to develop a canonical Web language for rules using XML markup and transformations from and to other rule standards and systems.

12.2.12 Semantic Web Rules Language (SWRL)

The Semantic Web Rule Language (SWRL) [312] is based on a combination of the OWL-DL and OWL Lite sublanguages of the OWL Web Ontology Language with the Unary/Binary Datalog RuleML sublanguages of the Rule Markup Language. SWRL includes a high-level abstract syntax for Horn-like rules in both the OWL-DL and OWL Lite sublanguages of OWL. A model-theoretic semantics is given to provide the formal meaning for OWL ontologies, including rules written in this abstract syntax, an XML syntax based on RuleML and the OWL XML. A presentation syntax and an RDF concrete syntax based on the OWL RDF/XML exchange syntax are also given.

12.2.13 Ontology Definition Metamodel (ODM)

The ODM specification [396] defines a family of independent metamodels, related profiles, and mappings among the metamodels corresponding to several international standards for ontology and topic maps definition, as well as capabilities supporting conventional modeling paradigms for capturing conceptual knowledge, such as entity-relationship modeling. In 2006, it was adopted as a standard by the Object Management Group (OMG) and is applicable to knowledge representation, conceptual modeling, formal taxonomy development and ontology definition, and enables the use of a variety of enterprise models as starting points for ontology development through mappings to UML and MOF. ODM-based ontologies can be used to support interchange of knowledge among heterogeneous computer systems, representation of knowledge in ontologies and knowledge bases, and specification of expressions that are the input to or output from inference engines.

12.2.14 Unified Modeling Language (UML)

The Unified Modeling Language (UML) [40] is a standardized specification language for object modeling and includes a graphical notation used to create an abstract model of a system, referred to as a UML model. UML is officially defined at the Object Management Group (OMG) by the UML metamodel, a Meta-Object Facility (MOF) metamodel. UML was designed to specify, visualize, construct, and document software-intensive systems. UML is also used for business process modeling, systems engineering modeling, and representing organizational structures. It is also being used for modeling ontologies.

12.2.15 Knowledge Interchange Format (KIF)

Knowledge Interchange Format (KIF) [50] is a computer-oriented language for the interchange of knowledge among disparate programs. It has declarative semantics (i.e., the meaning of expressions in the representation can be understood without an interpreter for manipulating those expressions); it is logically comprehensive (i.e., it provides for the expression of arbitrary sentences in the first-order predicate calculus); it provides for the representation of knowledge about the representation of knowledge; it provides for the representation of non-monotonic reasoning rules; and it provides for the definition of objects, functions, and relations.

12.2.16 Open Knowledge Base Connectivity Protocol (OKBC)

OKBC [397] is an API and reference implementation that allows representation-systemplatform- and language-independent knowledge-level communication. It enables knowledge application authors to write representation-system-independent tools, and to publish their knowledge easily.

12.2.17 DIG Description Logics Interface

The DIG Interface [398] is a standardized XML interface to Description Logics systems developed by the DL Implementation Group (DIG). DIG 2.0 from November 2006 provides a specification for a standardized interface across DL Reasoners supporting the W3C OWL Recommendation.

12.2.18 OWL API

OWL API [399] is a Java interface and implementation for the W3C Web Ontology Language (OWL), used to represent Semantic Web ontologies. The API is focussed toward OWL Lite, OWL-DL and OWL 1.1 and offers an interface to inference engines and validation functionality.

12.2.19 Standardized Vocabularies and Ontologies

A list of well known and standardized vocabularies and ontologies are provided below.

CIM/DMTF

The Common Information Model (CIM) [91] provides a common definition of management information for systems, networks, applications and services, and allows for vendor extensions. CIM's common definitions enable vendors to exchange semantically rich management information between systems throughout the network. CIM is composed of a specification and a schema. The schema provides the actual model descriptions, while the specification defines the details for integration with other management models.

SNOMED

SNOMED (Systematized Nomenclature of Medicine) [7] is a systematically organized computer processable collection of medical terminology covering most areas of clinical information such as diseases, findings, procedures, microorganisms, and pharmaceuticals. It allows a consistent way to index, store, retrieve, and aggregate clinical data across specialties and sites of care. It also helps in organizing the content of medical records, and in reducing the variability in the way data is captured, encoded and used for clinical care of patients and research.

Gene Ontology

The Gene Ontology project, or GO [10], provides a controlled vocabulary to describe gene and gene product attributes in any organism. It can be broadly split into two parts. The first is the ontology itself - actually three ontologies, each representing a key concept in molecular biology: the molecular function of gene products; their role in multi-step biological processes; and their localization to cellular components. The second part is annotation, the characterization of gene products using terms from the ontology. The members of the GO consortium submit their data and it is made publicly available through the GO Web site.

International Classification of Diseases (ICD-9)

The International Statistical Classification of Diseases and Related Health Problems 9th Revision (ICD-9) [8] is a coding of diseases and signs, symptoms, abnormal findings, complaints, social circumstances and external causes of injury or diseases, as classified by the World Health Organization (WHO).

Medical Subject Headings (MeSH)

MeSH [15] is the National Library of Medicine's controlled vocabulary thesaurus. It consists of sets of terms naming descriptors in a hierarchical structure that permits searching at various levels of specificity. MeSH descriptors are arranged in both an alphabetic and a hierarchical structure. At the most general level of the hierarchical structure are very broad headings such as “Anatomy” or “Mental Disorders.” More specific headings are found at more narrow levels of the eleven-level hierarchy, such as “Ankle” and “Conduct Disorder.” There are 22,997 descriptors in MeSH. In addition to these headings, there are more than 151,000 headings called Supplementary Concept Records (formerly Supplementary Chemical Records) within a separate thesaurus. There are also thousands of cross-references that assist in finding the most appropriate MeSH heading, for example, Ascorbic Acid for Vitamin C. These additional entries include 24,050 printed “see” references and 112,012 other entry points.

BioPax

BioPAX Level 2 [47] covers metabolic pathways, molecular interactions and protein post-translational modifications and is backward compatible with Level 1. Future levels will expand support for signaling pathways, gene regulatory networks and genetic interactions.

Cyc

The Cyc knowledge base (KB) [51] is a formalized representation of a vast quantity of fundamental human knowledge: facts, rules of thumb, and heuristics for reasoning about the objects and events of everyday life. The medium of representation is the formal language CycL, described below. The KB consists of terms, which constitute the vocabulary of CycL, and assertions, which relate those terms. These assertions include both simple ground assertions and rules.

Basic Formal Ontology (BFO)

BFO [93] is narrowly focussed on the task of providing a genuine upper ontology which can be used to support domain ontologies developed for scientific research, as for example in biomedicine.

IEEE Suggested Upper Merged Ontology (SUMO)

The Suggested Upper Merged Ontology (SUMO) [49] and its domain ontologies form the largest formal public ontology in existence today. They are being used for research and applications in search, linguistics and reasoning. SUMO is the only formal ontology that has been mapped to all of the WordNet lexicon. SUMO is written in the SUO-KIF language.

Unified Medical Language System (UMLS)

The Unified Medical Language System (UMLS) [16] is a compendium of many controlled vocabularies in the biomedical sciences. It provides a mapping structure between these vocabularies and thus allows us to translate between the various terminology systems; it may also be viewed as a comprehensive ontology of biomedical concepts. UMLS consists of the following components:

- Metathesaurus, the core database of the UMLS, a collection of concepts and terms from the various controlled vocabularies and their relationships;
- Semantic Network, a set of categories and relationships that are being used to classify and relate the entries in the Metathesaurus;
- SPECIALIST Lexicon, a database of lexicographic information for use in natural language processing;

12.3 Semantic Web Services Standardization Efforts

This section outlines the major standards activities in the Semantic Web Services area. Table 12.1 provides an overview of all standards activities that are discussed in the following.

Table 12.1. Standards overview

Organ-ization	Standard	Candidate Recomm-endation	Member Submission	Interest Group	Guide-lines	Proposed Architecture or Framework
ISO/IEC	PSL (ISO 18629)					
OASIS	SOA RM				SEE	

Table 12.1. Standards overview

Organ-ization	Standard	Candidate Recomm-endation	Member Submission	Interest Group	Guide-lines	Proposed Architecture or Framework
W3C		SAWSDL	OWL-S	SWSIG		
			WSMO			
			SWSF			
			WSDL-S			
SWSI (Non-affil-iated)						SWSA

Notably, PSL is already an international ISO standard. Other efforts like OWL-S, SWSL, WSDL-S and WSMO have been accepted as member submissions to the W3C Web Service Activity group. Architecturally, a working group has been formed in the context of OASIS and there are some additional relevant activities in this area.

12.3.1 ISO-18629 Process Specification Language (PSL)

PSL was developed by NIST as a neutral representation for manufacturing processes. It allows process information to be shared between autonomous applications that each use heterogeneous representations. The NIST homepage for PSL[18] describes how its rationale is to provide a language that is "common to all manufacturing applications, generic enough to be decoupled from any given application and robust enough to be able to represent process information from any given application". In August 2006, PSL was published as an international standard by ISO, giving the family of specifications the identifier ISO-18629.

PSL is organized as a layered ontology with the concepts defined formally using the Knowledge Interchange Format (KIF) [285]. The purpose of KIF is to provide a formal language for the interchange of knowledge rather than as an internal representation of knowledge in a system (although nothing prevents its being used in this way). The PSL ontology is organized into PSL-Core and a partially ordered set of extensions. All extensions must be consistent with PSL-Core but they need not be necessarily consistent with each other. The core ontology consists of four disjoint classes:

- **Activities** may have zero or more occurrences
- **Activity occurrences** begin and end at timepoints
- **Timepoints** constitute a linearly ordered set with endpoints at infinity
- **Objects** all elements that are not activities, occurrences or timepoints

18. http://www.mel.nist.gov/psl/rationale.html (Accessed September 10, 2007).

To supplement the PSL-Core, additional concepts are introduced in the PSL-Outer-Core theory. These include subactivities, occurrence trees, discrete states, atomic activities, and activity occurrences. Remaining core theories in the PSL ontology include: subactivity occurrence ordering, iterated occurrence ordering, duration and resource requirements. There is a distinction made between core theories and definitional extensions. Core theories introduce primitive concepts while all terminology introduced in definitional extensions are defined using the terminology of the core theories.

Section 11.3.2 describes how the Semantic Web Services Framework (SWSF) defines a conceptual model for Semantic Web Services called the Semantic Web Services Ontology (SWSO). The first-order axiomatization of SWSO is called First-Order Logic Ontology for Web Services (FLOWS) where FLOWS is based on PSL.

12.3.2 W3C Semantic Annotations for the Web Services Description Language (SAWSDL)

Section 11.3.3 introduced SAWSDL as a bottom-up approach to semantic markup of WSDL 2.0 Web Service descriptions. As it is possible for two services, offering different functionality, to be described using very similar WSDL, SAWSDL is intended as a means for allowing ambiguities to be resolved. It defines extensions for WSDL 2.0, enabling elements of WSDL 2.0 documents to be associated with ontological definitions. SAWSDL is agnostic to the choice of conceptual model. It simply provides the means by which such models can be associated with service descriptions.

As of January 2007, SAWSDL is a W3C Candidate Recommendation. The W3C policy guidelines state that a working group can only advance its specification to Candidate Recommendation level once the W3C is satisfied that the technical report is stable and appropriate for implementation. To reach the status of W3C Recommendation, the working group must gather implementation experience and update the specification accordingly. Once this has been completed, the specification may be put forward to the W3C Advisory Committee. If it is satisfied that the specification fulfills its mandate and any objections have been overcome, SAWSDL will become a W3C Recommendation.

SAWSDL came about as a result of the W3C Workshop on Frameworks for Semantics in Web Services,[19] held in Innsbruck, Austria in June 2005. A number of W3C member submissions were presented and debated. The aim of the workshop was to determine if the level of activity and maturity in the area of Semantic Web Services suggested the establishment of a W3C Working Group or some other activity. There was limited consensus at the end of the meeting between the more

---

19. http://www.w3.org/2005/04/FSWS/workshop-report.html (Accessed September 10, 2007).

comprehensive Semantic Web Service ontologies defined by WSMO, OWL-S and SWSF. On the other hand there was more agreement that the bottom-up approach of WSDL-S provided a common ground on which a W3C activity could progress. Ultimately, this led to the establishment of the SAWSDL Working Group.

12.3.3 OWL-S

A detailed description of OWL-S is provided in Section 11.3.1. Here, we examine the history of OWL-S and its status with respect to standardization.

OWL-S has roots in research communities on both sides of the Atlantic. The Ontology Inference Language (OIL) [286] took RDFS [287] as its starting point and added more intuitive semantics to enrich the language into one that was suitable as a Web ontology language. Specifically, OIL had well-defined formal semantics and reasoning support, and was intended to be intuitive to human readers and be linked properly with the Web languages XML and RDF. At the same time, in the US, a very similar parallel initiative was being undertaken called the DARPA Agent Markup Language (DAML), resulting in an ontology called DAML-ONT [289]. As both approaches had considerable overlap, in October 2000, the Joint US/EU ad hoc Agent Markup Language Committee was established to work on a DAML+OIL joint specification. The DAML-S ontology took over from DAML+OIL and, after a number of version iterations, DAML-S was renamed OWL-S to reflect its relationship to the Web Ontology Language (OWL).

OWL-S was submitted as a member submission to the W3C in 2005. It was a significant input to the W3C Workshop on Frameworks for Semantics in Web Services, mentioned in the previous section. In terms of standards OWL-S has no official status outside its own community of users, although this is a substantial community, particularly in the US.

12.3.4 Web Services Modeling Ontology (WSMO)

A detailed description of WSMO is included in Section 11.3.5. As with OWL-S, in this section we look at the status of WSMO with respect to standardization.

The WSMO is an initiative started in 2004, through EU funding, as an effort to address perceived fundamental problems in the OWL-S ontological model for Semantic Web Services (described in Section 11.3.5). It built on earlier research, notably the Web Services Modeling Framework [264]. As with OWL-S, WSMO was submitted to the W3C as a member submission and was presented and debated at the W3C Workshop on Frameworks for Semantics in Web Services. At the time of this publishing, the WSMO community is very active and evolving (in contrast to OWL-S which is relatively static), and the ontology, corresponding language, Web Service Modeling Language (WSML) and reference architecture, and Web Service Model Execution (WSMX) are foundational blocks in several current large-scale EU research projects, e.g., DIP,[20] SUPER,[21] SWING.[22]

12.3.5 Semantic Web Services Framework (SWSF)

SWSF is described in Section 11.3.2 and was also a W3C member submission for Semantic Web Services activity. Analogously to the motivation for WSMO, SWSF was established on the basis of perceived shortcomings of the OWL-S ontology, particularly with respect to the modeling of processes. We explained in Section 12.3.1 how SWSF is built on PSL, which itself has been promoted as an international process language standard by ISO.

12.3.6 WSDL-S

WSDL-S is described in Section 11.3.3 and also was a W3C member submission. We have already described how the bottom-up approach of WSDL-S was adopted by the W3C as the basis for the SAWSDL Working Group.

12.3.7 OASIS Semantic Execution Environment (SEE)

In November 2005, the Semantic Execution Environment Technical Committee (SEE TC) was formed at the OASIS eBusiness standards consortium to develop guidelines, justifications, and implementation directions for deploying Semantic Web Services in service-oriented architectures. A foundational input for the SEE TC is the open source work on WSMX as a reference architecture and prototype implementation for WSMO. The intent of the SEE TC is to:

- Provide a reference architecture for a Semantic Web Services execution environment
- Formally describe execution semantics for the SEE reference architecture
- Relate the SEE reference architecture to the OASIS Standard for a Service-Oriented Architecture Reference Model (SOA RM) as described in Section 12.3.8.

Service-oriented architectures anticipate a large number of ambient heterogeneous computational services which may be utilized in various combinations. However, composing a set of services to meet arbitrary goals is often an attempt to coordinate disparate resources with independent heterogeneous data and process models. The services in the composition may not know, or fully understand, the information models of each other in advance. So some interpretation, mediation or common understanding is essential for any significant deployment. The SEE TC aims to provide a reference architecture identifying the functional components, and the methods required of them, to overcome this challenge.

---

20. http://dip.semanticweb.org/ (Accessed September 10, 2007).
21. http://www.ip-super.org/ (Accessed September 10, 2007).
22. http://www.swing-project.org/ (Accessed September 10, 2007).

The TC is also defining a formal description of the necessary execution semantics of such a system. Taking advantage of the ontological models available for Web Service description, the TC will additionally define a generic and open mechanism, using metadata, to allow components to be plugged into the system and made available dynamically.

Corresponding to the works described in the previous paragraphs, there are three draft deliverables being developed by the SEE TC. A fourth document is available at the committee's Web page summarizing related work and background information. The three draft deliverables are:

1. Semantic Web Services Architecture and Information Model
2. SEE Execution Scenarios
3. Semantic Service-Oriented Architecture Reference Model

12.3.8 OASIS Service-Oriented Architecture Reference Model (SOA RM)

The OASIS SOA Reference Model (SOA RM) [288] specification defines a reference model as an "abstract framework for understanding significant relationships among the entities of some environment. It enables the development of specific reference or concrete architectures using consistent standards or specifications supporting that environment. A reference model consists of a minimal set of unifying concepts, axioms and relationships within a particular problem domain, and is independent of specific standards, technologies, implementations, or other concrete details."

The SOA RM aims to provide a single source of terminology for designers of SOAs. It is separated from architecture implementations by several layers of abstraction. The envisaged abstraction layers, starting from the top are:

- **SOA Reference Model.** Defines the concepts common across reference architectures for SOA, e.g., service, service description, visibility, interaction, contract and policy, execution context and effect.
- **SOA Reference Architecture.** In any domain a reference architecture identifies the specific elements, connections and patterns that are required for the architecture to achieve its aims. There may be many reference architectures in a given domain. The SOA RM provides an example from housing architecture.
- **SOA Concrete Architecture.** Adds specific additional elements that are required to realize the architecture as a working system. Concrete architectures have to take account of the deployment environment and the technologies that are available.
- **SOA Implementation.** The executable code that provides the functionality defined by the architecture.

From the above, it can be seen that the SOA RM is three levels of abstraction above the actual code that implements an SOA. At this level of abstraction, it is manageable to define all concepts that are required for the building of reference architectures.

Although the SOA RM defines the concepts, it falls short of providing a formal definition of the conceptual model which may result in different interpretations of the English descriptions of the concepts in different reference architectures. Nevertheless, the SOA RM provides a useful starting point and, in terms of OASIS, serves as important reference material to other technical committees. For example, the SEE TC (Section 12.3.7) relates the conceptual model used by it (WSMO) to the OASIS SOA RM. As of October 2006, the SOA RM specification became an official OASIS Standard.

12.3.9 Semantic Web Services Architecture (SWSA)

The Semantic Web Services Architecture (SWSA) is a product of the Semantic Web Services Initiative (SWSI),[23] "an ad hoc initiative of academic and industrial researchers, many of which are involved in DARPA and EU funded research projects". SWSI had two aims: the creation of a Semantic Web Services Language (SWSL) which we discussed as part of Section 11.3.2 on the Semantic Web Services Framework (a W3C member submission), and the development of a Semantic Web Services Architecture (SWSA) [290]. The SWSA framework addresses five categories of Semantic Web Services requirements. These are:

- **Dynamic service discovery**. The distributed search for services that can satisfy some part of a service requester's goal.
- **Service engagement**. Interpret candidate service enactment constraints (assuming they are published with the service descriptions) and then negotiate with the potential services until agreement, on how engagement with service can be carried out, is reached.
- **Service process enactment and management**. The process of completing the mutually agreed enactment between the service requester and provider. In particular, the various published communication protocols of requester and provider must be followed. When the process's primary goal is not achieved, some form of compensation activity may be necessary.
- **Community support services**. These include services for ontology lookup, mapping and version control, security and privacy services, confidentiality and trust services, reliability services, policy and protocol management services, and lifecycle management services.
- **Quality of service**. Requesters and providers of services in real-world business exchanges agree on contractual levels of service provision. An SWSA must provide services that can provide metrics on service provision and monitor if negotiated QoS levels are maintained.

23. http://www.swsi.org/ (Accessed September 10, 2007).

The SWSA was not directly submitted to a standardization body but is an influential body of work with respect to the WSMX architecture submitted to W3C and subsequently the OASIS SEE Technical Committee. Although the ad-hoc SWSI group has not officially disbanded, they are not active at this point in time. It is likely that the publication of the SWSA and SWSL specifications mark the end of their contribution to this area as a group.

12.3.10 Semantic Web Services Interest Group (SWS-IG)

The Semantic Web Services Interest Group (SWSIG)[24] is a W3C forum for collaboration, exchange of ideas and dissemination of information regarding ongoing research in the area of Semantic Web Services. It is part of the W3C Web Services Activity and is the coordinator of member submissions on Semantic Web Services to the W3C.

12.4 Summary

In this chapter, we provided a review of the standardization activities around Semantic Web Services. Of these, PSL as an ISO international standard is the most mature. The W3C is taking a bottom-up approach with SAWSDL on its way to becoming a W3C Recommendation. In terms of architecture and usage, OASIS is focussing on a reference model for service-oriented architecture which already is attracting notice and may be extended with formal semantics; and the Semantic Execution Environment Technical Committee aims at a specification for a Semantic Web Services reference architecture and relates its conceptual model to that described in the OASIS SOA RM standard.

---

24. http://www.w3.org/2002/ws/swsig/ (Accessed September 10, 2007).

Part V
Putting it All Together and Perspective

13 A Solution Approach to the Clinical Use Case

We now revisit functional requirements presented in Chapter 2 and illustrate them with examples based on the use case. We begin with a simplified clinical workflow in Figure 13.1 (which is an adaptation of Figure 2.1). This process is used as a (simplified) entry point into the translational medicine use case discussed earlier.

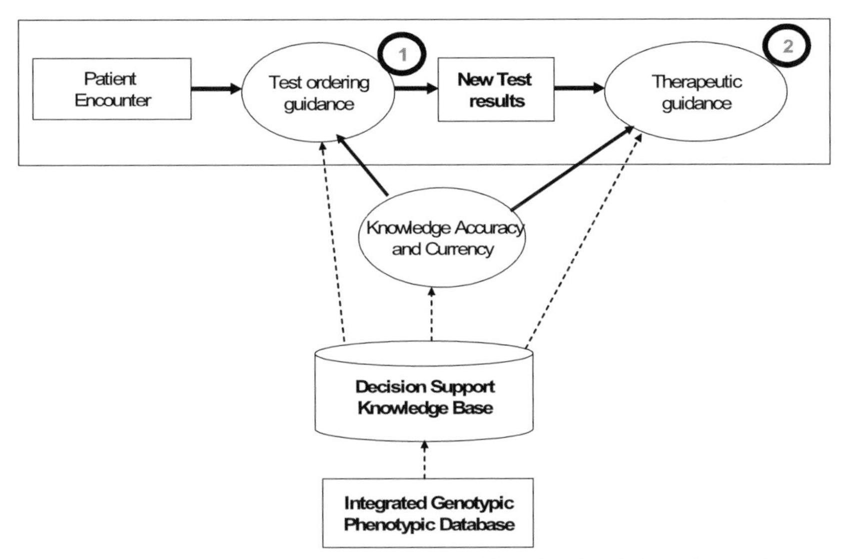

Fig. 13.1. Clinical Vignette for Translational Medicine: simplified clinical workflow

The process illustrated in Figure 13.1 above is an example of a clinical process and involves the basic patient-physician interactions in conjunction with a support process that maintains the accuracy and currency of the knowledge used in the clinical process.

- The patient encounter is the initial step where the patient is examined and observations about him and his observable state of health are recorded. Also, demographic information about the patient, including pertinent information about his family members, may be recorded in an integrated electronic medical record (EMR) at this time in case this has not been recorded yet in earlier visits.
- Based on the information available the physician orders clinical or genomic tests in order to validate or invalidate certain hypotheses about the potential disease a patient may be suffering from. In this he may be assisted by a test order-

ing guidance service. This test ordering guidance service internally uses decision support knowledge bases and other databases in order to provide its functionality.

- After receiving new test results, the physician typically identifies therapies in the form of drugs or clinical procedures for the patient. In this the physician is assisted by a therapeutic decision support service that internally uses knowledge bases and databases as in the previous service.
- Finally, there is a critical business requirement that the information and knowledge that is made available to the physicians in the clinical workflow is current with the latest discoveries and medicine and at the same time consistent with other pieces of knowledge and information. This function is supported by an ongoing knowledge accuracy and currency service that keeps the internal databases and knowledge bases current and consistent with each other.

One may view the clinical workflow as a business or "healthcare" process which can provide the framework for service discovery, composition and choreography. The support process for knowledge accuracy and currency can be viewed as a generic, domain-neutral, knowledge management process that aims to meet business requirements. In the following the various elements of Semantic Web Services are introduced for this example in more detail using the WSMO family of Semantic Web Service technologies.

Any of the Semantic Web Service approaches described in Chapter 11 could have been selected to flesh out our running example. WSMO is chosen as it offers a conceptual model allowing independent modeling of the perspectives of both the service requester and provider through the separate elements of Goals and Web Services. Additionally, WSMO anticpates that services within a single domain will likely use different models to describe their respective behaviour and data and, consequently, strong support for mediation is one of WSMO's fundamental principles. It is also arguable that WSML, as a frame-based language, provides a more readable syntax than OWL-S, for example. Finally, an RDF syntax is available for WSML[25] which can be used to incorporate the RDF examples (focussing on knowledge modeling) and the WSML example (focussing on service modeling) through this book.

13.1 Service Discovery, Composition and Choreography

We now illustrate how Semantic-Web-related technologies can be used to support annotations of services to facilitate service discovery, composition and choreography. As discussed in Chapter 2, there are two kinds of services:

25. http://www.wsmo.org/TR/d32/v0.1/ (accessed May 28, 2008).

- **Business/Clinical Services**. These are services that either are directly provisioned as compositions of services to implement a business process such as the clinical workflow illustrated in Figure 13.1 or could be implemented as an individual service to support certain business requirements. Business or clinical services may provide domain-specific functionality like the two in the example, as they return data that are used by a physician, or could be domain-independent services geared to meeting business requirements. The business services invoked are a *test ordering guidance service* and a *therapeutic guidance service*. Semantic Web technologies can be used to enable automatic discovery of these services and their subsequent composition/orchestration to implement the given business or clinical process.
- **Technology/Infrastructure Services**. Business services in turn may be implemented by technology services at run time to implement the advertised functionality. For example, decision support services (test ordering and therapeutic) may invoke technology services such as rules engine and ontology engine services, classification and inferencing services and data integration services. These services provide functionality independent of a specific application domain. Semantic Web technologies can be used to support dynamic orchestration of these services at run time. Knowledge management services either are invoked in response to certain events or can be executed as a monitoring process to support the accuracy and currency of various knowledge artifacts (e.g., decision support rules) used in the various services.

In the following sections, the two business services are defined, as are many infrastructure services. The discussion will clearly state when a business service and when an infrastructure service is used.

13.1.1 Specification of Clinical Workflow using WSMO

In this section, we model the workflow for the translational medicine example in terms of Semantic Web Service technology. We use WSMO for our conceptual model because of: (a) support for mediators; (b) ontological separation of requesters and providers; and (c) support for behavioral and process semantics. In other words, our model can use WSMO Goals to describe what the service requester wants and WSMO Web Service descriptions to describe what is available in the system. Where semantic mismatches arise at the data, process or protocol level, mediators can be defined to handle the heterogeneity.

From a modeling perspective, the translational medicine clinical workflow described above and depicted in Figure 13.1 can be represented as a single Goal to treat a patient. This high-level Goal can be modeled as a composition, or orchestration, of subgoals where each step in the orchestration is described by a Goal. Composing services into an orchestration is not a trivial task. Although it is arguable that if sufficient functional descriptions of services are available, then it is possible for them to be automatically composed, often this is blocked for reasons such as incomplete descriptions, or for more subjective reasons. Later in this chapter, we look at the value Semantic Web Service descriptions can add in assisting domain experts create service orchestrations in the context of the translational medicine example.

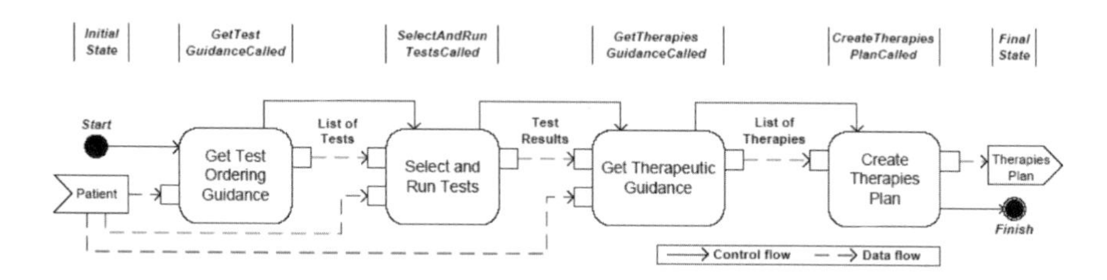

Fig. 13.2. Overview of translational medicine workflow

Figure 13.2 shows this orchestration as an UML 2 activity diagram. The solid lines represent control flow between each of the activities while the dashed lines represent data flow. The type of data being exchanged at each flow is labelled above it. The labels at the top of the figure indicate the state of the process as the goals are executed. This is referred to later in the WSML listing for this orchestration. It may be noted that each of these subgoals may be viewed as clinical/business services and may be further decomposed into technological/infrastructure services such as rules engine services, ontology services and data integration services.

The process starts when patient information is made available. The first step is to get guidance on relevant tests to order for the patient. A list of candidate tests is returned and those that are deemed suitable are selected (this may require expert input or be automated). Based on the selected tests and the patient information, a request is made for candidate therapies for treatment. As with the tests, suitable therapies are identified and a plan established for treatment of the patient. Figure 13.3 shows the relevant WSML definition of the WSMO Goal description containing the orchestration depicted in Figure 13.2. The orchestration is defined in terms of subgoals. As each Goal corresponds to a desired service, rather than a pre-selected one, the concepts defined as necessary inputs and outputs do not need to be grounded in specific messages in WSDL documents. Not binding the Goals to WSDL operations, allows for more flexibility in the run-time matching og Goals to Web Services.

```
wsmlVariant _"http://www.wsmo.org/wsml/wsml-syntax/wsml-rule" ...
interface TreatPatientInterface
orchestration TreatPatientOrchestration
  stateSignature TreatPatientStateSignature

  /* Concepts used as input and output to the orchestration */
  in PastMedicalHistory
  out MedicationOrders
```

```
  /* Concept used to maintain the control flow */
  controlled ControlState

  /* transition rules govern the control flow of the orchestration */
  transitionRules
  /* If initial state and available patient history, request test
guidance */
  if (
     ?patient memberOf Patient and
     ?cs[value hasValue InitialState] memberOf ControlState)
  then
     call getTestOrderingGuidance
            (?patient)
                 : _# memberOf ListTests
     update(?cs[value hasValue GetTestGuidanceCalled])
  endif

  /* Select and run subset of tests recommended by test guidance */
  if
     ?listPossibleTests memberOf ListTests and
     ?patient memberOf Patient and
     ?cs[value hasValue GetTestGuidanceCalled] memberOf ControlState
  then
     call selectAndRunTests
            (?listPossibleTests, ?patient)
                 : _# memberOf ListTestResults
     update(?cs[value hasValue SelectAndRunTestsCalled])
  endif

  /* Based on test results, request guidance on available therapies */
  if (
     ?listTestResults memberOf ListTestResults and
     ?patient memberOf Patient and
     ?cs[value hasValue SelectAndRunTestsCalled] memberOf Control-
State
  then
     call getTherapeuticGuidance
            (?listTestResults, ?patient)
                  : _# memberOf ListTherapies
     update(?cs[value hasValue GetTherapeuticGuidanceCalled])
  endif

  /* Establish a plan for the fulfillment of the therapies */
  if
     ?listTherapies memberOf ListTherapies and
     ?cs[value hasValue GetTherapeuticGuidanceCalled] memberOf con-
trolState
  then
     call TherapiesPlan(?listTherapies) : _# memberOf TherapiesPlan
     update(?cs[value hasValue CreateTherapiesPlanCalled])
  endif

  /* End orchestration once the therapies plan is created */
  if
```

```
      ?therapiesPlan memberOf TherapiesPlan and
      ?cs[value hasValue CreateTherapiesPlanCalled] memberOf control-
State
  then
     update(?cs[value hasValue FinalState])
  endif
```

Fig. 13.3. WSMO goal description for clinical workflow

We explain the syntax and meaning of WSMO orchestrations, including details on grounding, in the context of this example, later in the chapter in Section 13.1.6. Here we provide a brief overview. The main point to understand at this point is that the steps of the orchestration are declared in terms of Goals to be achieved. The orchestration says what the designer of the overall process wants to achieve without specifying with which specific services that process will be implemented. This illustrates the powerful abstraction that using Semantic Web Services allows.

The concepts involved in WSMO orchestrations take different modes. For our example, we are interested in three of these. Concepts of modes *in* and *out* are related to the data flow of the orchestration. Mode *in* means instances of the concept can only be read by the orchestration. Mode *out* means that instances of the concept can be created in the course of the orchestration and be available to the calling environment. Concepts with mode *controlled* are used as part of the control flow definition of the orchestration.

In Figure 13.3, each time a step of the orchestration needs to be executed, the *call* statement is used to indicate that a WSMO goal needs to be achieved so that the functionality of the step can be provided. The structure of the call statement is similar to a function call in software programming. The call keyword is followed by an identifier for the Goal. The variables inside the round brackets are similar to input parameters, indicating the flow of data from the orchestration instance to the execution of the goals. Finally, the expression after the closing bracket and the colon indicates the ontological concept defining the meaning of the data that the goal will return once it has been achieved. For example, the statement

```
call getTestOrderingGuidance
       (?patientHistory)
            : _# memberOf ListTests
```

has an execution semantics equivalent to achieving the Goal with ID `getTestOrderingGuidance`, making the data stored in the variable `?patientHistory`, available to the Goal execution mechanism, and expecting an instance of the concept `ListTests` to be provided to the orchestration on successful execution of the goal.

13.1.2 Data Structures in Data Flow

The data flow in the clinical workflow ensures that results of one activity are passed on to the next activity. This approach ensures that the correct data is available without the activities having to know about each other directly. The data structures used in data flows are defined in simplified form using WSML in the following bullet points.

- **Patient**. Patient is the data structure that initiates the data flow in the clinical workflow. The physician collects a patient history record, including current readings for vital signs and any family medical history information. This data is captured in an instance of the patient concept. The attribute for specifying the blood pressure details is typed using the medical information model known as CEN 13606 openEHR.

```
wsmlVariant _"http://www.wsmo.org/wsml/wsml-syntax/wsml-rule"
nameSpace {_"http://www.example.org/patientHistory#",
        ehr_bp _"http://www.example.org/ehr-bp-archetype#"}

importsOntology {_"http://www.example.org/ehr-bp-archetype}

concept Patient
   personalDetails ofType (1) PatientPersonalDetails
   ...
   familyHistory ofType FamilyHistory
   bloodPressureDetails ofType (1) ehr_bp#EHR_ComBP
   vitalSigns ofType (1) VitalSigns
   ...

concept PatientPersonalDetails
   firstName ofType _string
   lastName ofType string
   SSN ofType SSN
   ...
```

The cardinality of attributes is not constrained unless this is stated explicitly using the (n m) construct where n is the minimum cardinality and m is the maximum. Some examples of this are: (0 *) means zero or many values allowed, (0 1) means either zero or one values allowed, (1) means exactly one value allowed. If multiple values are allowed, they are provided in a space-separated list. This data structure of Patient is simplified; however, in the context of this example it is sufficient to show how the Semantic Web concepts are applied.

- **List of Tests**. The outcome of the first goal (or activity in the workflow) is a list of tests that have to be taken.

```
concept Test
   testId ofType _IRI
   input ofType TestInput
   output ofType TestOutput

concept ListTests
   patient ofType (1) Patient
   individualtest ofType Test
```

```
concept LiverTest subConceptOf Test
   ...

concept MolecularDiagnosticTest subConceptOf Test
   ...
```

- **Test Results**. After the tests are done the test results are provided by the second goal that is achieved. The definition of the test results is

```
concept TestResult
   resultId ofType (1) _IRI
   test ofType (1) Test

concept ListTestResults
   patient ofType (1) Patient
   individualResult ofType TestResult

concept LiverTestResult subConceptOf TestResult
   ...

concept MolecularDiagnosticTestResult subConceptOf TestResult
   ...
```

- **List of Therapies**. Based on the outcome of the tests the therapies that have to be conducted are determined. This is defined as follows

```
concept Therapy
   therapyId ofType (1) _IRI
   ...

concept ListTherapies
   patient ofType (1) Patient
   individualTherapy ofType Therapy

concept FibrateTherapy subConceptOf Therapy
   ...
```

- **Therapies Plan**. Finally, a therapy plan is established and in the use case this is the final outcome:

```
concept TherapiesPlan
   ...
   patient ofType (1) Patient
   therapiesPlanID ofType _IRI
   therapy ofType Therapy
   ...
```

These various data structures are represented as concepts in ontologies that are used in the scope of the clinical workflow. However, each individual activity may use different concepts internally for various reasons. One reason could be that underlying repositories and databases might have existed for a while and their ontologies been defined a long time back. Another reason could be that their definition might possibly have taken place before the clinical workflow came into existence, as they might have been used in different contexts long before it.

In order to ensure that the data structures in the data flow match with the data structures or ontologies in the activities, data mappings may be necessary so that data mediation can take place whenever a data structure from the data flow is given to the activity or whenever an activity's result is put into the data flow. The concrete data mediation that is discussed next is needed for the clinical workflow.

13.1.3 Data Mediation

There are multiple data standard specifications for the healthcare industry that overlap to greater or lesser extents. This results in typical problems of data sharing that are in no way unique to healthcare informatics. Adding to the complexity, there are different levels of interoperability problems that need to be tackled before healthcare systems can really communicate with each other. These include levels for message protocol and message content. In our example, we will assume that message level interoperability is guaranteed by all services adopting the HL7 protocol. While HL7 is by no means universally adopted, it is the strongest candidate at this level. However, there are numerous possible specifications for the description of message content. Two of these include CEN 13606 and HL7-CDA for the definition of the structure of the messages. The content of the messages may be defined by medical vocabularies such as SNOMED, ICD-10 and LOINC amongst others. As our example is based on translational medicine, it is likely that services from the life sciences domain will also be required with corresponding vocabulary specifications including BIOPAX and the Gene Ontology.

In Figure 13.2, each step of the orchestration represents a WSMO Goal that resolves to a WSMO Web Service at run time. Each of these WSMO Web Services, in turn, may represent a single Web Service or a further orchestration of Goals and/or services. To illustrate how data heterogeneity impacts the process in our use case, we focus on the Goal represented by the *GetTestOrderingGuidance* activity. We assume that this Goal uses the CEN 13606 data standard for electronic health records. The Goal's capability matches that of a service providing the required guidance on tests but this service uses the HL7 CDA data model, meaning that data mediation is required to enable the interaction between the requester (represented by the Goal) and the provider of the service.

In particular, we look at the attribute of the Patient concept for recording a blood pressure measurement. The Goal states that this information is available in terms of the CEN 13606 information model. However, let us assume that the service providing the functionality for the *GetTestOrderingGuidance* step requires the blood pressure information to be specified using HL7-CDA. Both models allow the same information to be specified but use different datatype definitions. At the instance level, these overlaps are not immediately obvious. However, looking at a conceptual view on the models and using Semantic Web modeling tools to identify the underlying base concepts allow the overlaps to be highlighted. Once this has been achieved, mappings at the ontological level between the models can be created. These mappings can be applied to instance data at run time to achieve the desired data mediation. The example we use here is fully based on the example described in [366] with the difference that WSMO rather than OWL is used as the model for specifying the domain ontologies. The language for WSMO is less verbose than OWL and has greater tool support for data mediation. The example is intended to illustrate how semantic annotation facilitates service discovery and data mediation to resolve interoperability issues. We do not imply that any additional semantics are added to the information models or that we extend the work described in the paper at [366]. A full discussion on the three underlying information models and the algorithm for determining mappings between them is presented and analyzed in the context of health informatics in that paper.

Snapshot of CEN 13606 and HL7-CDA

HL7 is a non-profit standards organization whose primary goal is to provide standards for the exchange of information between healthcare systems. HL Version 3 is based on a data model called the Reference Information Model (RIM), consisting of six backbone classes shown in Figure 13.4.

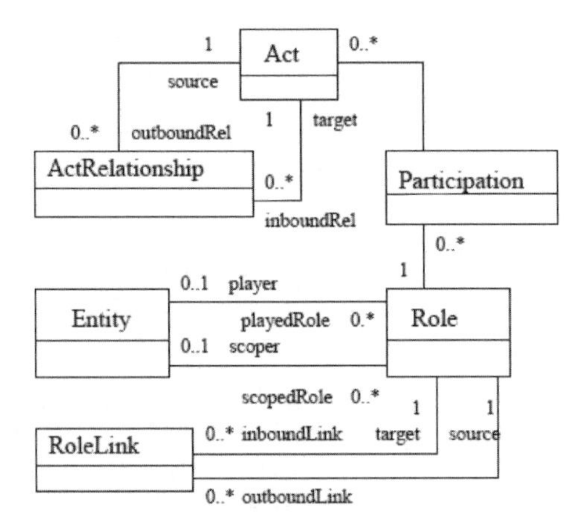

Fig. 13.4. HL7 RIM backbone classes

A fragment of this model is represented as the WSMO ontology in Figure 13.5. Both the CEN 13606 and the HL7-CDA models derive their machine processable meaning or semantics from the HL7 RIM. In other words both of these specifications refine the RIM to more specialized Refine Message Information Models (R-MIMs). WSMO ontology fragments for these R-MIMs are shown in Figure 13.6 and Figure 13.7. In each case the ontologies import the WSMO ontology for the RIM using its concepts to create more refined models.

```
wsmlVariant _"http://www.wsmo.org/wsml/wsml-syntax/wsml-rule"
namespace { _"http://www.example.org/hl7-rim#" }

ontology hl7rim

concept Act
    code ofType CD
    outBoundRelationship ofType ActRelationship

concept Observation subConceptOf Act
    value ofType ANY

concept ActRelationship
    _target ofType Act

concept ANY

concept CD subConceptOf ANY
    CD_code ofType _string
    CD_codeSystem ofType _string
    CD_codeSystemName ofType _string
    CD_displayName ofType _string
```

Fig. 13.5. WSMO ontology fragment for HL7 RIM

```
wsmlVariant _"http://www.wsmo.org/wsml/wsml-syntax/wsml-rule"

namespace { _"http://www.example.org/hl7-cda-rmim#",
     rim _"http://www.wsmo.org/ontologies/hl7-rim#"
}

importsOntology {_"http://www.wsmo.org/ontologies/hl7-rim"}

ontology hl7rmim

concept EntryRelationship subConceptOf rim#ActRelationship
    clinicalStatement ofType Observation

concept Observation subConceptOf rim#Observation
      entryRelationship ofType EntryRelationship
```

Fig. 13.6. WSMO ontology fragment for HL7 CDA R-MIM

```
wsmlVariant _"http://www.wsmo.org/wsml/wsml-syntax/wsml-rule"
namespace { _"http://www.example.org/ehr-rmim#",
     rim _"http://www.wsmo.org/ontologies/hl7-rim#"
}

importsOntology {_"http://www.wsmo.org/ontologies/hl7rim"}

ontology ehr_rmim
```

```
concept Element subConceptOf rim#Observation
    value ofType _integer

concept Entry subConceptOf rim#Act
    component1 ofType Component1

concept Component1 subConceptOf rim#ActRelationship
    element ofType Element
```

Fig. 13.7. WSMO ontology fragment for CEN 13606 R-MIM

Both HL7 CDA and CEN 13606 use archetypes as a means for modeling entities within specific subdomains. In our example, blood pressure information is a typical example of an archetype that is useful. It can be thought of as a template specifying a number of concepts, relationships between them and the constraints that are relevant for the specific entity being described.

The blood pressure archetypes are modeled as an additional layer building on the concepts defined in the R-MIMs of the respective models. Figure 13.8 and Figure 13.9 show the WSMO ontology fragments for the blood pressure archetypes in each of the two models. We are interested in how the concepts defined for the archetype in the ontology for one R-MIM map to the corresponding R-MIM in the second ontology. Creating such a mapping remains a job for domain experts but the semantic description of the information model greatly helps this task. We use the mediation tool available with the Web Services Modeling Toolkit (WSMT).

```
wsmlVariant _"http://www.wsmo.org/wsml/wsml-syntax/wsml-rule"
namespace { _"http://www.example.org/hl7-cda#",
     cda_rmim _"http://www.example.org/hl7-cda-rmim#",
     rim _"http://www.wsmo.org/ontologies/hl7-rim#"
}

importsOntology {_"http://www.example.org/hl7-cda-rmim",
     _"http://www.wsmo.org/ontologies/hl7-rim"}

ontology hl7_cda_bp_archetype

concept CDADP subConceptOf cda_rmim#Observation
    cda_rmim#entryRelationship ofType
    CDASystolicBPER_or_CDADiastolicBPER

concept CDASystolicBPER subConceptOf
     cda_rmim#EntryRelationship
    _target ofType CDASystolicBP

concept CDADiastolicBPER subConceptOf
     cda_rmim#EntryRelationship
    _target ofType CDADiastolicBP

concept CDASystolicBP subConceptOf cda_rmim#Observation

concept CDADiastolicBP subConceptOf cda_rmim#Observation
```

```
  moodCode hasValue "EVN"
  codeBlock hasValue codeBlock1
  entryRelationship hasValue systolicBPER diastolicBPER

instance codeBlock1 memberOf CodeBlock
  code hasValue 251076008
  codeSystem hasValue "2.16.840.1.113883.6.96"
  codeSystemValue hasValue "SNOMED CT"
  displayName hasValue "Cuff Blood Pressure"

instance systolicBPER memberOf CDASystolicBPER
  typeCode hasValue "COMP"
  observation hasValue CDASystolicBP_1_2234

instance CDASystolicBP_1_2234 memberOf CDASystolicBP
  codeBlock hasValue codeBlock2
  classCode hasValue "OBS"
  moodCode hasValue "EVN"
  valueType hasValue "PQ"
  value hasValue 132
  valueUnit hasValue "mm[Hg]"

instance codeBlock2 memberOf codeBlock
  ...
  displayName hasValue "Systolic Blood Pressure"

instance DiastolicBPComponent memberOf EHRComDiastolicBPComp
  typeCode hasValue "COMP"
  observation hasValue CDADiastolicBP_1_4524

instance CDADiastolicBP_1_4524 memberOf CDADiastolicBP
  codeValues hasValue codeBlock3
  classCode hasValue "OBS"
  moodCode hasValue "EVN"
  valueType hasValue "PQ"
  value hasValue 86
  valueUnit hasValue "mm[Hg]"

instance codeBlock3 memberOf codeBlock
  ...
    displayName hasValue "Diastolic Blood Pressure"
```

Fig. 13.13. WSMO instances for a HL7-CDA 13606 blood pressure artefact

As the definition shows, not all of the activities needed data mediation in this case.

13.1.4 Goal Definition

In the clinical workflow described in Section 13.1.1, each activity is represented as a Goal that can be resolved using Semantic Web Service discovery and mediation techniques at run time or by using Semantic Web Service composition and orchestration techniques at design time. The designers of such a system could restrict the pool of candidate services in such a way that only known and trusted services are available for discovery, or they could leave this more open and allow possibly new services to be offered as candidates, leaving a decision on suitability to an expert user. The richer a Goal description can be in terms of specifying what a service requester wants and the constraints that must be in place, the greater the possibility for accuracy in locating suitable services. Nevertheless, if a Goal is created to heal a patient showing the symptoms of a rash across the back and a headache, it is likely that a physician will use an SWS-based discovery mechanism to identify probable tests and therapies as guidance rather than allow these to be decided automatically.

In this section, we will show some examples of how the Goals for our workflow can be defined using WSMO as the conceptual model and WSML as the corresponding formal language. We will also identify the potential of using Semantic Web Services techniques to enable service discovery, composition and orchestration.

- **TreatPatient.** The topmost Goal is to treat a patient. This is the goal the physician has when a patient comes into his office. The goal might resolve into many different workflows once the physician enters it into the clinical information system. The clinical workflow shown in Figure 13.2 is one orchestration that can fulfill the goal and we assume in this example that the physician selects it. We show a WSML fragment of the capability for this goal below. The capability states that the owner of this goal wishes to get an instance of the TherapiesPlan concept (postcondition) for a specific patient based on the provision of an instance of the Patient concept (precondition).

```
goal _"http://www.example.org/TreatPatient"
...
capability _"http://www.example.org/TreatPatient#capability"

   precondition
       definedBy
           ?patient memberOf Patient.

   postcondition
       definedBy
           ?therapiesPlan memberOf TherapiesPlan and
           ?therapiesPlan[patient hasValue ?patient].
...
```

The TreatPatient Goal can be decomposed into a number of subgoals that form a composition which can satisfy the overall request. We look at each subgoal corresponding to the orchestration in Figure 13.2 in the next bullet points. Functional decomposition is a mature research area in the Artificial Intelligence (AI) and Multi-Agent System (MAS) communities. Some examples are the use of Hierarchical Task Network (HTN) planning using OWL-S service descriptions from Sirin et al. [343] and McIlraith and Son [342]. In the latter, the authors acknowledge that automatic decomposition is difficult precisely because it is difficult to predict what knowledge necessary for the decomposed process will be available at run time. The approach allows plans to be constructed that can be customized and executed by agents operating on behalf of individuals. For example, the orchestration of Figure 13.2 could act as a template for matching the TreatPatient Goal but the services identified to match the GetTestOrderingGuidance and GetTherapiesGuidance subgoals may require additional information about allergies not present in the initial Patient data. This knowledge could be retrieved either from preferences known to the agent or by including an additional knowledge retrieval service in the orchestration.

- **GetTestOrderingGuidance**. The first step in satisfying the overall TreatPatient goal is to determine what tests should be ordered based on the symptoms and medical history presented in the instance of the patient concept. This is represented by the GetTestOrderingGuidance Goal whose WSML fragment is given here:

```
goal _"http://www.example.org/GetTestOrderingGuidance"
...
capability  _"http://www.example.org/GetTestOrderingGuidance#capa-
bility"

   precondition
       definedBy
           ?patient memberOf Patient.

   postcondition
       definedBy
           ?listTests memberOf ListTests and
          ?listTests[patient hasValue ?patient].
...
```

- **SelectAndRunTests**. Once a list of suitable tests has been obtained from a test ordering guidance service, it is necessary for the most suitable tests to be identified, selected and run. This is the aim of the SelectAndRunTests Goal. Services that match this Goal do not have to be restricted to only traditional Web Services. The service may be implemented by posting a list of candidate tests to a human expert's monitoring system. The expert then selects tests and is responsible for their execution. The results of these tests are then returned to the overall composition. The point here is that Semantic Web Service descriptions may ground to human interaction systems as well as to completely machine-based service implementations. An example of how the capability of this goal would be expressed in WSML is:

```
goal _"http://www.example.org/SelectAndRunTests"
...
capability _"http://www.example.org/SelectAndRunTests#capability"
   precondition
      definedBy
         ?patient memberOf Patient and
         ?listTests memberOf ListTests and
         ?listTests[patient hasValue ?patient].

   postcondition
      definedBy
         ?listTestResults memberOf ListTestResults and
         ?listTestResults[patient hasValue ?patient].
...
```

- **GetTherapeuticGuidance.** Once the tests have been run and the results made available, the next step is to use those results to get guidance on what therapies are available and suitable for the given patient. As before the specification of the Goal makes no assumption as to how any of the services matching it are implemented. The WSML fragment for the capability is:

```
goal _"http://www.example.org/GetTherapeuticGuidance"
...
capability _"http://.../GetTherapeuticGuidance#capability"

   precondition
      definedBy
         ?patient memberOf Patient and
         ?listTestResults memberOf ListResultsTests and
         ?listTestResults[patient hasValue ?patient].

   postcondition
      definedBy
         ?listTherapies memberOf ListTherapies and
         ?listTherapies[patient hasValue ?patient].
...
```

- **CreateTherapiesPlan.**

```
goal _"http://www.example.org/CreateTherapiesPlan"
...
capability    _"http://www.example.org/CreateTherapiesPlan#capabil-
ity"

   precondition
      definedBy
         ?listTherapies memberOf ListTherapies and
         ?listTherapies[?patient ofType Patient].
```

```
   ...

webService _"http://www.example.org/TherapyUnlimited"
   ...
   nonFunctionalProperties
    description hasValue "Therapy Guidance Service"
    provider hasValue "Therapy Unlimited, US."
    implementationMechanism hasValue "Oracle 10 RDBMS"
    USStateAppr hasValue hasValue "ALL"
    SecurityProtocol hasValue {"SSL", "PKI"}
   endNonFunctionalProperties
   ...
webService _"http://www.example.org/Therapy4Us"
   ...
   nonFunctionalProperties
    description hasValue "Therapy Guidance Service"
    provider hasValue "Therapy4Us Inc, US."
    implementationMechanism hasValue "Rules + DL Reasoner"
    USStateAppr hasValue hasValue {"TX", "CA", "MI", "GA"}
    SecurityProtocol hasValue {"PKI"}
   endNonFunctionalProperties
   ...
```

13.1.6 Orchestration/Service Composition

Each of the top-level Goals (GetTestOrderingGuidance, SelectAndRunTests, GetTherapeuticGuidance) in the orchestration of Goals in Figure 13.2 may be achieved by a single service or by a further orchestration of services. In WSMO, the description of how a service uses other services (or Goals) to achieve its capability is defined by the service orchestration. If Goals are used, then an execution environment, such as WSMX, enables the Goals to be resolved to service invocations at run time using discovery and mediation mechanisms. This gives an additional layer of flexibility when service orchestrations are being created, allowing the orchestration designers to focus on what they want to achieve rather than having them be aware of all potential services that may be available.

Orchestrations are typically kept private by the service provider if they reflect private business logic. From an external perspective, whether a service has an orchestration or not is usually not visible to clients. They see only the publicly described capability and interface. For the client, this means that discovering and interacting with a service is the same regardless of its internal design. In other words, if the orchestration of a service is predefined, then identifying an orchestration for use in a system becomes a service discovery problem. A WSMO Web Service description, for example, may represent a composition of services that is described in the orchestration part of the service description. It is up to the service provider whether this description is made public or not.

WSMO orchestrations allow the process implemented by services to be declaratively described using ontologized Abstract State Machines (ASMs). ASMs are defined in terms of states and transition rules that model how the system, represented by the ASM, transitions from state to state. ASM was chosen as a general model as it provides a minimal set of modeling primitives that is sufficiently expressive to rigorously model dynamic semantics. If the provider wishes to change this process, it can be done declaratively with minimal or no recoding. There is an analogy with a business process described with a language such as WSBPEL that is abstracted behind a Web Service interface. One question that arises is how such orchestrations are created and whether the creation can be automated.

Fensel and Bussler point out three possibilities for the definition of orchestrations of Web Services in WSMF [264]. The first is provider-based orchestration where the result, from a requester's perspective, is itself a Web Service; the requester does not see details of the orchestration. The second is client-based composition where the requester combines services as necessary and the provider has no knowledge on how the services are being composed. The third is where the provider declares additional information about how the service should be used and the requester is free to combine services as long as any provider constraints are not violated. It is this third way that we propose in this section.

Semantic Web Service descriptions allow providers to formally state constraints on the usage of their services. Automated orchestration of services, particularly in a specialist domain such as medicine, faces practical difficulties as knowledge beyond technical suitability is often taken into account when combining services. Services may be available that are incompletely described, but a physician, based on his own information, must be able to use such services if he wishes.

The added value of Semantic Web Service descriptions over the use of WSDL in WSBPEL orchestrations is that constraints on the usage of each test can be specified by that test's providers using rules made up of ontologized logical expressions. The tests (represented by services) can possibly be composed in multiple different sequences, using a graphical orchestration tool, but none of these compositions may violate the constraints specified by other services. A reasoning engine built into the tool ensures this.

We take the SelectAndRunTests Goal as an initial simple example. Assume the GetTestGuidance goal has been matched to a service and has returned a set of candidate tests. The next step is to select the most suitable tests and request that they be performed. Each service has been semantically described using WSMO. Based on this set of descriptions, a physician (through tool support) may identify three genomic tests (G1, G2, G3) looking for genetic markers for specific diseases, and three clinical tests (C1, C2, C3) looking for evidence of disease in blood and tissue samples. An orchestration may be constructed specifying that these tests should be carried out in the sequence (G1, G2, C1, G3, C2, C3). However local hospital policy dictates that test G2 only be carried out after the other five as it involves a high cost and may be unnecessary depending on results of other tests. This constraint is encoded in the service description. The SWS-assisted orchestration engine guides the creation of the orchestration so that this constraint is met. Although the orchestration is not automated, the expert is presented with rich semantic information to assist with the task and any violations of the constraints on the services will be picked up by the semantically enabled tool.

WSML Example: GetTestOrderingGuidance Orchestration

We now take a look at the WSML representation of the orchestration defined for the GetTestOrderingGuidance subgoal of the TreatPatient Goal shown in Figure 13.2. The subgoals correspond to technological/infrastructural services such as a data integration service and a rules engine service. As mentioned, WSMO uses a rule-based approach to declare orchestrations using ontologized Abstract State Machines (ASMs). Each orchestration has a state signature which is a state ontology used by the service together with the definition of the modes the concepts and relations within that ontology may have. The input and output messages of the orchestration are associated with concepts having modes in and out respectively. An additional mode controlled is used for concepts that are internal to the orchestration and used to define sequential control flow. This is necessary as when the condition of a transition rule is met, any update rules in the body of that rule will fire in parallel by default.

Grounding provides the link between the semantic and syntactic descriptions of a Web Service provided by WSMO and WSDL respectively. WSDL provides a mechanism by which each element of a WSDL document can be uniquely identified using a URI. How this mechanism is used in WSMO grounding is explained by Kopecký et al. in [376]. The WSML *withGrounding* keyword is used to declare the groundings. Concepts with mode in or shared may be grounded in input messages of WSDL operations and concepts with mode out or shared may be grounded in output messages of WSDL operations. It is possible to have multiple groundings declared for each concept (also explained in [376]). Transformations are necessary between WSML and the instances of XML types defined for messages in WSDL. We assume that this transformation, called lowering (WSML to XML) or, correspondingly, lifting (XML to WSML), is available but details of how this is implemented is out of scope of this section. The following WSML listing shows the orchestration of a service to match the GetTestOrderingGuidance Goal.

```
wsmlVariant _"http://www.wsmo.org/wsml/wsml-syntax/wsml-rule"
...
interface GetTestOrderingGuidanceInterface
orchestration TestOrderingGuidanceOrchestration
  stateSignature GetTestOrderingGuidanceSignature

  /* Concepts used as input and output to the orchestration */
  in Patient withGrounding {
       _"http://.../RulesEngineService.wsdl#wsdl.interfaceMessageRef-
erence
            (GetTestOrderingGuidance/GetTests/In)"}
  in Patient withGrounding {
```

```
      _"http://.../DataIntegration.wsdl#wsdl.interfaceMessageRefer-
ence
            (PatientHistory/GetCardiacHistory/In)"}
  shared PatientCardiacHistory withGrounding {
      _"http://.../RulesEngineService.wsdl#wsdl.interfaceMessageRef-
erence
            (GetTestOrderingGuidance/GetTests/In)"}
  shared PatientCardiacHistory withGrounding {
      _"http://.../DataIntegration.wsdl#wsdl.interfaceMessageRefer-
ence
            (PatientHistory/GetPatientCardiacHistory/Out)"}
  out ListTests withGrounding {
      _"http://.../RulesEngineService.wsdl#wsdl.interfaceMessageRef-
erence
            (GetTestOrderingGuidance/GetTests/Out)"}
  /* Concept used to define sequential control */
  controlled ControlState

  /* transition rules define the state changes of the orchestration */
  transitionRules

  if (
     ?patient memberOf Patient and
     ?patientCardiacHistory memberOf PatientCardiacHistory and
  then
     add(_# memberOf ListTests)
     update(?cs[value hasValue RulesEngineServiceCalled])
  endif

  if (
     ?patient memberOf Patient and
     ?cs[value hasValue initialState] memberOf ControlState)
  then
     add(_# memberOf PatientCardiacHistory)
     update(?cs[value hasValue PatientCardiacHistoryAvailable])
  endif
```

The first part of the orchestration declares the state signature. This consists of the concepts that are used by the transition rules within the orchestration. We assume that all the concepts are from the same namespace. The listing includes concepts marked with modes in, out, shared and controlled as explained earlier. Each concept unless marked as controlled has a grounding which links the concept to the in or out message of an operation in a WSDL document. For example:

```
in Patient withGrounding {
_"http://.../RulesEngineService.wsdl#wsdl.interfaceMessageReference
          (GetTestOrderingGuidance/GetTests/In)"}
```

The fragment identifier for Interface Message References used with the withGrounding keyword is defined as

```
wsdl.interfaceMessageReference(interface/operation/message)
```

with the three parts in parentheses replaced with the following:

- **message** is the message label property of the Interface Message Reference component. It can be either In or Out.
- **operation** is the local name of the operation that contains the message.
- **interface** is the local name of the interface owning the operation.

The *Patient* and *PatientCardiacHistory* concepts in this example have two grounding declarations each. In both cases the correct grounding to use depends on the transition rule being executed and this can be determined by the orchestration engine. Two rules are declared.

The first states that if instances of *Patient* and *PatientCardiacHistory* and are available in the ontology, defined by the state signature of the orchestration, an instance of the *ListTests* concept will be added to that ontology. In terms of execution, this means that if the header part of the rule is true, the engine will look for a concept of type *ListTests* with mode *out* in the state signature and will use its grounding information to determine which service needs to be invoked to provide the data. In this case it is the service defined by *RulesEngineService.wsdl*. The grounding information in the state signature is sufficient to determine the operation to invoke and the data required for the input is identified by the concepts in the header of the rule.

The second rule is used if an instance of *PatientCardiacHistory* is missing, meaning that the first rule cannot fire. An instance of the *PatientCardiacHistory* for the *Patient* can be retrieved using the *DataIntegration* service. The header of the second rule is checked in the same way (and at the same time) as the first. An instance of the concept *ControlState* of mode *controlled* is used to ensure that the *DataIntegration* service is only invoked (if at all) when the orchestration is in its initial state. The grounding information in the state signature is used, as before, to identify the correct WSDL service endpoint and operation to invoke. Once the *PatientCardiacHistory* instance has been retrieved using the service, the ontology of the orchestration is updated and the conditions of both transition rules are checked again. Now all the information required for the first rule is available; it fires, causing the *RulesEngineService* to be invoked and an instance of *ListTests* to be returned. As no more rules can fire, the orchestration execution ends.

As with the GetTestOrderingGuidance Goal of the TreatPatient orchestration, the clinical Goal, GetTherapeuticGuidance, may also be further decomposed into subgoals, which correspond to technological/infrastructural services such as a data integration service and a rules engine service. This is similar to the GetTestOrderingGuidance service; only the input and out parameters are different. We will discuss further technological implementations of these services in later parts of the chapter.

Steps Required To Create an Orchestration

Encapsulating a process definition as an orchestration of services with its own service interface and description enables processes to be discovered and used as building blocks for other orchestrations in the same way as any other service. Usually, the description of a service's orchestration is not visible to the requesting client. The following steps provide an example of what is required to create an orchestration of Web Services in the context of WSMO.

Identify the tasks that go into making up the orchestration. The overall goal to be achieved needs to be decomposed into individual tasks. Each task is then matched to an existing goal description or a new goal description is created.

Discover services to realize the goals. Semantic Web Service discovery is used to find candidate Web Services. Constraints may be specified in the descriptions of the goals for each task and only Web Services that do not violate these constraints will be included as candidates for the orchestration.

Select a service for each task. Using the information provided in the semantic descriptions of the services, the tool assists the expert in selecting the services whose non-functional characteristics best match those required for the task.

Define the concepts used for input and output data. The selected Web Services will have choreographies that describe their public interfaces. With tool assistance this information can be used to declare the concepts representing messages for the orchestration with modes *in*, *out* or *shared*. For example, if a concept is declared with mode *out* in the choreography of a selected service then this means that it will be made available to the orchestration from the Web Service and, therefore, should be marked with mode *in* for the orchestration.

Identify and include any mediation that may be necessary. Mediation is necessary where the data or process descriptions of service choreographies, required to cooperate, do not match. Services that provide this capability need to be discovered. As all information for this discovery is available, it should be possible for the tool to automatically carry out this step.

Define the control flow between the selected services. The first step is to identify the sequence in which services should be invoked. Preconditions and postconditions in the service capability define the data that the service will consume and provide. It needs to be verified that preconditions of all services in the sequence can be met. For example, in the GetTestOrderingGuidance example earlier, the RulesEngineService requires an instance of PatientCardiacHistory for one of its inputs. This precondition cannot be satisfied by the information available in the instance of the Patient concept. This is a case of missing data which can be obtained by using an additional service to get this information. The DataIntegration service is identified for this and must precede the RulesEngineService in the invocation sequence so that the pre- and postconditions are not violated. This verification can be carried out by an orchestration tool. As the orchestration is being built, the designer can be warned where sequence violations occur and guided in resolving them.

Another type of violation is one where constraints on non-functional aspects of services are included in the service descriptions. In Section 13.1.5, we gave some examples of these, including USStateAppr (the US states in which the service was approved) and SecurityProtocol (the security protocol required by the Web Service implementation). As services are composed into an orchestration, these constraints may be checked by the tool to ensure that they are not being violated. For example, the expert using the tool may add configuration information about issues such as security and which geographic region the orchestration will have to be deployed to. As with functional conditions, the tool using logical reasoning support can monitor the orchestration as it is being constructed and notify the user if a constraint is being violated and possibly suggest alternative action.

As services are organized into the orchestration in the desired sequence, the tool assists the individual operations required on the services to be selected. The final result is the generation of the abstract state machine representation of the orchestration which consists of the state signature (the ontology defining all the data that the orchestration uses) and the set of transition rules that define the sequence in which the services will be invoked.

Change is inevitable and after an orchestration has been designed, there is no guarantee that the behavior and data defined at the interfaces of services will remain the same. There is a requirement that executions of orchestrations be monitored so that changes that may occur can be handled. For example, the RulesEngineService may add an additional input to allow selection of rules engine implementations with differing characteristics. A new precondition in the service description reflects this change. A simple approach would be for the orchestration engine to notify that this new precondition was not being satisfied the next time an instance of the orchestration is requested. This could prompt the expert to modify the orchestration to take the new condition into account. We assume that the execution engine takes care of any consistency issues in migrating from the old orchestration instance to the new one. A more intelligent approach would be for the engine to use the knowledge available to it to reason over what rules engine would be most suitable and to carry out the modification and migration automatically.

An alternative to the last two steps above is to create the orchestration using only goal descriptions. Data and control flow would still need to be specified but the resolution of the goals to individual services and the handling of any mediation could be delegated to an SWS execution environment at run time.

13.1.7 Process and Protocol Mediation

In the last Section, we looked at how the GetTestGuidance sub-goal could be resolved using the RulesEngineService (with some assistance from the DataIntegration service). We described in Section 11.3.5 how WSMO conceptually separates the concerns of service requesters and providers into goals and Web Service descriptions respectively. The choreography of a goal describes the information and behavior that a service requester would like a matching service to have at its interface. The Web Service choreography description describes the interface that it actually offers. Ideally, where a service capability matches that of a goal, the choreographies match each other exactly as well. In reality this is less likely to be the case. Three categories of mismatches listed in Section 11.3.5 are:

- **Data mediation.** Handle mismatches between datatypes used in the choreogarphies.
- **Protocol mediation.** Handle mismatches between message exchange protocols. This relates to the messaging and communication protocols used, e.g., SOAP, REST, HTTP, and FTP.
- **Process mediation.** Handle mismatches between interaction patterns defined for interface behavior. These can be considered as public processes supported by services or goals and may correspond to international standards such as those defined by the RosettaNet[26] or ebXML[27] for eCommerce.

A fourth type of mismatch is identified by Fensel and Bussler in WSMF [264]. This is the *mediation of dynamic service invocation*. It relates to how orchestrations may be flexibly designed as compositions of goals which need to be resolved to a concrete orchestration instance at run time. In this section, we focus on process and protocol mediation.

Where Web Services are used to represent trading partners participating in an interaction, the invocation of operations on each Web Service interface will have to follow a defined sequence. It must be possible for this sequence to be defined in such a way that other partners can determine the sequence itself and, if applicable, the standard being used. If two services using different invocation sequences wish to interact in a process, then process mediation is required to align the invocation sequences. It is possible for this type of mismatch to be resolved at run time if sufficient machine-understandable semantics of the public processes supported by the Web Services is made available. Bussler points out in [372] that semantic definitions, in addition to data and message sequencing, are required to support B2B interactions between Web Services. These include: the intent of the messages, whether messages are asynchronous or synchronous, trading-partner-specific configurations, message definitions, syntax of the messages, security, reliability and communication protocols.

The terms protocol mediation and process mediation are often used interchangeably, e.g., in Williams et al. [373]. Williams points out that this probably reflects research that dates back to the 1980s such as [374] [375] on how systems supporting differing communication protocols could be enabled to communicate with each other. In the context of this section, protocol mediation deals with handling interoperability problems that the communication of messages between two parties. These include the communication protocol used, the security and reliability

26. http://www.rosettanet.org/ (Accessed September 10, 2007).

27. http://www.ebxml.org/ (Accessed September 10, 2007).

constraints at the communication level, the binding of the messages to transportation protocols such as HTTP or FTP, the structure of messages, the syntax of the language used in messages. Process mediation is concerned with the mismatches at the process level and is a superset of protocol mediation.

In WSMF, the authors point out that sequence mismatches can be separated into those that can and cannot be resolved. An unresolvable sequence mismatch, for example, is one where a requester expects that a blood test be conducted, followed by a liver test and a genomic test, in that order, while a provider expects the genomic test first followed by the blood test and then the liver test. These sequences cannot be matched. A resolvable mismatch is one in which the requester does specify in which sequence the tests should be carried out and the sequence can be matched by a mediator to that expected by the provider. Cimpian and Mocan in [350] identify types of process sequence mismatches that may be resolved by the process mediator component of the WSMX execution environment [330] [332].

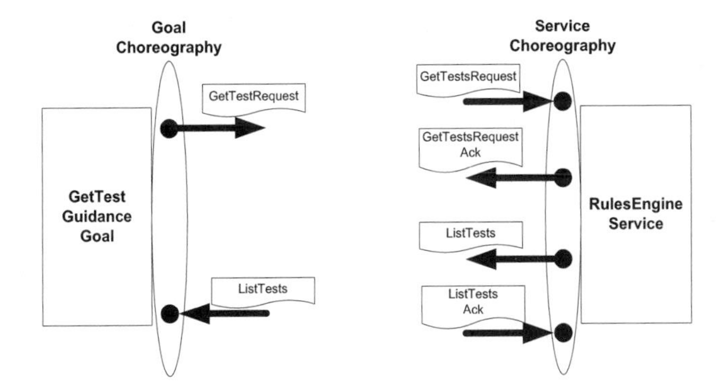

Fig. 13.14. Choreography mismatch between goal and Web Service

We return to the example of the GetTestGuidance goal and the RulesEngineService. The RulesEngineService is used heavily by a large number of medical diagnostic facilities. There is no guarantee on the speed of the response to queries, as some may require considerable processing, and resources are limited.

Consequently the service's choreography specifies that for each request it will synchronously provide an acknowledgement message and the response message containing the list of recommended tests will be sent in a later message for which the RulesEngineService expects a return acknowledgement message. Additionally as the messages contain potentially confidential patient data they must be encoded using the SSL security mechanism. The choreography for the GetTestGuidance goal expects only one message to be sent out with the request and one to be received with the list of recommended tests. It does not specify anything about acknowledgement messages. Additionally it does not specify any requirement with regard to the security mechanism for the service request. Both choreographies are illustrated in Figure 13.14.

Process mediation is required to handle the acknowledgement messages and protocol mediation is required to handle the SSL. Both are provided through a ProcessMediator which has access to the choreographies of the goal and the service and which acts as an intermediary. In Figure 13.15, the request message is sent from the requester. The ProcessMediator adds SSL security and passes the message to the RulesEngineService. The RulesEngineService replies synchronously with an acknowledgement message which is consumed by the process mediator. As the goal choreography does not expect this message, it is discarded. Later the RulesEngineService sends a message with the list of recommended tests. This is expected by the goal choreography and passes through the process mediator unchanged with only the SSL security removed. The RulesEngineService now expects an acknowledgement message in return. The information required for this message is available from earlier information and is generated by the ProcessMediator which sends it back to the RulesEngineServices with the appropriate SSL security.

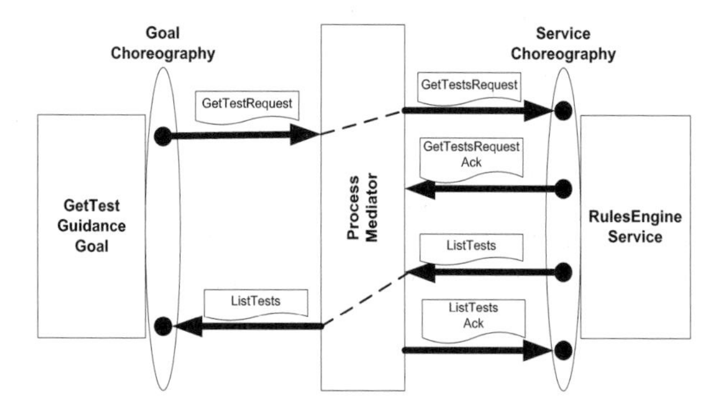

Fig. 13.15. Choreography mismatch between goal and Web Service

13.2 Data and Knowledge Integration

Up to now, in this chapter, we have examined how an overall process in translational medicine can be modeled using Semantic Web Service technology. In particular, we have used the WSMO model and corresponding WSML language to focus on the description of a composition and the goals and services that could be used to realize it. We highlighted how heterogeneity between the interfaces of a goal (what a service requester wants) and a service (what is being provided) could be resolved using ontologized mappings.

This section takes a data-centric view on integration issues in translational medicine, focussing on the creation and mapping or merging of data models that span the clinical and genomic domains. RDF and OWL are used to specify the ontologies in the following examples as they are the most common Semantic Web languages for knowledge representation. The Semantic Web Service models can be linked to these examples through the RDF syntax for WSML which allows for bi-driectional translation. In particular the RDF knowledge examples in the followiug sections can be translated and applied to the WSML Goal and Web Service listings used so far and in the remaining sections of this Chapter.

This component of the data integration architecture enables integration of genotypic and phenotypic patient data, and reference information data. This integrated data could be used for enabling clinical care transactions, knowledge acquisition of clinical guidelines and decision support rules, and for hypothesis discovery for identifying promising drug targets. We now describe with the help of an example our implementation approach for data integration based on Semantic Web specifications such as RDF and OWL to bridge clinical data obtained from an EMR and genomic data obtained from a LIMS. The data integration approach consists of the following steps:

1. Creation of a domain ontology identifying key concepts across the clinical and genomic domains.
2. Design and creation of wrappers that expose the data in a given data repository in an RDF view.
3. Specification of mapping rules that provide linkages across data retrieved from different data repositories.
4. A user interface for: (a) specification of data linkage mappings; and (b) visualization of the integrated information.

We first present a WSMO/WSML specification of Data Integration Services, followed by a discussion of the semantic data integration architecture underlying the implementation of a Data Integration Service. This is followed by a discussion of the various steps enumerated above.

13.2.1 Data Integration Services: WSMO/WSML Specification

We now present a WSML specification of Data Integration Services in the context of the GetTherapeuticGuidance clinical service. The orchestration of the data integration service is as follows:

- GetRelevantDataSources takes as input pointers to clinical data such as family history, test results and contraindications and identifies the relevant data sources
- DistributedQueryProcessing takes as input the list of data sources and a query, typically specified as a join of the various types of clinical data for a given patient, and returns a list of RDF models/graphs from various data sources
- GraphMerge takes as input the set of RDF graphs and merges them based on mapping rules which may also be input. In case the mapping rules are not input, the default mapping rule for mapping URIs is used for the graph merge operation.

```
wsmlVariant _"http://www.wsmo.org/wsml/wsml-syntax/wsml-rule" ...
```

```
interface DataIntegrationInterface
orchestration DataIntegrationOrchestration
  stateSignature DataIntegrationSignature

  /* Concepts used as input and output to the orchestration */
  in FamilyHistory, ListTestResults, ListContraindiations
  out ListTherapies
  /* Concept used to maintain the control flow */
  controlled ControlState

  /* Transition rules govern the control flow of the orchestration */
  transitionRules

  if (?familyHistory memberOf FamilyHistory and
     ?listTestResults memberOf ListTestResult and
     ?contraindication memberOf Contraindication
    ?cs[value hasValue InitialState] memberOf ControlState)
  then
     call GetRelevantDataSources
        (?familyHistory ?listTestResults ?contraindication)
             : _# memberOf DataSourceList
     update(?cs[value hasValue GetRelevantDataSourcesCalled])
  endif

  if (?dataSourceList memberOf DataSourceList and
     ?query = "join(FamilyHistory, ListTestResults, Contraindica-
tion)") and
     ?cs[value hasValue GetRelevantDataSourcesCalled] memberOf Con-
trolState)
  then
    call DistributedQueryProcessing
        (?dataSourceList ?query)
             : _# memberOF RDFGraphList
    update(?cs[value hasValue DistributedQueryProcessingCalled])
  endif

  if (?rdfGraphList memberOf RDFGraphList and
     ?mappingRules memberOf MappingList)
     ?cs[value hasValue DistributedQueryProcessingCalled] memberOf
ControlState)
  then
      call GraphMerge
           (?rdfGraphList ?mappingRules)
                : _# member Of IntegratedPatientRDFGraph
      update(?cs[value hasValue GraphMergeCalled])
  endif
```

13.2.2 Semantic Data Integration Architecture

The semantic data integration architecture is a federation of data repositories as illustrated in Figure 13.5 below and has the following components:

Data Repositories: Data repositories that participate in the federation offer access to all or some portion of the data. In the translational medicine context, these repositories could contain clinical data stored in the EMR system or genomic data stored in the LIMS system. Data remain in their native repositories in a native format and are not moved to a centralized location, as would be the case in data-warehouse-based approach.

Domain Ontologies: Ontologies contain a collection of concepts and relationships that characterize the knowledge in the clinical and genomic domains. They provide a common reference point that supports the semantic integration and interoperation of data.

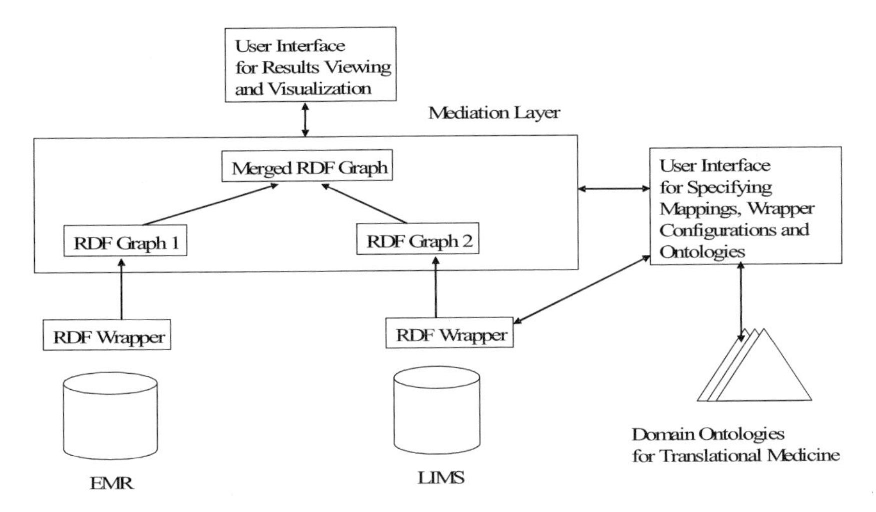

Fig. 13.16. Semantic Data Integration Architecture

RDF Wrappers: Wrappers are data-repository-specific software modules that map internal database tables or other data structures to concepts and relationships in the domain ontologies. Data in a repository is now exposed as RDF graphs for use by the other components in the system.

Mediation Layer: The mediation layer takes as input mapping rules that may be specified between various RDF graphs and computes the merged RDF graphs based on those mapping rules. In Section 13.1.3, we looked at how mediation between services could be established using ontologized mappings applied at run time. Here we focus on providing a global view on merged data models abstracting from the underlying heterogeneous data sources.

User Interfaces: User interfaces support: (a) visualization of integration results; (b) design and creation of domain ontologies; (c) configuration of RDF wrappers; and (d) specification of mapping rules to merge RDF graphs.

The main advantage of the approach is that one or more data sources can be added in an incremental manner. According to the current state of the art, data integration is implemented via one-off programs or scripts where the semantics of the data is hard-coded. Adding more data sources typically involves rewriting these programs and scripts. In our approach, the semantics are made explicit in RDF graphs and the integration is implemented via declarative specification of mappings and rules (this is analogous to the ontologized mapping examples of Section 13.1.3). These can be configured to incorporate new data sources via appropriate configurations of mappings, rules and RDF wrappers, leading to a cost- and time-effective solution.

13.2.3 A Domain Ontology for Translational Medicine

A key first step in semantic data integration is the definition of a domain ontology spanning both the clinical and genomic domains. A portion of the domain ontology is illustrated in Figure 13.17 and contains the following key concepts and relationships.

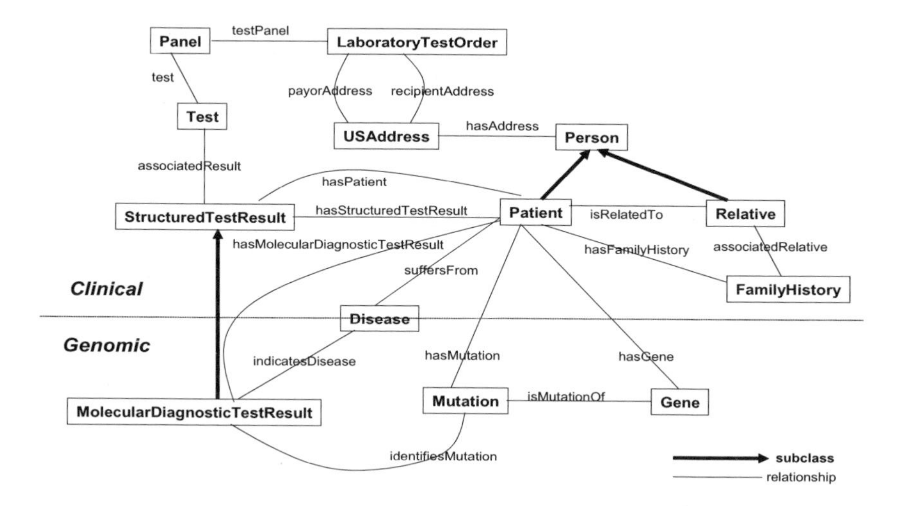

Fig. 13.17. A domain ontology for translational medicine

- The class `LaboratoryTestOrder` represents the order for a laboratory test for a patient. The order may be for a panel of tests represented in the class `Panel` which may contain one or more tests represented in the class `Test`. The order may have a `recipientAddress` and a `payorAddress` represented by the class USAddress, representing the set of addresses in the US.
- The class `Patient` is a core concept that characterizes patient state information, such as values of various patient state parameters, the results of diagnostic tests and his family and lineage information. It is related to the class `Person` through the subclass relationship. Information about a patient's relatives is represented using the `is_related` relationship and the `Relative` concept. The class `FamilyHistory` captures information of family members who may have had the dis-

ease for which the patient is being evaluated, and is related to the Patient concept via the `hasFamilyHistory` relationship.

- The `StructuredTestResult` captures results of laboratory tests and is related to the `Patient` class via the `hasStructuredTest` relationship and the `Test` class via the `associatedResult` relationship. Various types of test results such as LDL, AST, ALT, TotalBilirubin, etc. can be represented as subclasses of this class. The `MolecularDiagnosticTestResult` class represents the results of a molecular diagnostic test result, a type of structured test result (represented using the `subclass` relationship). Molecular diagnostics identify mutations (represented using the `identifiesMutation` relationship) and indicates diseases (represented using the `indicatesDisease` relationship) in a patient.
- The class `Gene` represents information about genes and is linked to the `Patient` class via the hasGene relationship. Genetic variants or mutations of a given gene are represented using the `Mutation` class which is linked to the `Patient` class via the `hasMutation` relationship. The relationship between a gene and mutation is represented using the `isMutationOf` relationship.
- The `Disease` class characterizes the disease states which can be diagnosed about a patient, and is related to the `Patient` class via the `suffersFrom` relationship and to the molecular diagnostic test results concept via the `indicatesDisease` relationship.

OWL Representation of Domain Ontology

The OWL representation of the domain ontology dicussed above is as follows.

```
<owl:Class rdf:ID="Person"/>
<owl:Class rdf:ID="Patient">
  <rdfs:subClassOf rdf:resource="#Person"/>
</owl:Class>
<owl:Class rdf:ID="Relative">
  <rdfs:subClassOf rdf:resource="#Person"/>
</owl:Class>
<owl:Class rdf:ID="StructuredTestResult"/>
<owl:Class rdf:ID="MolecularDiagnosticTestResult"/>
<owl:Class rdf:ID="FamilyHistory"/>
<owl:Class rdf:ID="Disease"/>
<owl:Class rdf:ID="Gene"/>
<owl:Class rdf:ID="Mutation"/>
<owl:Class rdf:ID="LaboratoryTestOrder"/>
<owl:Class rdf:ID="Panel"/>
<owl:Class rdf:ID="Test"/>
<owl:Class rdf:ID="USAddress"/>

<owl:ObjectProperty rdf:ID="isRelatedTo">
  <rdf:type rdf:resource="&owl;TransitiveProperty" />
  <rdfs:domain rdf:resource="#Patient"/>
  <rdfs:range rdf:resource="#Relative"/>
</owl:ObjectProperty>
```

```
<owl:ObjectProperty rdf:ID="hasFamilyHistory">
  <rdfs:domain rdf:resource="#Patient"/>
  <rdfs:range rdf:resource="#FamilyHistory"/>
</owl:ObjectProperty>
<owl:ObjectProperty rdf:ID="associatedRelative">
  <rdfs:domain rdf:resource="#FamilyHistory"/>
  <rdfs:range rdf:resource="#Relative"/>
</owl:ObjectProperty>
<owl:ObjectProperty rdf:ID="hasStructuredTestResult">
  <rdfs:domain rdf:resource="#Patient"/>
  <rdfs:range rdf:resource="#StructuredTestResult"/>
</owl:ObjectProperty>
<owl:ObjectProperty rdf:ID="hasStructuredTestResult">
   <owl:inverseOf rdf:resource="#hasPatient"/>
</owl:ObjectProperty>
<owl:ObjectProperty rdf:ID="hasMolecularDiagnosticTestResult">
    <rdfs:subPropertyOf rdf:resource="#hasStructuredTestResult" />
    <rdfs:range rdf:resource="#MolecularDiagnosticTestResult" />
</owl:ObjectProperty>
<owl:ObjectProperty rdf:ID="identifiesMutation"/>
  <rdfs:domain rdf:resource="#MolecularDiagnosticTestResult"/>
  <rdfs:range rdf:resource="#Mutation"/>
</owl:ObjectProperty>
<owl:ObjectProperty rdf:ID="indicatesDisease">
  <rdfs:domain rdf:resource="#MolecularDiagnosticTestResult"/>
  <rdfs:range rdf:resource="#Disease"/>
</owl:ObjectProperty>
<owl:ObjectProperty rdf:ID="suffersFrom">
  <rdfs:domain rdf:resource="#Patient"/>
  <rdfs:range rdf:resource="#Disease"/>
</owl:ObjectProperty>
<owl:ObjectProperty rdf:ID="hasMutation">
  <rdfs:domain rdf:resource="#Patient"/>
  <rdfs:range rdf:resource="#Mutation"/>
</owl:ObjectProperty>
<owl:ObjectProperty rdf:ID="hasGene">
  <rdfs:domain rdf:resource="#Patient"/>
  <rdfs:range rdf:resource="#Gene"/>
</owl:ObjectProperty>
<owl:ObjectProperty rdf:ID="isMutationOf">
  <rdfs:domain rdf:resource="#Mutation"/>
  <rdfs:range rdf:resource="#Gene"/>
</owl:ObjectProperty>
<owl:ObjectProperty> rdf:ID="hasAddress">
  <rdfs:domain rdf:resource="#Person"/>
  <rdfs:range rdf:resource="#USAddress"/>
</owl:ObjectProperty>
<owl:ObjectProperty rdf:ID="recipientAddress">
  <rdfs:domain rdf:resource="#LaboratoryTestOrder"/>
  <rdfs:range rdf:resource="#USAddress"/>
</owl:ObjectProperty>
<owl:ObjectProperty rdf:ID="payorAddress">
  <rdfs:domain rdf:resource="#LaboratoryTestOrder"/>
```

```
  <rdfs:range rdf:resource="#USAddress"/>
</owl:ObjectProperty>
<owl:ObjectProperty rdf:ID="testPanel">
  <rdfs:domain rdf:resource="#LaboratoryTestOrder"/>
  <rdfs:range rdf:resource="#Panel"/>
</owl:ObjectProperty>
<owl:ObjectProperty rdf:ID="test">
  <rdfs:domain rdf:resource="#Panel"/>
  <rdfs:range rdf:resource="#Test"/>
</owl:ObjectProperty>
<owl:ObjectProperty rdf:ID="associatedResult">
  <rdfs:domain rdf:resource="#Test"/>
  <rdfs:range rdf:resource="#StructuredTestResult"/>
</owl:ObjectProperty>
<owl:DataTypeProperty rdf:ID="orderDateTime">
    <rdf:type rdf:resource="&owl;FunctionalProperty"/>
    <rdfs:domain rdf:resource="#LaboratoryTestOrder"/>
    <rdfs:range  rdf:resource="&xsd;datetime" />
</owl:DataTypeProperty>
<owl:Class rdf:ID="Patient">
  <rdfs:subClassOf>
    <owl:Restriction>
      <owl:onProperty rdf:resource="#isRelatedTo"/>
      <owl:allValuesFrom rdf:resource="#Relative"/>
    </owl:Restriction>
  </rdfs:subClassOf>
</owl:Class>
<owl:Class rdf:ID="Mutation">
   <rdfs:subClassOf>
    <owl:Restriction>
      <owl:onProperty rdf:resource="#isMutationOf"/>
      <owl:someValuesFrom rdf:resource="#Gene"/>
    </owl:Restriction>
  </rdfs:subClassOf>
</owl:Class>
<owl:Class rdf:ID="StructuredTestResult">
  <rdfs:subClassOf>
    <owl:Restriction>
      <owl:onProperty rdf:resource="#hasPatient"/>
      <owl:cardinality rdf:datatype="&xsd;nonNegativeInteger">
        1
     </owl:cardinality>
    </owl:Restriction>
  </rdfs:subClassOf>
</owl:Class>
<owl:Class rdf:ID="PatientWithMYH7Gene">
  <rdfs:subClassOf>
    <owl:Restriction>
      <owl:onProperty rdf:resource="#hasGene"/>
      <owl:hasValue rdf:resource="#MYH7"/>
    </owl:Restriction>
  </rdfs:subClassOf>
</owl:Class>
```

```
<owl:Class rdf:ID="DiabeticPatient">
  <owl:intersectionOf rdf:parseType="Collection">
    <owl:Class rdf:about="#Patient"/>
    <owl:Restriction>
      <owl:onProperty rdf:resource="#suffersFrom"/>
      <owl:someValuesFrom rdf:resource="#Diabetes"/>
    </owl:Restriction>
  </owl:intersectionOf>
</owl:Class>
<owl:Class rdf:ID="StructuredTestResult">
  <owl:unionOf rdf:parseType="Collection">
    <owl:Class rdf:resource="#NormalStructuredTestResult"/>
    <owl:Class rdf:resource="#AbnormalStructuredTestResult"/>
  </owl:unionOf>
</owl:Class>
<owl:Class rdf:ID="NormalStructuredTestResult">
  <rdfs:subClassOf rdf:resource="#StructuredTestResult"/>
  <owl:disjointWith rdf:resource="#AbnormalStructuredTestResult/>
</owl:Class>
<owl:Class rdf:ID="AbnormalStructuredTestResult">
  <rdfs:subClassOf rdf:resource="#StructuredTestResult"/>
  <owl:disjointWith rdf:resource="#NormalStructuredTestResult"/>
</owl:Class>
```

Advantages of OWL-based Semantic Web specifications

The OWL Specifications above illustrate some of the key features, based on which one may chose OWL as opposed to other alternatives:

- OWL seeks to model the semantics of the information through constructs such as `owl:Class`, `owl:ObjectProperty` and `owl:DatatypeProperty`.
- OWL supports the ability to iteratively add descriptions as more knowledge and information becomes available. For instance, in the ontology below, the declaration of the class `Mutation` could be added first. The property restriction (`<owl:onProperty rdf:resource="#isMutationOf"> <owl:someValuesFrom rdf:resource="#Gene">`) could be added later independently by another domain expert.
- Unlike RDF Schema, OWL supports the ability to locally restrict the values of a particular property, e.g., the values of the property `suffersFrom` are restricted to instances of the class `Diabetes` when applied to instances of the class `DiabetesPatient`. Other classes of patients may be restricted to instances of other diseases.
- In contrast with RDF Schema, OWL supports the ability to support complex classes (`StructuredTestResult` is the union of `NormalStructuredTestResult` and `AbnormalStructuredTestResult`), disjoint classes (`NormalStructuredTestResult` and `AbnormalStructuredTestResult`) and cardinality constraints (e.g., each instance of `StructuredTestResult` has exactly 1 value for the `hasPatient` property).

13.2.4 Use of RDF to represent Genomic and Clinical Data

As discussed in Section 13.2.1, RDF wrappers perform the function of transforming information as stored in internal data structures in LIMS and EMR systems into RDF-based graph representations. We illustrate with examples (Figure 13.18), the RDF representation of clinical and genomic data in our implementation.

Clinical data related to a patient with a family history of `SuddenDeath` is illustrated. Nodes corresponding Patient ID and Person ID are connected by an edge labeled `isRelatedTo`, modeling the relationship between a patient and his father. The name of the patient ("Mr. X") is modeled as another node, and is linked to the patient node via an edge labeled `name`. Properties of the relationship between the patient ID and person ID nodes are represented by reification of the edge labeled `isRelatedTo` and by attaching labeled edges for properties such as the `type` of relationship (paternal) and the `degree` of the relationship (1).

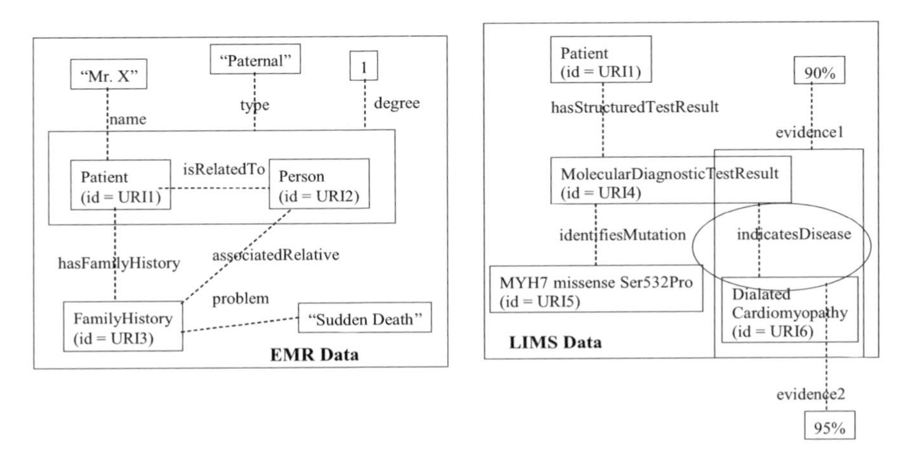

Fig. 13.18. RDF Representation of clinical and genomic data

Genomic data related to a patient evaluated for a given mutation (`MYH7 missense Ser532Pro`) is illustrated. Nodes corresponding to Patient ID and Molecular Diagnostic Test Result ID are connected by an edge labeled `hasStructuredTestResult` modeling the relationship between a patient and his molecular diagnostic test result. Nodes are created for the genetic mutation `MYH7 missense Ser532Pro` and the disease `Dialated Cardiomyopathy`. The relationship of the test result to the genetic mutation and disease is modeled using the labeled edges `identifiesMutation` and `indicatesDisease` respectively. The degree of evidence for the dialated cardiomyopathy is represented by reification of the `indicatesDisease` relationship and attaching labeled edges `evidence1` and `evidence2` to the reified edge. Multiple confidence values expressed by different experts can be represented by reifying the edge multiple times.

RDF Representation

The RDF representations of these graphs using the triples syntax is as follows:

```
# Available at http://www.hospital.org/EMR
PREFIX skos: <http://www.w3.org/2004/02/skos/core#>

URI1 name "X" ;
URI1 isRelatedTo URI2 .

_:stmt1 rdf:type rdf:Statement .
_:stmt1 rdf:subject URI1 .
_:stmt1 rdf:predicate isRelatedTo .
_:stmt1 rdf:object URI2 .
_:stmt1 type "Paternal"@en .
_:stmt1 degree 1 .

URI1 hasFamilyHistory URI3 .
URI3 associatedRelative URI2 .
URI3 problem URI7 .

URI7 rdf:type skos:Concept .
URI7 skos:preflabel "Sudden Death"@en .
URI7 skos:inScheme URI8 .
URI8 rdf:type skos:ConceptScheme
URI8 dc:title "Systematized Nomenclature of Medicine (SNOMED)" .

# Available at http://www.laboratory.com/LIMS
PREFIX skos: <http://www.w3.org/2004/02/skos/core#>

URI1 hasStructuredTestResult URI4 .
URI4 identifiesMutation URI5 .
URI4 indicatesDisease URI6

_:stmt2 rdf:type rdf:Statement .
_:stmt2 rdf:subject URI4 .
_:stmt2 rdf:predicate identifiesMutation .
_:stmt2 rdf:object URI6 .
_:stmt2 evidence1 "90%" .
_:stmt2 evidence2 "95%" .

URI5 rdf:type skos:Concept .
URI5 skos:preflabel "MYH7 missense Ser532Pro"@en .
URI5 skos:inScheme URI9 .
URI9 rdf:type skos:ConceptScheme
URI9 dc:title "Human Genome Nomenclature"@en .

URI6 rdf:type skos:Concept .
URI6 skos:preflabel "Dialated Cardiomyopathy"@en .
URI6 skos:inScheme URI10 .
URI10 rdf:type skos:ConcpetScheme .
URI10 dc:title "NCI Thesaurus" .
```

Advantages of RDF Specification

The RDF representations of example EMR and LIMS data above illustrate the following key features that are enabled by Semantic Web specifications:

- As discussed earlier, one of the key aspects of the RDF specification is the ability to uniquely identify resources using URIs which are available. In the example above, the same URI (e.g., `URI1`) is used to identify a patient in the EMR and the LIMS data-set. This is an important feature from the point of view of achieving web-scale data linkage and integration.
- The ability to "reify" an edge in the RDF graph (e.g., `URI1 isRelatedTo URI2`) and attaching additional properties and values (e.g., `type` and `"Paternal"`).
- There are multiple standardized vocabularies in use in the healthcare and life sciences. Some examples referenced in the above example are NCI Thesaurus, SNOMED and Human Genome Nomenclature. The Semantic Web specification through the Simple Knowledge Organization Scheme (SKOS) [403] provides a standardized way to link to concepts from these standardized vocabularies. For e.g., the RDF graph refers to a standardized vocabulary code for "Sudden Death" from the SNOMED controlled vocabulary by using the following RDF triples.

```
URI3 problem URI7 .
URI7 rdf:type skos:Concept .
URI7 skos:preflabel "Sudden Death"@en .
URI7 skos:inScheme URI8 .
URI8 rdf:type skos:ConceptScheme
URI8 dc:title "Systematized Nomenclature of Medicine (SNOMED)" .
```

13.2.5 The Integration Process

The data integration process is an interactive one and involves the end user, who in our case might be a clinical or genomic researcher. RDF graphs from different data sources are displayed. The steps in the process that lead to the final integrated result are enumerated below.

1. RDF graphs are displayed in an intuitive and understandable manner to the end user in a graphical user interface.
2. The end user previews them and specifies a set of rules for linking nodes across different RDF models. Some examples of simple rules that are implemented in our system are:

   (A) Merge nodes that have the same IDs or URIs.

   (B) Merge nodes that have matching IDs, per a lookup on the Enterprise Master Patient Index (EMPI).

   (C) If there are three nodes in the merged graph, `Node1`, `Node2` and `Node3` such that `Node1` and `Node2` are linked with an edge labeled `hasStructuredTestResult`, and `Node2` and `Node3` are linked with an edge labeled `indicatesDisease`, then introduce a new edge labeled `suffersFrom` that links `Node1` and `Node3`.

3. Merged RDF graphs that are generated based on these rules are displayed to the user, who may then decide to activate or deactivate some of the rules displayed.
4. New edges (e.g., `suffersFrom`) that are inferred from these rules may be added back to the system based on the results of the integration. Sophisticated data mining that determines the confidence and support for these new relationships might be invoked.

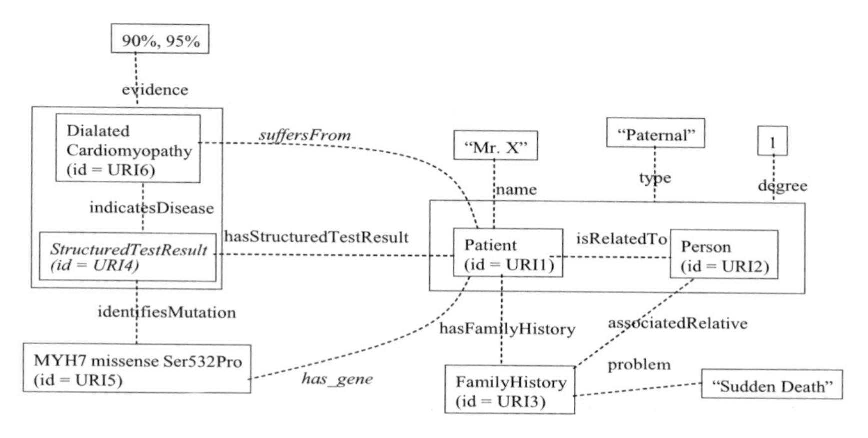

Fig. 13.19. RDF representation of the integrated result

Linking Data From Multiple Sources

A key construct supported in the SPARQL specification is the ability to define a graph data-set containing a default graph and other default graphs. For instance the RDF graph describing the EMR data can be specified as a default graph and the RDF graph describing the LIMS data can be specified as the named graph. An interesting capability is that these graphs can be distributed and available at different URIs as illustrated below.

```
# Default graph <http://www.hospital.org/EMR>
/* ... RDF representation as illustrated in Section 4.4.1 above ... */

# Named graph: <http://www.laboratory.com/LIMS>
/* ... RDF representation in Section 4.4.1 above ... */
```

Consider the merged RDF graph illustrating the integration of clinical and genomic data as illustrated in Figure 13.19 above. A default mapping is the ability to match on URIs, a mappping we get for free by using the Semantic Web infrastructure. The merged graph based on ID matching is available based on the graph merge operation discussed above. However, the merged graph in Figure 13.19 above illustrates two new edges, `suffersFrom` and `hasMutation` which identify the association between the patient and a disease and gene based on the results of the molecular diagnostic test result. The mappings required for enabling this are:

- If a structured test result for a patient indicates a disease with higher then 90% probability, then the patient may suffer from that disease. This can be represented using SPARQL `CONSTRUCT` expression as follows:

```
CONSTRUCT { ?s suffersFrom ?o}
WHERE {
          ?s hasStructuredTestResult ?result .
          ?result indicatesDisease ?o
      }
```

- If the structured test result for a patient indicates a mutation, then the patient has the mutation is part of the patient's genome. This can be represented using the following SPARQL `CONSTRUCT` expression.

```
CONSTRUCT { ?s hasMutation ?o}
 WHERE {
        ?s hasStructuredTestResult ?result .
        ?result identifiesMutation ?o .
       }
```

The merged graph can then be created by appropriately constructing the SPARQL query, where the `CONSTRUCT` part will contain the new predicates and the `WHERE` clauses can be appropriately combined.

An alternate approach by which these mappings can be represented is using a rule based formalism. A representation of the same rules using the N3 rules syntax is as follows.

- ```
  {?s hasStructuredTestResult ?result .
   ?result indicatesDisease ?o}
  => {?s suffersFrom ?o}
  ```
- ```
  {?s hasStructuredTestResult ?result .
   ?result identifiesMutation ?o}
  => {?s suffersFrom ?o}
  ```

Advantages of using Semantic Web Specifications

The solution approach proposed above supports an incremental approach for data integration. Furthermore, the solution leverages the underlying web infrastructure to uniquely identify resources referred to in RDF graphs. This enables easy linking and integration of data across multiple RDF data sources, without the need implement costly data value mapping techniques. The most valuable aspect of this approach is that it enables a flexible approach to ground data representing in RDF graphs to concepts in standardized concepts and vocabularies. The SKOS standard enables the association of semantics with the data in consistent manner. Finally, the key advantage of semantic web specifications is the ability to specify mapping rules at the information level using the SPARQL CONSTRUCT expression or using rules expressed in the N3 syntax. The enables integration and linking at the "information" level in contrast with the current state of art with one of java and perl scripts that implement one-off integration solutions. The externalization and representation of mappings using semantic web specifications enable re-use and configuration of mappings that can be leveraged to implement data linking and integration.

13.3 Decision Support

As illustrated in the clinical use case, there is a need for providing guidance to a clinician for ordering the right molecular diagnostic tests in the context of phenotypic observations about a patient and for ordering appropriate therapies in response to molecular diagnostic test results. The decision support functionality spans both the clinical and biological domains and depends on effective integration of knowledge and data across data repositories containing clinical and biological data and knowledge.

In order to maintain the currency and consistency of decision support knowledge across all clinical information systems and applications, a rules-based approach for representing and executing decision support knowledge has been adopted [367] [368]. We present an approach and architecture for implementing scalable and maintainable clinical decision support using a business rules engine and an OWL-based ontology engine. Various clinical applications will invoke this clinical decision support service for their decision support needs.

The architecture integrates a business rules engine that executes declarative if-then rules stored in a rule base referencing objects and methods in a business object model. The rules engine executes object methods by invoking services implemented on the EMR. Specialized inferences that support classification of data and instances into classes are identified and an approach to implementing these inferences using an OWL (Web Ontology Language) based ontology engine is presented. Architectural alternatives for integration of clinical decision support functionality with the invoking application and the underlying clinical data repository and their associated trade-offs are also discussed. Consider the following decision support rule:

```
IF the patient's LDL > 120
AND the patient has a contraindication to Fibric Acid
THEN Prescribe the Zetia Lipid Management Protocol
```

In Section 5.4.2, we presented a solution approach using the SWRL specification. However, SWRL is currently a W3C member submission and is likely to be a while before it is standardized. In this section, we present a solution approach based on a commerical BRMS such as ILOG [405]. The steps for implementing the above clinical guideline are:

1. Create the business object model that defines patient-related classes and methods.
2. Specify rules to encode decision support logic.
3. Delineate definitions characterizing patient states and classes and represent them in an ontology.

We begin by presenting our clinical decision support architecture and then illustrating with the example given above the steps for creation of appropriate ontologies and rule bases.

13.3.1 Decision Support Services: WSMO/WSML Specification

We now present a WSML specification of Decision Support Service in the context of the GetTherapeuticGuidance clinical service. The orchestration of the decision support service is as follows:

- SelectAndLoadRuleBase takes as input the integrated clinical and genomic data associated with a patient and identifies the appropriate rulebase to be loaded into the rules engine.
- AssertFactsInRuleEngine asserts all the patient data in the rules engine.
- ExecuteInferences initiates the execution of the rules engine and returns the facts that are inferred in the process, in this case resulting in the therapeutic recommendations appropriate for the patient.

```
wsmlVariant _"http://www.wsmo.org/wsml/wsml-syntax/wsml-rule" ...
interface DataIntegrationInterface
orchestration DataIntegrationOrchestration
  stateSignature DataIntegrationSignature

  /* Concepts used as input and output to the orchestration */
  in IntegratedPatientData, ListContraindications, ListDiseases
  out ListTherapies
  /* Concept used to maintain the control flow */
  controlled ControlState

  /* transition rules govern the control flow of the orchestration */
  transitionRules
  if (?integratedPatientData memberOf Patient and
     ?listContraindications memberOf ListContraindications and
     ?listDiseases memberOf ListDiseases
     ?cs[value hasValue InitialState] memberOf ControlState)
  then
     call SelectandLoadRuleBase
        (?integratedPatientData ?listContraindications ?listDiseases)
             : _# memberOf RuleEngine
     update(?cs[value hasValue SelectAndLoadRuleBaseCalled])
  endif

  if (?ruleEngine memberOf RuleEngine and
```

```
    ?integratedPatientData memberOf Patient and
    ?listContraindications memberOf ListContraindications and
    ?listDiseases memberOf ListDiseases and
    ?cs[value hasValue SelectAndLoadRuleBaseCalled])
  then
    call AssertFactsIntoRuleEngine
     (?integratedPatientData ?listContraindications ?listDiseases
?ruleEngine)
          : _# memberOf RuleEngine
   update(?cs[value hasValue AssertFactsIntoRuleEngineCalled])
  endif

  if (?ruleEngine memberOf RuleEngine and
     ?cs[value hasValue AssertFactsIntoRuleEngineCalled])
  then
     call ExecuteInferences(?ruleEngine): _# memberOf ListTherapies
     update(?cs[value hasValue ExecuteInferencesCalled])
  endif
```

13.3.2 Architecture

The architecture for implementing clinical decision support is illustrated in Figure 13.20 below and consists of the following components:

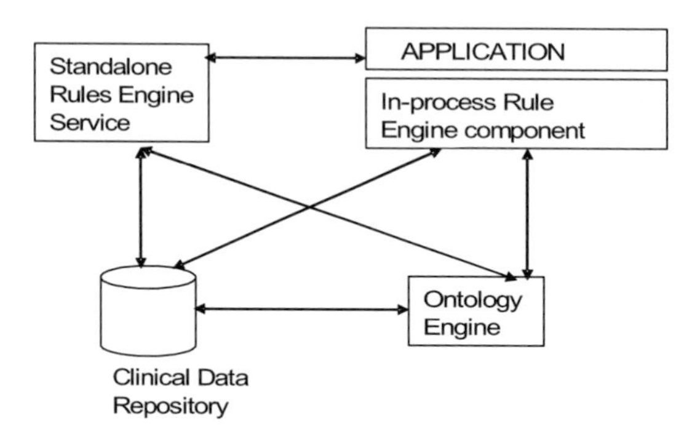

Fig. 13.20. Clinical decision support architecture

Clinical Data Repository: The clinical data repository stores patient-related clinical data. External applications, the rules engine (via methods defined in the business object model) and the ontology engine retrieve patient data by invoking services implemented by the clinical data repository.

Standalone Rules Engine Service: A standalone rules engine service is implemented using a business rules engine. On receiving a request, the service initializes a rules engine instance, and loads the rule base and business object model. The rules engine service then executes methods in the business object model and performs rule-based inferences. The results obtained are then returned to the invoking application.

In-Process Rules Engine Component: This provides functionality similar to that of the rules engine service, except that the rules engine component is loaded in the same process space in which the application is executing.

Ontology Engine: This will be implemented using an OWL-based ontology engine. On receiving a request, the ontology engine performs classification inferences on patient data to determine if a patient belongs to a particular category, e.g., patients with contraindication to fibric acid.

13.3.3 Business Object Model Design

The business object model for the above clinical decision support rule could be specified as follows.

```
Class Patient: Person
method getName(): string;
method hasMolecularDiagnosticTestResult(): StructuredTestResult;
method hasLiverPanel(): LiverPanelResult;
method hasLDL(): real;
method hasContraindication(): set of string;
method hasMutation(): string;
method recommendedTherapy(): set of string;
method setRecommendedTherapy(string): void;
method isAllergicTo(): set of string;

Class StructuredTestResult
method getPatient(): Patient;
method indicatesDisease(): Disease;
method identifiesMutation(): set of string;
method evidenceOfMutation(string): real;

Class LiverPanelResult
method getPatient(): Patient;
method getALP(): real;
method getALT(): real;
method getAST(): real;
method getTotalBilirubin(): real;
method getCreatinine(): real;
```

The model describes patient state information by providing a class and set of methods that make patient state information, e.g., results of various tests, therapies, allergies and contraindications, available to the rules engine. The model also contains classes corresponding to complex tests such as a liver panel result and methods that retrieve information specific to those tests, e.g., methods for retrieving creatinine clearance and total bilirubin. The methods defined in the object model are executed by the rules engine, which results in invocation of services in the clinical data repository for retrieval of patient data.

13.3.4 Rule Base Design

The business object model defined in the previous section provides the vocabulary for specifying various clinical decision support rules. Consider the following specification of the clinical decision support rule discussed earlier.

```
IF the_patient.hasLDL() > 120
AND ((the_patient.hasLiverPanel().getALP() > <Normal>
AND the_patient.hasLiverPanel().getALT() > <Normal>
AND the_patient.hasLiverPanel().getAST() > <Normal>
AND the_patient.hasLiverPanel().getTotalBilirubin() > <Normal>
AND the_patient.hasLiverPanel().getCreatinine() > <Normal>)
OR "Fibric Acid Allergy" memberOf the_patient.isAllergicTo())
THEN the_patient.setRecommendedTherapy("Zetia Lipid Management Ther-
apy")
```

The above rule represents the various conditions that need to be specified (the `IF` part) so that the system can recommend a particular therapy for a patient (the `THEN` part). The following conditions are represented on the `IF` part of the rule:

1. The first condition is a simple check on the value of the LDL test result for a patient.
2. The second condition is a complex combination of conditions that check whether a patient has contraindication to fibric acid. This is done by checking whether the patient has an abnormal liver panel or an allergy to fibric acid.

13.3.5 Definitions vs. Actions: Ontology Design

Our implementation of the clinical decision support service using a business rules engine involved encoding decision support logic across a wide variety of applications using rule sets and business object models. An interesting design pattern that emerged is described below:

- Rule-based specifications of conditions that describe patient states and classes, for instance, “Patient with contraindication to fibric acid”. They also involve characterization of normal or abnormal physiological patient states, for instance, “Patients with abnormal liver panel”. These specifications are also called definitions.
- Rule-based specifications that propose therapies, medications and referrals, for instance, prescribing lipid management therapy for a patient with a contraindication to fibric acid. These specifications are called actions.

The rule sets are modularized by separating the definition of a “*Patient with a contraindication to fibric acid*” from the decisions that are recommended once a patient is identified as belonging to that category. The definitions of various patient states and classes can be represented as axioms in an ontology that could be executed by an OWL ontology engine. At execution time, the business rules engine can invoke a service that interacts with the ontology engine to infer whether a particular patient belongs to a given class of patients, in this case, whether a patient has a contraindication to fibric acid. The ontology of patient states and classes is represented as follows and is illustrated in Figure 13.21.

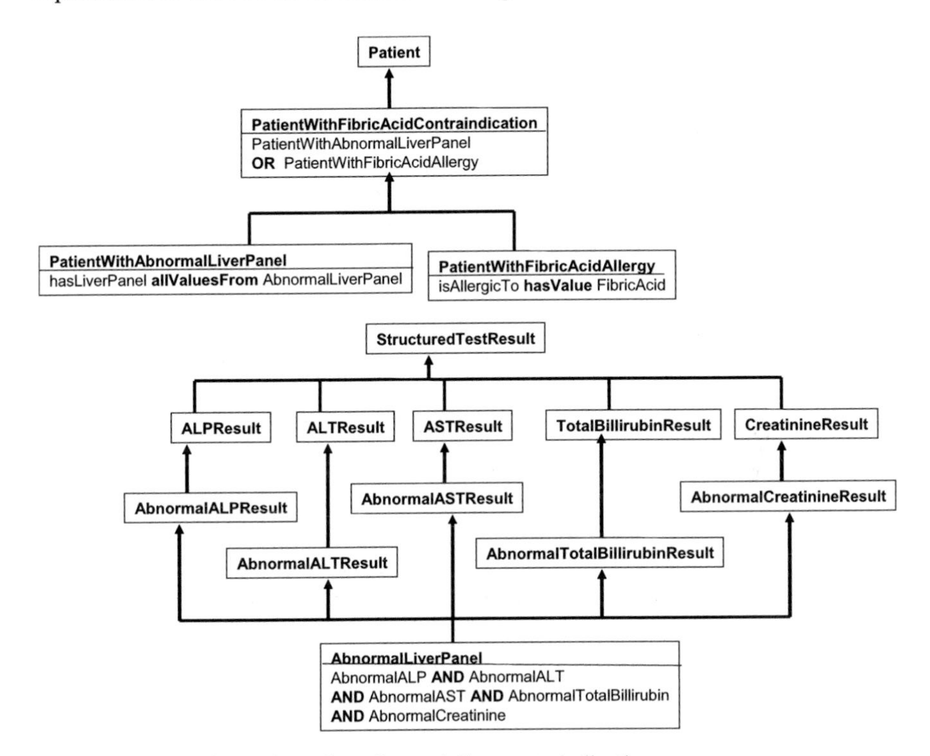

Fig. 13.21. An enhanced ontology for modeling contraindications

```
<owl:ObjectProperty hasLiverPanel>
  <rdfs:subPropertyOf rdf:resource="#hasStructuredTestResult"/>
</owl:ObjectProperty>
<owl:ObjectProperty hasALP>
  <rdfs:subPropertyOf rdf:resource="#hasStructuredTestResult"/>
</owl:ObjectProperty>
<owl:ObjectProperty hasALT>
  <rdfs:subPropertyOf rdf:resource="#hasStructuredTestResult"/>
</owl:ObjectProperty>
<owl:ObjectProperty hasAST>
  <rdfs:subPropertyOf rdf:resource="#hasStructuredTestResult"/>
</owl:ObjectProperty>
<owl:ObjectProperty hasCreatinine>
  <rdfs:subPropertyOf rdf:resource="#hasStructuredTestResult"/>
</owl:ObjectProperty>
<owl:ObjectProperty hasTotalBilirubin>
  <rdfs:subPropertyOf rdf:resource="#hasStructuredTestResult"/>
</owl:ObjectProperty>
```

```
<owl:ObjectProperty rdf:ID="isAllergicTo">
  <rdfs:domain rdf:resource="#Patient"/>
  <rdfs:range rdf:resource="#Allergen"/>
</owl:ObjectProperty>
<owl:ObjectProperty rdf:ID="recommendedTherapy">
  <rdfs:domain rdf:resource="#Patient"/>
  <rdfs:range rdf:resource="#Therapy"/>
</owl:ObjectProperty>

<owl:Class rdf:ID="Allergen"/>
<Allergen rdf:ID="FibricAcid"/>

<owl:Class rdf:ID="Therapy"/>
<Therapy rdf:ID="ZetiaLipidManagementTherapy"/>
<owl:Class rdf:ID="LiverPanelResult">
  <owl:unionOf rdf:parseType="Collection">
    <owl:Class rdf:resource="#NormalLiverPanelResult"/>
    <owl:Class rdf:resource="#AbnormalLiverPanelResult"/>
  </owl:unionOf>
</owl:Class>
<owl:Class rdf:ID="AbnormalLiverPanelResult"/>
  <rdfs:subClassOf rdf:resource="#LiverPanelResult"/>
  <owl:disjointWith rdf:resource="#NormalLiverPanelResult"/>
</owl:Class>
   <owl:Class rdf:ID="NormalLiverPanelResult"/>
  <rdfs:subClassOf rdf:resource="#LiverPanelResult"/>
  <owl:disjointWith rdf:resource="#AbnormalLiverPanelResult"/>
</owl:Class>
/* Similar definitions for ALPResult, ALTResult, */
/* CreatinineResult and TotalBilirubinResult */

<owl:DatatypeProperty hasALPValue>
  <rdfs:domain rdf:resource="#ALPResult"/>
  <rdfs:range rdf:datatype="&xsd;float"/>
</owl:DatatypeProperty>
/* Similar properties for hasALTValue, hasASTValue, */
/* hasCreatinineValue, hasTotalBilirubinValue */
```

Consider the definition of a FibricAcidContraindication represented using OWL as follows and illustrated in Figure 13.21.

```
<owl:Class rdf:ID="PatientContraindicatedToFibricAcid">
  <owl:unionOf rdf:parseType="Collection">
    <owl:Class rdf:resource="#Patient"/>
    <owl:Restriction>
      <owl:onProperty="#isAllergicTo"/>
      <owl:hasValue="#FibricAcid"/>
    </owl:Restriction>
    <owl:Restriction>
      <owl:onProperty="#hasLiverPanel"/>
      <owl:allValuesFrom="#AbnormalLiverPanelResult"/>
    </owl:Restriction>
  </owl:unionOf>
</owl:Class>
```

The above OWL class defines patients with contraindication to Fibric Acid as patients having an abnormal liver panel and having an allergy to Fibric Acid. Abnormal Live Panel is further defined as:

```
  <owl:Class rdf:ID="AbnormalLiverPanel">
 <owl:intersectionOF rdf:parseType="Collection">
    <owl:Restriction>
      <owl:onProperty="#hasALP"/>
      <owl:allValuesFrom="#AbnormalALPResult"/>
    </owl:Restriction>
    <owl:Restriction>
      <owl:onProperty="#hasALT"/>
      <owl:allValuesFrom="#AbnormalALTResult"/>
    </owl:Restriction>
    <owl:Restriction>
      <owl:onProperty="#hasAST"/>
<owl:allValuesFrom="#AbnormalASTResult"/>
    </owl:Restriction>
    <owl:Restriction>
      <owl:onProperty="#hasCreatinine"/>
      <owl:allValuesFrom="#AbnormalCreatinine"/>
    </owl:Restriction>
    <owl:Restriction>
      <owl:onProperty="#hasTotalBilrubin"/>
      <owl:allValuesFrom="#AbnormalTotalBilirubin"/>
    </owl:Restriction>
  </owl:intersectionOf>
</owl:Class>
```

Based on the above definitions, a highly simplified version of the clinical decision rule discussed above can now be implemented as follows:

```
IF the_patient.hasLDL() > 120
AND the_patient.hasContraindiction() contains "Fibric Acid Contrain-
dication"
THEN the_patient.setRecommendedTherapy("Zetia Lipid Management Proto-
col")
```

The class `Patient` and properties `isAllergicTo`, `hasLiverPanel` and others provide a framework for describing the patient. The class `PatientContraindicatedToFibricAcid` is a subclass of all patients that are known to have contraindication to fibric acid. This is expressed using an OWL axiom. The class `Allergen` represents various diseases allergens of interest including `FibricAcid`. The classes `AbnormalALPResult`, `AbnormalALTResult`, `AbnormalASTResult`, `AbnormalTotalBilirubinResult` and `AbnormalCreatinineResult` represent ranges of values of abnormal ALP, ALT, AST, total bilirubin and creatinine results respectively.

Custom datatypes based on the OWL specifications provide the ability to map XML Schema datatypes to OWL classes. The class `AbnormalLiverPanel` is defined using an axiom to characterize the collection of abnormal values of various component test results (e.g., ALP, ALT, AST) that belong to a liver panel.

The representation of an axiom specifying the definition of `PatientContraindicatedToFibricAcid` enables the knowledge engineer to simplify the rule base significantly. The separation of definitions from actions and their implementation in an ontology engine reduces the complexity of the rule base maintenance significantly. It may be noted that the conditions that comprise a definition may appear multiple times in multiple rules in a rule base. Our approach enables the encapsulation of these conditions in a definition, e.g., `PatientContraindicatedToFibricAcid`. Thus all rules can now reference the class `PatientContraindicatedToFibricAcid` which is defined and maintained in the ontology engine. Whenever the definition of `PatientContraindicatedToFibricAcid` changes, the changes can be isolated within the ontology engine and the rules that reference this definition can be easily identified. Issues related to Knowledge change, maintenance and Provenance are discussed next.

13.4 Knowledge Maintenance and Provenance

All the functional requirements identified above (service composition, data integration and decision support) critically depend on domain-specific knowledge that could be represented as ontologies, rule bases, semantic mappings (between data and ontological concepts), and bridge ontology mappings (between concepts in different ontologies). The healthcare and life sciences domains are experiencing a rapid rate of new knowledge discovery and change. A knowledge change "event" has the potential of introducing inconsistencies and changes in the current knowledge bases that inform semantic data integration and decision support functions. There is a critical need to keep knowledge bases current with the latest knowledge discovery and changes in the healthcare and life sciences domains. Some requirements for knowledge maintenance and provenance that characterize these challenges are:

- Knowledge Management (KM) systems should have the ability to manage knowledge change at different levels of granularity.
- The impact of knowledge change at one level of granularity should be propagated to related knowledge at multiple levels of granularity.
- The impact of knowledge change of one type, e.g., definition of a contraindication, should be propagated to knowledge of another type, e.g., clinical decision support rules, containing references to that definition.

- The impact of knowledge on the data stored in the EMR should be accounted for. For instance, changes in the logic of clinical decision support, may invalidate earlier patient states that might have been inferred, or add new information to the EMR.

We now present a WSML specification of the KnowledgeAccuracyAndCurrencty service as an orchestration of the following goals/services:

- IdentifyRelevantKnowledgeBase identifies the potential knowledge bases that are relevant to a knowledge change event, such as the change in the normal values of a blood test result or the discovery of a new biomarker.
- IdentifyKnowledgeImpact identifies the ontology definitions or rules that are likely to be impacted. This could potentially be implemented as a technological service by an underlying rule or ontology engine and results in notifications being sent to the relevant knowledge engineers.
- IdentifyDataImpacts identifies the past inferences about a patient that might be invalidated due to the changes in the knowledge and results in notifications being sent to the appropriate patients or physicians.

```
wsmlVariant _"http://www.wsmo.org/wsml/wsml-syntax/wsml-rule" ...
interface KnowledgeAccuracyAndCurrencyInterface
orchestration KnowledgeAccuracyandCurrencyOrchestration
  stateSignature KnowledgeAccuracyandCurrencySignature

  /* Concepts used as input and output to the orchestration */
  in ListKnowledgeChangeEvent
  out ListKnowledgeEngineerNotifications, ListPhysicianPatientNotifi-
cations
  /* Concept used to maintain the control flow */
  controlled ControlState

  /* transition rules govern the control flow of the orchestration */
  transitionRules
  if (?listKnowledgeChangeEvent memberOf ListKnowledgeChangeEvent and
      ?cs[value hasValue InitialState] memberOf ControlState)
  then
     call IdentifyRelevantKnowledgeBase
           (?listKnowledgeChangeEvent)
               : _# memberOf KBList
      update(?cs[value hasValue IdentifyRelevantKnowledgeBaseCalled])
  endif

  if (?kbList memberOf KBList and
      ?listKnowledgeChangeEvent memberOf ListKnowledgeChangeEvent and
      ?cs[value hasValue IdentifyRelevantKnowledgeBaseCalled]
            memberOf ControlState)
  then
      call IdentifyKnowledgeImpacts
            (?kbList ?listKnowledgeChangeEvent)
                : _# memberOf ListKnowledgeElements
      update(?cs[value hasValue IdentifyKnowledgeImpactsCalled])
  endif
```

```
  if (?listKnowledgeElements memberOf ListKnowledgeElements and
     ?cs[value hasValue IdentifyKnowledgeImpactsCalled] memberOf Con-
trolState)
  then
      call NotifyKnowledgeEngineers
           (?listKnowledgeElements)
               : _# memberOf ListKnowledgeEngineerNotifications
      call IdentifyDataImpacts
            (?listKnowledgeElement)
               : _# memberOf ListDataItems
      update(?cs[value hasValue KnowledgeEngineersNotified-DataImpac-
tIdentified])
  endif

  if (?listDataItems memberOf ListDataItems and
    ?cs[value hasValue KnowledgeEngineersNotified-DataImpactIdenti-
fied]
            memberOf ControlState)
  then
    call NotifyPhysiciansAndPatients
           (?listDataItems):
               _# memberOf ListPhysicianPatientNotifications
    update(?cs[value hasValue NotifyPhysiciansAndPatientsCalled])
  endif
```

There is a close relationship between knowledge change, the core issue in the context of knowledge maintenance; and provenance. Issues related to when and by whom the change was effected are issues related to knowledge provenance, and provide useful information for maintaining knowledge. The issue of representing the rationale behind the knowledge change involves both knowledge change and provenance. On the one hand, the rationale behind the change could be that a knowledge engineer did it, which is an aspect of provenance. On the other hand, if the change in knowledge is due to the propagation of a change in either a knowledge component or related knowledge, it is knowledge change propagation invoked in the context of knowledge provenance.

We address the important issue of knowledge change propagation in this section. Consider the definition in natural language of fibric acid contraindication:

```
A patient is contraindicated for fibric acid if he or she has an
allergy to fibric acid or has an abnormal liver panel.
```

Suppose there is a new (hypothetical) biomarker for fibric acid contraindication for which a new molecular diagnostic test is introduced in the market. This leads to a redefinition of a fibric acid contraindication as follows.

```
The patient is contraindicated for fibric acid if he has an allergy to
fibric acid or has elevated Liver Panel or has a genetic mutation.
```

Let us also assume that there is a change in a clinically normal range of values for the lab test AST which is a part of the liver panel lab test. This leads to a knowledge change and propagation across various knowledge objects that are sub-components and associated with the fibric acid contraindication concept. A diagrammatic representation of the OWL representation of the new fibric contraindication with the changes marked in red ovals is illustrated below. The definition of "fibric acid contraindication" changes, triggered by changes at various levels of granularity.

A potential sequence of change propagation steps are enumerated below:

1. The clinically normal range of values for the AST lab test changes.
2. This leads to a change in the abnormal value ranges for the AST lab test.
3. This leads to a change in the definition of an abnormal liver panel.
4. This leads to a change in what it means to be a patient with an abnormal liver panel.
5. The definition of fibric acid contraindication changes due to the following changes.

   (A) The change in the definition of a patient with an abnormal liver panel as enumerated in steps 1-4 above.

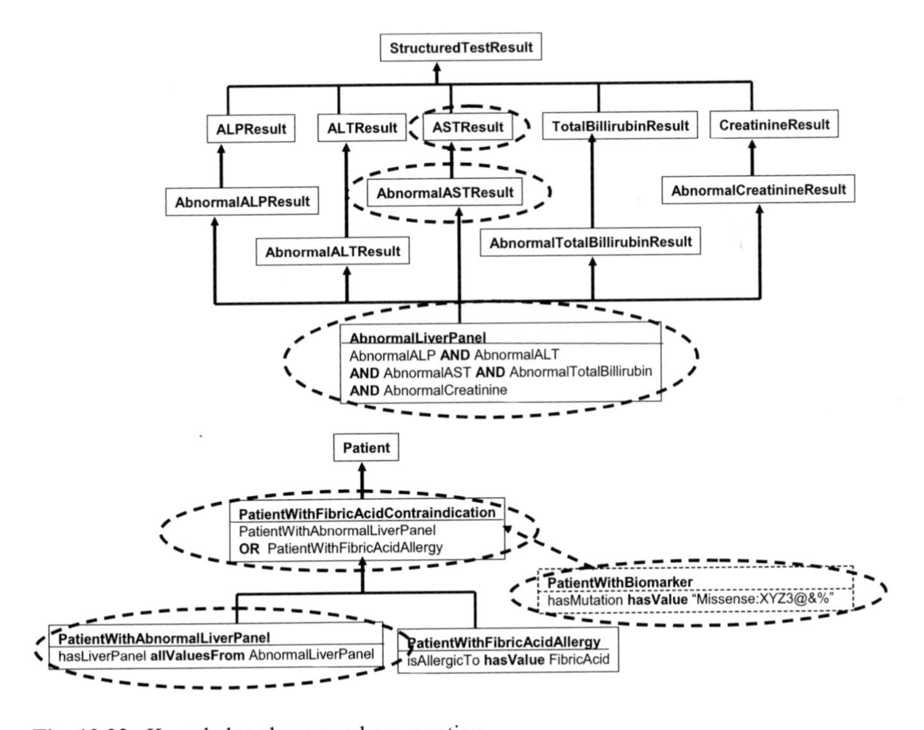

Fig. 13.22. Knowledge change and propagation

(B) Introduction of a new condition: a patient having a mutation: "Missense: XYZ3@&%" (hypothetical). This is a new condition which could lead to a change in what it means to be a patient with a contraindication to fibric acid.

It may be noted that in our discussion in Section 5.3.1, none of the ontology editors and tools today support versioning and change management functionality. In our solution approach, we propose to load these ontologies as data into a rules engine and write specialized rules to identify the impacts of a change operation.

14 Outlook: The Good, the Bad and the Ugly?

In the introduction of this book we wrote: "The Semantic Web is a vision: the idea of having data on the Web defined and linked in a way that it can be used by machines not just for display purposes, but for automation, integration and reuse of data across various applications. The goal of the Semantic Web initiative is as broad as that of the Web: to create a universal medium for the exchange of data. It is envisaged to smoothly interconnect personal information management, enterprise application integration, and the global sharing of commercial, scientific and cultural data."

So it is only appropriate in the outlook chapter to ask: "Where are we in the time line for Semantic Web and Semantic Web Services?"

There is the notion of a time line for new technologies. One of these is the 5-5-5 rule from [384]. A fundamentally new concept seems to move from initial concept to the mass market in 15 years. The first five years are for research, the second five for refining products based on the research, and the last five are where the concept reaches saturation in the market. This would mean that we are in the middle of the second 5 years, meaning that the pure research is done and product refinements are currently taking place.

At the same time this also means that the field is not done yet and a lot of questions and issues remain that we discuss next as The Good, The Bad and The Ugly.

14.1 The Good - Progress and Impact

A lot of positive and constructive work happened and is ongoing in the area of the Semantic Web and Semantic Web Services. The most important are the following.

- **Move from information to knowledge.** The most important commodity an organization can acquire is information. There is a significant move to the mainstream for knowledge representation being driven to a growing extent by social networking e.g. FOAF, RSS, tagging and tag clouds, see [380] or [381].
- **Everything is connected.** The ever increasing capability of mobile devices means a move away from interaction models with central control and central systems toward P2P-based communication without the need for a middle man connecting the various devices. Every internet-enabled device has a unique identifier and therefore can be a node in a network (e.g. Microsoft follows this trend through the Peer Name Resolution Protocol (PNRP)[28]). Devices that can

connect (using technologies such as HTTP) will interact and this is where semantic interoperability will be more and more relevant in making heterogeneity transparent.

- **Push for openly verifiable benchmarking.** Efforts like the SWS Challenge [385] push toward a public benchmark of claims, following traditional technologies like EDI. This ensures that semantic interoperability claims are verified and form the basis of building trust into this new technology.
- **Commercialisation.** Companies like Ontoprise[29], TopBraid Composer[30], Cycorp[31] and Siderean[32] are commercializing Semantic Web technology by developing software products. Companies with significant semantics-based products are being acquired by mature players in the market such as WebMethods (acquired Cerebra) and Google (acquired Applied Semantics) . Established companies like Oracle now include Semantic Web technology in their established products e.g. RDF supported in the heart of the Oracle database architecture.
- **Semantics for business processes.** Joint industrial and academic research is building on how Semantic Web principles can be applied to Web Services to how they can be applied intrinsically in business process description and management technologies e.g. through projects such as SUPER [383]. While the research for the lower layers of the architecture stack has progressed a lot, research on the higher layers is moving into the active research phase.
- **Standardization.** SAWSDL is a W3C recommendation that extends WSDL to support a minimal but useful set of semantic annotations. Given the popularity of WSDL, there is a good chance for traction with SAWSDL. Another standard that has become a W3C Recommendation is GRDDL which seeks to transform the current web, i.e., XML, Microformats and RDFa documents to RDF.

All in all, the technology around the Semantic Web and Semantic Web Services is being taken up not only in research, but also in product development and standardization. What remains is the clear identification of some "killer" applications that will take over a specific area completely.

---

28. http://technet.microsoft.com/en-us/library/bb726971.aspx (Accessed March 21, 2008).
29. http://www.ontoprise.de/ (Accessed March 21, 2008).
30. http://www.topquadrant.com/topbraid/composer/ (Accessed March 21, 2008).
31. http://www.cyc.com/ (Accessed March 21, 2008).
32. http://www.siderean.com/ (Accessed March 21, 2008).

14.2 The Bad - Major Obstacles to Overcome

Not all is good and in the area of Semantic Web and Semantic Web Services some open issues remain and some tough problems are still to be addressed and resolved. Some of these are

- **Efficiency, speed, security.** Semantic Web Service expectations are high but there are some elephants lurking in the corners. Speed in the case of the SOAP/ HTTP model; lack of support for process models in the REST HTTP/XML model (and possibly reliability and security at the message level rather than the transport protocol level); the focus on grounding is to WSDL but the focus of the WSDL binding is more or less RPC [379]. These are all infrastructure issues and so questions needs to be resolved on where the improvement in efficiency will come from what has been promised?
- **Integration problem.** Regardless of how services find each other, the processes at the interfaces of services wishing to interact need to overcome their data and process heterogeneity and have to match for a successful communication. If a solvable mismatch exists, the communication between the services needs to be mediated. A lot of the focus stays on discovery and composition for services to find each other, but data and process mediation remains largely an unsolved problem. Is the cart before the horse?
- **Absence of systems that solve problems on a real-world scale.** As is natural for any new technology, the smaller problems are solved first. Bigger and more demanding problems remain open and there is still no solution in sight for the "Proposed Challenge for Measuring the Success of SWS" [382]. As the technology matures more relevant and more complex problems require solutions.
- **Agreeing terms is hard.** The fundamental aspect of the Semantic Web is in achieving the agreement of meanings of terms declared in ontologies; and this is hard. Its likely that the most popular concepts are the ones whose definition may change the most [381] and comprehensive agreement, in any specific domain, across many communities or countries remains a big challenge.
- **Humans are still needed.** While the vision puts forward the idea that software agents find each other dynamically and figure out any mediation needed for interoperation, currently we are far from achieving this. Furthermore, for fuzzy problems humans are still in the loop and probably will for some time.
- **Learning curve:** The ability to design and understand RDF and OWL requires an understanding of logic and is akin to learning a different paradigm of programming. The description of logical expressions is not trivial and tool support is still limited. Any shortage of intuitive and user-friendly tools for computer scientists, or domain modelers, not well versed in logic-based formalisms makes the problem worse. In order to achieve a major shift that enables software engineers to easily deal with this different type of approach would also require that the curricula in educational institutions changes significantly. And this will take

time as we have seen, courses on operating systems can still be found in many universities.

While a lot of progress has been made and the Semantic Technologies are being taken up, there are major challenges remaining, as just discussed. This is not really a big problem as this is a property of a new and evolving field. At the same time it shows that a lot more work has to be accomplished in order to make this technology the basis for day-to-day software engineering.

14.3 The Ugly - Possible Prohibitors

Not surprising, there are forces that push against the progress of the new technologies and are inhibitors. In the long run it will be seen if they are effective inhibitors or just a temporal appearance.

- **Why should I share my knowledge?** Socio-economic factors dictate reality. Who will develop the common ontologies and will they invest the effort and then allow them to be used for free? Commercial systems such as RosettaNet and EDI operate a different model as they address a very specific need in very specific industries. Existing environments also are not attempting to solve a general problem, but a very specific one.
- **Logical reasoners are slow.** Reasoners are slow and this is unlikely to change dramatically soon. If performance is a real and hard requirement in many situations then the use of reasoners is not possible *now*. This is somewhat similar to the situation of logic programming that never became mainstream as the performance and the tooling was not useful in day-to-day software engineering (in addition to the lack of education).
- **The "dirty" Semantic Web.** The Semantic Web is evolving and co-exists with the current Web. It will not be a a clean formalized Semantic Web as expressed in the vision and as can be seen with Swoogle. In fact, Semantic Web content will be a mixture of HTML, XML, RDF and OWL markup expressions in addition to many documents containing proprietary formalizations. This is similar to the situation where it takes a long time for legacy systems to die out and be replaced by new technologies (and no end in sight here, either).
- **The Data as opposed to the Semantic Web.** The Semantic Web is more likely to be a data web rather than a Semantic Web, where high levels of data standardization will likely be achieved within the context of a domain or business vertical, not across all areas of human knowledge. The data web will not be formal or clean, but in fact characterized by errors and inconsistencies as the current web is characterized today, for e.g., Error 404.
- **Co-existing Semantic Web islands.** There will be multiple Semantic Web islands created in the context of multiple domains, industries and areas of activity where the value proposition is clearly understood, articulated and communicated. These domains will have standardized vocabularies, information models

and ontologies developed in a bottom up manner. Interoperation across these islands will be achieved but in a hard-coded one-off manner driven by specific business needs and consumer demands which will result in cross-maps across the vocabularies generated in these islands.

- **Dominance of textual and unstructured content.** The majority of the Semantic Web content will still be textual and unstructured with a thin metadata layer, if at all. The situation will be similar to the current web where a significant portion of the data is unstructured. However, as the value proposition is realized, the percentage of structured metadata descriptions will increase with the benefit that can be realized.
- **The prevalence of "uncertainty".** In contrast to the "crisp" and formal notion of relationships proposed in the current Semantic Web vision, data and relationships between the data will be characterized by uncertainty. This uncertainty will be due to the inconsistent and incomplete nature of information present on the web and also due to the lack of knowledge as seen in rapidly evolving domains such as the Healthcare and the Life Sciences.
- **Business as usual**. Probably one of the biggest inhibitors is the human nature of doing business as usual and rejecting change. It can be currently observed that Semantic Technology starts moving into specialized areas and specific technologies. Not broad application across all areas of computer science can be seen. This has many reasons, like the human nature of rejecting change, but also the behavior in research and industry to take up new developments only in their specific fields. So we can see Semantic Databases, and Semantic Languages, etc., but not the formation of a community around Semantic Computing that fundamentally revisits all aspects of computer science.

Time will tell how the whole area of the Semantic Web and Semantic Web Services is going to succeed. There is a lot of progress being made based on a very powerful vision. Still, problems have to be solved and some potentially powerful inhibitors exist. For sure, specific areas and domains will greatly benefit from the new technology and are already doing it. So partial success is achieved. However, based on the 5-5-5 rule, it'll take quite a number of years to come to have Semantic Technology be omnipresent in computer science.

Part VI
References and Index

References

1. Kashyap, V., and A. Sheth. Semantics-based Information Brokering. Proceedings of the Third International Conference on Information and Knowledge Management, 1994
2. Berners-Lee, T., J. Hendler and O. Lassila. The Semantic Web. Scientific American, May 2001
3. http://www.translational-medicine.com/info/about (Accessed September 10, 2007)
4. http://webster.com/dictionary/semantics (Accessed September 10, 2007)
5. Kashyap, V., E. Neumann and T. Hongsermeier. Tutorial on Semantics in the Healthcare and LifeSciences. The 15th International World Wide Conference (WWW 2006), 2006. http://lists.w3.org/Archives/Public/www-archive/2006Jun/att-0010/Semantics_for_HCLS.pdf (Accessed September 10, 2007)
6. American College of Cardiology. http://www.acc.org (Accessed September 10, 2007)
7. SNOMED International. http://www.ihtsdo.org/our-standards/snomed-ct/ (Accessed September 10, 2007)
8. International Classification of Diseases. http://www.who.int/classifications/icd/en/ (Accessed September 10, 2007)
9. BioPAX. http://www.biopax.org (Accessed September 10, 2007)
10. The Gene Ontology. http://www.geneontology.org (Accessed September 10, 2007)
11. Lawrence, S., and C. L. Giles. Accessibility of Information on the Web. Nature Magazine. Vol. 400, No. 107, 1999
12. Kashyap, V. Information Modeling on the Web. The Role of Metadata, Semantics and Ontologies. Practical Handbook of Internet Computing. CRC Press, 2004
13. Boll, S., W. Klas, and A. Sheth. Overview on Using Metadata to manage Multimedia Data. A. Sheth and W. Klas (editors): Multimedia Data Management. McGraw-Hill, 1998
14. The Standard Generalized Markup Language. http://www.w3.org/MarkUp/SGML (Accessed September 10, 2007)

15. Nelson, S. J., W. D. Johnston, and B. L. Humphreys. Relationships in Medical Subject Headings (MeSH). C. A. Bean and R. Green (editors): Relationships in the Organization of Knowledge. Kluwer Academic Publishers, 2001
16. Lindbergh, D., B. Humphreys, and A. McCray. The Unified Medical Language System. Methods Inf. Med., Vol. 32, No. 4, 1993. http://umlsks.nlm.nih.gov (Accessed September 10, 2007)
17. Goldberg, H., M. Vashevko, A. Postilnik, K. Smith, N. Plaks, B. Blumenfeld. Evaluation of a Commercial Rules Engine as a basis for a Clinical Decision Support Service. Proceedings of the Annual Symposium on Biomedical and Health Informatics, AMIA, 2006
18. Kashyap, V., A. Morales and T. Hongsermeier. Implementing Clinical Decision Support: Achieving Scalability and Maintainability by combining Business Rules with Ontologies. Proceedings of the Annual Symposium on Biomedical and Health Informatics, AMIA, 2006
19. Bohm, K., and T. Rakow. Metadata for Multimedia Documents. In [27]
20. Chen, F., M. Hearst, J. Kupiec, J. Pederson, and L. Wilcox. Metadata for Mixed-Media Access. In [27]
21. Collet, C., M. Huhns, and W. Shen. Resource Integration using a Large Knowledge Base in Carnot. IEEE Computer, December 1991
22. Deerwester, S., S. Dumais, G. Furnas, T. Landauer, and R. Hashman. Indexing by Latent Semantic Indexing. Journal of the American Society for Information Science. Vol. 41, No. 6, 1990
23. Glavitsch, U., P. Schauble, and M. Wechsler. Metadata for Integrating Speech Documents in a Text Retrieval System. In [27]
24. Jain, R., and A. Hampapur. Representations of Video Databases. In [27]
25. Kahle, B., and A. Medlar. An Information System for Corporate Users: Wide Area Information Servers. Connexions — The Interoperability Report. Vol. 5, No. 11, November 1991
26. Kiyoki, Y., T. Kitagawa, and T. Hayama. A meta-database System for Semantic Image Search by a Mathematical Model of Meaning. In [27]
27. Klaus, W., and A. Sheth (editors): Metadata for digital media. SIGMOD Record, special issue on Metadata for Digital Media. Vol 23, No. 4, December 1994
28. The PubMed MEDLINE system. http://www.ncbi.nlm.nih.gov/entrez/query.fcgi?db=PubMed (Accessed September 10, 2007)
29. Ordille, J., and B. Miller. Distributed Active Catalogs and Meta-Data Caching in Descriptive Name Services. Proceedings of the 13th International Conference on Distributed Computing Systems, 1993
30. Anderson, J., and M. Stonebraker. Sequoia 2000 Metadata Schema for Satellite Images, In [27]
31. Shklar, L., A. Sheth, V. Kashyap, and K. Shah. Infoharness: Use of Automatically Generated Metadata for Search and Retrieval of Heterogeneous Information. Proceedings of CAiSE '95. Lecture Notes in Computer Science #932, 1995

32. Sciore, E., M. Siegel, and A. Rosenthal. Context Interchange using Meta-Attributes. Proceedings of the CIKM, 1992
33. Kashyap, V., and A. Sheth. Semantics-based Information Brokering. Proceedings of the Third International Conference on Information and Knowledge Management (CIKM), 1994
34. Mena, E., V. Kashyap, A. Sheth, and A. Illarramendi. OBSERVER: An approach for query processing in global information systems based on interoperation across pre-existing ontologies. Proceedings of the First IFCIS International Conference on Cooperative Information Systems (CoopIS '96), 1996
35. Sheth, A., and V. Kashyap. Media-independent Correlation of Information. What? How? Proceedings of the First IEEE Metadata Conference, 1996
36. Shoens, K., A. Luniewski, P. Schwartz, J. Stamos, and J. Thomas. The Rufus System: Information Organization for Semi-Structured Data. Proceedings of the 19th VLDB Conference, 1993
37. Gruber, T. A translation approach to portable ontology specifications. International Journal of Knowledge Acquisition for Knowledge-Based Systems, Vol. 5, No. 2, June 1993
38. Codd, E. F. A relational model of data for large shared data banks. Communications of the ACM, Vol. 13, No. 6, June 1970
39. Chen, P. P. The entity relationship model — toward a unified view of data. ACM Transactions on Database Systems. Vol. 1, No. 1, March 1976
40. Unified Modeling Language. http://www.uml.org (Accessed September 10, 2007)
41. Health Level 7 (HL7). http://www.hl7.org/library/standards_non1.htm#HL7%20Version%203 (Accessed September 10, 2007)
42. Resource Description Framework. http://www.w3.org/RDF (Accessed September 10, 2007)
43. Extensible Markup Language. http://www.w3.org/XML (Accessed September 10, 2007)
44. Kyoto Encyclopedia of Genes and Genomes. http://www.genome.jp/kegg (Accessed September 10, 2007)
45. OWL Web Ontology Language Overview. http://www.w3.org/TR/owl-features (Accessed September 10, 2007)
46. Goble, C. A., R. Stevens, G. Ng, S. Bechhofer, N. W. Paton, P. G. Baker, M. Peim, and A. Brass. Transparent Access to Multiple Bioinformatics, Information Sources. IBM Systems Journal Special Issue on Deep computing for the Life Sciences. Vol. 40, No. 2, 2001
47. Biological Pathways Exchange. BioPAX, http://www.biopax.org (Accessed September 10, 2007)
48. Spackman, K. A., R. Dionne R, E. Mays, and J. Weis. Role grouping as an extension to the description logic of Ontylog, motivated by concept modeling in SNOMED. Proceedings of the Annual Symposium on Biomedical Informatics, AMIA, 2002

49. IEEE Standard Upper Ontology. http://suo.ieee.org (Accessed September 10, 2007)

50. Knowledge Interchange Format (KIF). http://ksl.stanford.edu/knowledge-sharing/kif/ (Accessed September 10, 2007)

51. Cyc Ontology. http://research.cyc.com (Accessed September 10, 2007)

52. Patel-Schneider, P., and J. Simeon. The Yin/Yang Web: XML Syntax and RDF Semantics. Proceedings of the 11th International World Wide Web Conference (WWW 2002), 2002

53. Hayes, P. RDF Semantics. http://www.w3.org/TR/rdf-mt/ (Accessed September 10, 2007)

54. OWL Web Ontology Language Semantics and Abstract Syntax. http://www.w3.org/TR/owl-semantics/ (Accessed September 10, 2007)

55. Deutsch, A., M. Fernandez, D. Florescu, A. Levy, and D. Suciu. A Query Language for XML. Proceedings of the 8th International World Wide Web Conference (WWW 1999), 1999

56. Lassila, O., and R. Swick. Resource description framework (RDF) model and syntax specification. W3C Recommendation, 1999. http://www.w3.org/TR/REC-rdf-syntax/ (Accessed September 10, 2007)

57. Boley, H. A web data model unifying XML and RDF. Unpublished draft, 2001. http://www.dfki.uni-kl.de/~boley/xmlrdf.html (Accessed September 10, 2007)

58. Patel-Schneider, P. F., and I. Horrocks. Mapping RDF Graphs to OWL. http://www.w3.org/TR/2004/REC-owl-semantics-20040210/mapping.html (Accessed September 10, 2007)

59. OWL Web Ontology Language Reference. http://www.w3.org/TR/owl-ref/ (Accessed September 10, 2007)

60. Patel-Schneider, P. F., P. Hayes, and I. Horrocks. RDF-Compatible Model Theoretic Semantics. http://www.w3.org/TR/owl-semantics/rdfs.html (Accessed September 10, 2007)

61. Bonifati, A., and S. Ceri. Comparative Analysis of Five XML Query Languages. SIGMOD Record 29(1): 68-79, 2000

62. Abiteboul, S., D. Quass, J. McHugh, J. Widom, J. Wiener, and J. Widom. The Lorel Query Language for Semistructured Data. Intenational Journal on Digital Libraries (IJDL). Vol. 1, No. 1, April 1997

63. Ceri, S., S. Comai, E. Damiani, P. Fraternali, S. Paraboschi, and L. Tanca. XML-GL - a Graphical Language for Querying and Restructuring WWW Data. International World Wide Web Conference (WWW), 1999

64. Boag, S., D. Chamberlin, M. F. Fernandez, D. Florescu, J. Robie, and J. Simeon (editors): XQuery 1.0: An XML Query Language. W3C Recommendation, January 23, 2007. http://www.w3.org/TR/xquery/ (Accessed September 10, 2007)

65. Clark, J. XML Transformations (XSLT). Version 1.0. W3C Recommendation November, 1999. http://www.w3.org/TR/xslt (Accessed September 10, 2007)

66. Derksen, E., P. Fankhauser, E. Howland, G. Huck, I. Macherius, M. Murata, M. Resnick, and H. Schöning, J. Robie (editors): XQL (XML Query Language), 1999. http://www.ibiblio.org/xql/xql-proposal.html (Accessed September 10, 2007)
67. Clark, J., and S. DeRose (editors): XML Path Language (XPath). Version 1.0. http://www.w3.org/TR/xpath (Accessed September 10, 2007)
68. DeRose, S., E. Maler, and D. Orchard. XML Linking (XLink). Version 1.0. W3C Recommendation, 2001. http://www.w3.org/TR/xlink (Accessed September 10, 2007)
69. DeRose, S., R. Daniel Jr., P. Grosso, E. Maler, J. Marsh, and N. Walsh. XML Pointer Language (XPointer). W3C Working Draft, 2002. http://www.w3.org/TR/xptr (Accessed September 10, 2007)
70. Chamberlin, D. XQuery: An XML Query Language. IBM Systems Journal. Vol. 41, No. 4, 2002
71. Chamberlin, D., J. Robie, and D. Florescu. Quilt: An XML Query Language for Heterogeneous Data Sources. Lecture Notes in Computer Science, 2000. http://www.almaden.ibm.com/cs/people/chamberlin/quilt.html (Accessed September 10, 2007)
72. Atwood, T., D. Barry, J. Duhl, J. Eastman, G. Ferran, D. Jordan, M. Loomis, and D. Wade. The Object Database Standard: ODMG-93. Release 1.2. R. G. C. Catell (editor). Morgan Kaufmann Publishers, 1996
73. Information Technology-Database Language SQL. Standard No. ISO/IEC 9075. International Organization for Standardization (ISO), 1999
74. XML Query Use Cases. W3C Working Group Note, 23 March 2007. http://www.w3.org/TR/xmlquery-use-cases (Accessed September 10, 2007)
75. Karvounarakis, G., S. Alexaki, V. Christophides, D. Plexousakis, and M. Schol. RQL: A Declarative Query Language for RDF. Proceedings of the Eleventh International World Wide Web Conference (WWW'02), 2002
76. Broekstra, J., and A. Kampman. SeRQL: An RDF Query and Transformation Language. Proceedings of the SWAD-Europe workshop on Semantic Web Storage and Retrieval, 2003
77. Sintek, M., and S. Decker. TRIPLE - an RDF query, inference and transformation language. Deductive Databases and Knowledge Management (DDLP), 2001
78. Seaborne, A. RDQL: A Query Language for RDF. W3C Member Submission. http://www.w3.org/Submission/2004/SUBM-RDQL-20040109/ (Accessed September 10, 2007)
79. Berners-Lee, T. (editor): Notation 3 (N3), A Readable RDF Syntax. http://www.w3.org/DesignIssues/Notation3 (Accessed September 10, 2007)
80. Olson, M., and U. Obguji. Versa. http://copia.ogbuji.net/files/Versa.html (Accessed September 10, 2007)

81. Haase, P., J. Broekstra, A. Eberhart, and R. Volz. A Comparison of RDF Languages. Proceedings of the 3rd International Semantic Web Conference (ISWC), 2004
82. Kifer, M., G. Lausen, and J. Wu. Logical foundations of object-oriented and frame-based languages. Journal of the ACM, Vol. 42, 1995
83. Prud'hommeaux, E., and A. Seaborne (editors): SPARQL query language for RDF. W3C Proposed Recommendation, 12 November 2007. http://www.w3.org/TR/rdf-sparql-query (Accessed November 24, 2007)
84. Beckett, D., and J. Broekstra (editors): SPARQL Query Results Format. W3C Candidate Recommendation, 2006. http://www.w3.org/TR/rdf-sparql-XMLres/ (Accessed September 10, 2007)
85. Clark, K. G. (editor): SPARQL Protocol for RDF. W3C Candidate Recommendation, 2006. http://www.w3.org/TR/rdf-sparql-protocol/ (Accessed September 10, 2007)
86. Fikes, R., P. Hayes, and I. Horrocks. OWL-QL - A Language for Deductive Query Answering on the Semantic Web. Knowledge Systems Laboratory. Technical Report No. KSL-03-14, Department of Computer Science, Stanford University, 2003
87. W3C XML Schema. http://www.w3.org/XML/Schema (Accessed September 10, 2007)
88. RDF Vocabulary Description Language 1.0: RDF Schema. http://www.w3.org/TR/rdf-schema/ (Accessed September 10, 2007)
89. Gruber, T. R. A translation approach to portable ontologies. Knowledge Acquisition. Vol. 5, No. 2, 1993
90. de Keizer, N. F., A. Abu-Hanna, and J. H. Zwetsloot-Schonk. Understanding terminological systems. I: Terminology and Typology. Methods of Information in Medicine. Vol. 39, No. 1, 2000
91. Distributed Management Task Force — Common Information Model. http://www.dmtf.org/standards/cim/ (Accessed September 10, 2007)
92. DOLCE: A Descriptive Ontology for Linguistic and Cognitive Engineering. http://www.loa-cnr.it/DOLCE.html (Accessed September 10, 2007)
93. Basic Formal Ontology (BFO). http://www.ifomis.uni-saarland.de/bfo/home.php (Accessed September 10, 2007)
94. Wallside, D. C., and P. Walmsley. XML Schema Part 0: Primer. Second Edition. http://www.w3.org/TR/xmlschema-0 (Accessed September 10, 2007)
95. Thompson, H. S. Towards a logical foundation for XML Schema. Proceedings of XML Europe, 2004. http://www.ltg.ed.ac.uk/~ht/XML_Europe_2004.html (Accessed September 10, 2007)
96. Manola, F., and E. Miller (editors): RDF Primer. http://www.w3.org/TR/rdf-primer/ (Accessed September 10, 2007)
97. Smith, M. K., C. Welty, and D. L. McGuinness (editors): OWL Web Ontology Language Guide. http://www.w3.org/TR/owl-guide/ (Accessed September 10, 2007)

98. Gil, Y., and V. Ratnakar. A Comparison of (Semantic) Markup Languages. Proceedings of the 15th International FLAIRS Conference, 2002
99. Gomez Perez, A. (editor): Ontoweb Deliverable D1.3. A survey of ontology tools. http://www.deri.at/fileadmin/documents/deliverables/Ontoweb/D1.3.pdf (Accessed September 10, 2007)
100. Denny, M. Ontology Building: A survey of ontology tools. http://www.xml.com/pub/a/2002/11/06/ontologies.html (Accessed September 10, 2007)
101. The Apollo Ontology Editor. http://apollo.open.ac.uk (Accessed September 10, 2007)
102. The Link Factory Ontology Editor. http://www.landcglobal.com/pages/link-factory.php (Accessed September 10, 2007)
103. OntoStudio. http://www.ontoprise.de/content/e1171/e1249/index_eng.html (Accessed September 10, 2007)
104. Ontolingua. http://ontolingua.stanford.edu (Accessed September 10, 2007)
105. Ontosaurus. http://www.isi.edu/isd/ontosaurus.html (Accessed September 10, 2007)
106. Protege. http://protege.stanford.edu (Accessed September 10, 2007)
107. WebODE. http://webode.dia.fi.upm.es/ (Accessed September 10, 2007)
108. Arpírez, J. C., O. Corcho, M. Fernández-López, and A. Gómez-Pérez. WebODE: A Scalable Workbench for Ontological Engineering. First International Conference on Knowledge Capture (KCAP01), 2001
109. Fernández-López, M., A. Gómez-Pérez, A. Pazos, and J. Pazos. Building a Chemical Ontology Using Methontology and the Ontology Design Environment. IEEE Intelligent Systems and Their Applications. Vol. 4, No. 1, 1999
110. WebOnto. http://kmi.open.ac.uk/projects/webonto/ (Accessed September 10, 2007)
111. ICOM. http://www.inf.unibz.it/~franconi/icom/ (Accessed September 10, 2007)
112. IODE. http://www.ontologyworks.com/iode.php (Accessed September 10, 2007)
113. Visual Ontology Modeler. http://www.sandsoft.com/products.html (Accessed September 10, 2007)
114. Semtalk. http://www.semtalk.com (Accessed September 10, 2007)
115. CoBra. http://www.xspan.org/cobra/index.html (Accessed September 10, 2007)
116. Generic Knowledge Base Editor. http://www.ai.sri.com/~gkb/ (Accessed September 10, 2007)
117. SWOOP. http://code.google.com/p/swoop/ (Accessed September 10, 2007)
118. Fisher, D. H. Knowledge Acquisition via incremental conceptual clustering. Machine Learning, No. 2, 1987
119. Clerkin, P., P. Cunningham, and C. Hayes. Ontology Discovery for the Semantic Web using Hierarchical Clustering. Proceedings of the Semantic Web Mining Workshop, 2001

120. Cohen, W. W., and H. Hirsh. Learning the CLASSIC Description Logic: Theoretical and Experimental Results. Principles of Knowledge Representation and Reasoning. Proceedings of the Fourth International Conference, 1994
121. Suryanto, H. and P. Compton. Learning Classification taxonomies from a classification knowledge based system. Proceedings of Workshop on Ontology Learning, 2000
122. Maedche, A., G. Neumann, and S. Staab. Bootstrapping an Ontology Based Information Extraction System. Studies in Fuzziness and Soft Computing. Szczepaniak, J. Segovia, J. Kacprzyk, and L.A. Zadeh (editors): INTELLIGENT EXPLORATION OF THE WEB, 2003
123. Riloff, E., and J. Shepherd. A corpus-based approach for building semantic lexicons. Proceedings of the Second Conference on Empirical Methods in Natural Language Processing (EMNLP-97), 1997
124. Dill, S., N. Eiron, D. Gibson, D. Gruhl, R. V. Guha, A. Jhingran, T. Kanungo, S. Rajagopalan, A. Tomkins, J. A. Tomlin, J. Y. Zien. SemTag and Sem-Seeker: Bootstrapping the Semantic Web via automated semantic annotation. Proceedings of the 12th International WWW Conference (WWW 2003), 2003
125. Jacobs, P., and U. Zernik. Acquiring Lexical Knowledge from Text: A Case Study. Proceedings of the Seventh National Conference on Artificial Intelligence, 1988
126. Hastings, P., and S. Lytinen. The Ups and Downs of Lexical Acquisition. Proceedings of the Twelfth National Conference on Artificial Intelligence, 1994
127. Berwick, R. C. Learning Word Meanings from Examples. Semantic Structures: Advances in Natural Language Processing. Lawrence Erlbaum Associates, 1989
128. Cardie, C. A Case-based Approach to Knowledge Acquisition for Domain Specific Sentence Analysis. Proceedings of the Eleventh National Conference on Artificial Intelligence, 1993
129. Missikoff, M., P. Velardi, and P. Fabriani. Text Mining Techniques to automatically enrich a Domain Ontology. Applied Intelligence Vol. 18, 2003
130. Sanderson, M., and B. Croft. Deriving Concept Hierarchies from Text. International Conference on Research and Development in Information Retrieval (SIGIR 1999), 1999
131. Nazarenko, A., P. Zweigenbaum, J. Bouaud, and B. Habert. Corpus-based identification and refinement of semantic classes. Proceedings of the AMIA Annual Symposium, 1997
132. Lin, D. Automatic retrieval and clustering of similar words. Proceedings of COLING-ACL-98, 1998
133. Fiszman, M., T. C. Rindflesch, and H. Kilicoglu. Integrating a Hypernymic Preposition Interpreter into a Semantic Processor for Biomedical Texts. Proceedings of the AMIA Annual Symposium on Medical Informatics, 2003

134. Hearst, M. Automatic acquisition of hyponyms from large text corpora. Proceedings of the 14th International Conference on Computational Linguistics, 1992
135. Finkelstein-Landau, M., and E. Morin. Extracting Semantic Relationships between Terms: Supervised vs Unsupervised Methods. Proceedings of International Workshop on Ontological Engineering on the Global Information Infrastructure, 1999
136. Maedche, A., and S. Staab. Discovering conceptual relations from text. Technical Report 399, Institute AIFB, Karlsruhe University, 2000
137. Everitt, B. S., S. Landau, and M. Leese. Cluster Analysis. Edward Arnold. 4th Edition, May 2001
138. Zhang, Y., and G. Karypis. Criterion functions for Document Clustering. Technical Report, U. Minnesota, Dept. of Computer Science, #TR-01-40, 2002
139. Chakrabarti, S. Data Mining for Hypertext: A Tutorial Survey. ACM SIGKDD Explorations, Vol. 1, No. 2, 2000
140. Rasmussen, E. Clustering Algorithms. W. B. Frakes and R. Baeza-Yates (editors): Information Retrieval: Data Structures and Algorithms. Prentice Hall, 1992
141. Cutting, D. R., D. R. Karger, J. O. Pedersen, and J. W. Tukey. Scatter/Gather: A cluster-based approach to browsing large document collections. Annual International Conference on Research and Development on Information Retrieval, 1992
142. Zamir, O., and O. Etzioni. Web Document Clustering: A Feasibility Demonstration. Proceedings of ACM SIGIR Conference, 1998
143. Buckley, C., M. Mitra, J. Walz, and C. Cardie. Using clustering and superconcepts within SMART: TREC 6. Sixth Test Retrieval Conference (TREC-6), 1997
144. Maedche, A., and S. Staab. Ontology learning for the Semantic Web. IEEE Intelligent Systems, Vol. 16, 2001
145. Kashyap, V., C. Ramakrishnan, C. Thomas, and A. Sheth. TaxaMiner: An Experimental Framework for Automated Taxonomy Bootstrapping. International Journal of Web and Grid Services. Special Issue on Semantic Web and Mining Reasoning, 2005
146. Davulcu, H., S. Vadrevu, and S. Nagarajan. OntoMiner: Bootstrapping and Populating Ontologies from Domain Specific Websites. Proceedings of the First International Workshop on Semantic Web and Databases (SWDB 2003), 2003
147. Mena, E., A. Illarramendi, V. Kashyap, and A. Sheth. OBSERVER: An Approach for Global Query Processing in Global Information Systems based on Interoperation across Pre-existing Ontologies. Distributed and Parallel Databases. Vol. 8, No. 2, 2000
148. Mena, E., V. Kashyap, A. Illarramendi, and A. Sheth. Imprecise Answers in Distributed Environments: Estimation of Information Loss for Multiple

Ontology based Query Processing. International Journal of Cooperative Information Systems (IJCIS). H. Wache and D. Fensel (editors): Special Issue on Intelligent Integration of Information, Vol. 9, No. 4, 2000

149. Chimaera. http://ksl.stanford.edu/software/chimaera/ (Accessed September 10, 2007)
150. Noy, N. F., and M. A. Musen. PROMPT: Algorithm and Tool for Automated Ontology Merging and Alignment. Seventeenth National Conference on Artificial Intelligence (AAAI-2000), 2000
151. Ramos, J. A. Mezcla automática de ontologías y catálogos electrónicos. Final Year Project. Facultad de Informática de la Universidad Politécnica de Madrid. Spain, 2001
152. de Diego, R. Método de mezcla de catálogos electrónicos. Final Year Project. Facultad de Informática de la Universidad Politécnica de Madrid. Spain, 2001
153. Baader, F., C. Lutz, and B. Suntisrivaraporn. CEL - A Polynomial-time Reasoner for Life Science Ontologies. U. Furbach and N. Shankar (editors): Proceedings of the 3rd International Joint Conference on Automated Reasoning (IJCAR'06). Lecture Notes in Artificial Intelligence #4130, 2006
154. The GALEN Model. http://www.opengalen.org/themodel/ontology.html (Accessed September 10, 2007)
155. OWL: FaCT++. http://owl.man.ac.uk/factplusplus (Accessed September 10, 2007)
156. The FaCT System. http://www.cs.man.ac.uk/~horrocks/FaCT/ (Accessed September 10, 2007)
157. The fuzzy DL System. http://gaia.isti.cnr.it/~straccia/software/fuzzyDL/fuzzyDL.html (Accessed September 10, 2007)
158. KAON2: Ontology Management System for the Semantic Web. http://kaon2.semanticweb.org (Accessed September 10, 2007)
159. Pellet, An Open Source OWL-DL Reasoner in Java. http://pellet.owldl.com (Accessed September 10, 2007)
160. RacerPro. http://www.racer-systems.com/products/racerpro/index.phtml (Accessed September 10, 2007)
161. Grosof, B. N., I. Horrocks, R. Volz, and S. Decker. Description Logic Programs: Combining Logic Programs with Description Logic. Proceedings of the Twelfth International Conference World Wide Web Conference (WWW 2003), 2003
162. Borgida, A. On the relative expressiveness of description logics and predicate logics. Artificial Intelligence, Vol. 82, No. 1-2, 1996
163. Donini, F. M., M. Lenzerini, D. Nardi, and W. Nutt. The complexity of concept languages. Proceedings of KR '91, 1991
164. Vardi, M. Y. Why is modal logic so robustly decidable? N. Immerman and P. Kolaitis (editors): Descriptive Complexity and Finite Models. American Mathematical Society, 1997

165. Borgida, A., and P. F. Patel-Schneider. A semantics and complete algorithm for subsumption in the CLASSIC description logic. Journal of Artificial Intelligence Research, Vol. 1, 1994
166. Patel-Schneider, P. F., D. L. McGuinness, R. J. Brachman, L. A. Resnick, and A. Borgida. The CLASSIC Knowledge Representation System: Guiding principles and implementation rationale. SIGART Bulletin, Vol. 2, No. 3, 1991
167. Levy, A. Y., and M. C. Rousset. Combining Horn Rules and Description Logics in CARIN. Artificial Intelligence, Vol. 104, No. 1-2, 1998
168. Horrocks, I., and P. Patel-Schneider. A Proposal for an OWL Rules Language. Proceedings of the Thirteenth World Wide Web (WWW) Conference, 2004
169. Padgham, I., and P. Lambrix. A framework for part-of hierarchies in terminological logics. Proceedings of the 14th International Conference on the Principles of Knowledge Representation and Reasoning (KR '94), 1994
170. Rector, A., and I. Horrocks. Experience building a large re-usable medical ontology using a description logic with transitivity and concept inclusion. Proceedings of the Workshop on Ontological Engineering, AAAI Spring Symposium, 1997
171. Uren, V., P. Cimiano, J. Iria, S. Handschuh, M. Vargas-Vera, E. Motta, and F. Ciravegna. Semantic annotation for Knowledge Management: Requirements and a survey of the state of the art. Web Semantics: Science, Services and Agents on the World Wide Web. Vol. 4, No. 1, 2005
172. Kahan, J., M.-J. Koivunen, E. Prud'Hommeaux, and R. Swick. Annotea: An open RDF infrastructure for shared web annotations. Proceedings of the 10th International World Wide Web Conference (WWW 2001), 2001
173. Handschuh, S., and S. Staab. Authoring and annotation of web pages in CREAM. Proceedings of the 11th International World Wide Web Conference (WWW 2002), 2002
174. Quint, V., and I. Vatton. An Introduction to Amaya, W3C Note, 1997
175. McDowell, L., O. Etzioni, S. Gribble, A. Halevy, H. Levy, W. Pentney, D. Verma, and S. Vlasseva. Enticing ordinary people onto the Semantic Web via instant gratification. Proceedings of the 2nd International Semantic Web Conference (ISWC 2003), 2003
176. McDowell, L., O. Etzioni, and A. Halevy. Semantic email: theory and applications. Journal of Web Semantics. Vol. 2, No. 2, 2004
177. Schroeter, R., J. Hunter, and D. Kosovic. Vannotea, A collaborative video indexing, annotation and discussion system for broadband networks. Proceedings of the K-CAP 2003 Workshop on "Knowledge Markup and Semantic Annotation", 2003
178. Hunter, J., R. Schroeter, B. Koopman, and M. Henderson. Using the semantic grid to build bridges between museums and indigenous communities. Proceedings of the GGF11—Semantic Grid Applications Workshop, 2004
179. Annozilla annotator. http://annozilla.mozdev.org (Accessed September 10, 2007)

180. Teknowledge Annotation Applications. http://mr.teknowledge.com/DAML/ (Accessed September 10, 2007)

181. Handschuh, S., S. Staab, and R. Studer. Leveraging metadata creation for the Semantic Web with CREAM. Proceedings of the Annual German Conference on AI, 2003

182. Ciravegna, F., and Y. Wilks. Designing adaptive information extraction for the Semantic Web in amilcare. S. Handschuh and S. Staab (editors): Annotation for the Semantic Web. Frontiers in Artificial Intelligence and Applications, IOS Press, Amsterdam, 2003

183. Volz, R., S. Handschuh, S. Staab, L. Stojanovic, and N. Stojanovic. Unveiling the hidden bridge: deep annotation for mapping and migrating legacy data to the Semantic Web. Journal of Web Semantics. Vol. 1, No. 2, 2004

184. Bloehdorn, S., K. Petridis, C. Saathoff, N. Simou, V. Tzouaras, Y. Avrithis, S. Handschuh, Y. Kompatsiaris, S. Staab, and M.G. Strintzis. Semantic annotation of images and videos for multimedia analysis. Proceedings of the 2nd European Semantic Web Conference (ESWC 2005), 2005

185. SMORE: Semantic Markup, Ontology and RDF Editor. http://www.ece.umd.edu/~adityak/editor.html (Accessed September 10h, 2007)

186. Heflin, J., and J. Hendler. A portrait of the Semantic Web in action. IEEE Intelligent Systems. Vol. 16, No. 2, 2001

187. Collier, N., A. Kawazoe, A.A. Kitamoto, T. Wattarujeekrit, T.Y. Mizuta, and A. Mullen. Integrating deep and shallow semantic structures in open ontology forge. Proceedings of the Special Interest Group on Semantic Web and Ontology. JSAI (Japanese Society for Artificial Intelligence). Vol. SIG-SWO-A402-05, 2004

188. Bechhofer, S., and C. Goble. Towards annotation using DAML+ OIL. Proceedings of the Workshop on Semantic Markup and Annotation, 2001

189. Bechhofer, S., C. Goble, L. Carr, W. Hall, S. Kampa, and D. De Roure. COHSE: conceptual open hypermedia service, S. Handschuh and S. Staab (editors): Annotation for the Semantic Web, IOS Press, Amsterdam, 2003

190. Plessers, P., S. Casteleyn, Y. Yesilada, O. De Troyer, R. Stevens, S. Harper, and C. Goble. Accessibility: a web engineering approach. Proceedings of the 14th International World Wide Web Conference (WWW2005), 2005

191. Baumgartner, R., R. Flesca, and G. Gottlob. Visual web information extraction with Lixto. Proceedings of the International Conference on Very Large Data Bases (VLDB), 2001

192. Vargas-Vera M., E. Motta, J. Domingue, M. Lanzoni, A. Stutt, and F. Ciravegna. MnM: A tool for automatic support on semantic markup. KMi Technical Report Number 133, 2003

193. Ciravegna, F., A. Dingli, D. Petrelli, and Y. Wilks. User-system cooperation in document annotation based on information. Proceedings of the 13th International Conference on Knowledge Engineering and KM (EKAW02), 2002

194. Gilardoni, L., M. Biasuzzi, M. Ferraro, R. Fonti, and P. Slavazza. Machine learning for the Semantic Web: putting the user in the cycle. Proceedings of the Dagstuhl Seminar Machine Learning for the Semantic Web, 2005

195. Black, W. J., J. McNaught, A. Vasilakopoulos, K. Zervanou, B. Theodoulidis, and F. Rinaldi. CAFETIERE conceptual annotations for facts, events, terms, individual entities, and relations. Parmenides Technical Report, TR-U4.3.1, 2005

196. Vasilakopoulos, A., M. Bersani, and W.J. Black. A Suite of Tools for Marking Up Textual Data for Temporal Text Mining Scenarios. Proceedings of the 4th International Conference on Language Resources and Evaluation (LREC-2004), 2004

197. Siliopoulou, M., F. Rinaldi, W.J. Black, G.P. Zarri, R.M. Mueller, M. Brunzel, B. Theodoulidis, G. Orphanos, M. Hess, J. Dowdall, J. McNaught, M. King, A. Persidis, and L. Bernard. Coupling information extraction and data mining for ontology learning in PARMENIDES. Proceedings of the Recherche d'Information Assist´ee par Ordinateur (RIAO'2004), 2004

198. Ciravegna, F., S. Chapman, A. Dingli, and Y. Wilks. Learning to harvest information for the Semantic Web. Proceedings of the 1st European Semantic Web Symposium, 2004

199. Etzioni, O., M.J. Cafarella, D. Downey, A.-M. Popescu, T. Shaked, S. Soderland, D.S. Weld, and A. Yates. Unsupervised named-entity extraction from the Web: an experimental study. Artificial Intelligence. Vol. 165, No. 1, 2005

200. Buitelaar, P., and S. Ramaka. Unsupervised ontology based semantic tagging for knowledge markup. Proceedings of the Workshop on Learning in Web Search, 2005

201. Cimiano, P., S. Handschuh, and S. Staab. Towards the self-annotating web. Proceedings of the 13th International World Wide Web Conference (WWW 2004), 2004

202. Kogut, P., and W. Holmes. AeroDAML: applying information extraction to generate DAML annotations from web pages. Proceedings of the Workshop on Knowledge Markup and Semantic Annotation, 2001

203. Popov, B., A. Kiryakov, D. Ognyanoff, D. Manov, A. Kirilov, and M. Goranov. Towards Semantic Web information extraction. Proceedings of the Human Language Technologies Workshop, 2003

204. Popov, B., A. Kirayakov, D. Ognyanoff, D. Manov, and A. Kirilov. KIM—a semantic platform fo information extraction and retrieval. Natural Language Engineering. Vol. 10, No. 3-4, 2004

205. Svatek, V., M. Labsky, and M. Vacura. Knowledge modelling for deductive web mining. Proceedings of the 14th International Conference on Knowledge Engineering and Knowledge Management (EKAW 2004), 2004

206. Maynard, D., M. Yankova, A. Kourakis, and A. Kokossis. Ontology-based information extraction for market monitoring and technology watch. Proceedings of the Workshop on User Aspects of the Semantic Web (UserSWeb), 2005

207. Dowman, M., V. Tablan, H. Cunningham, and B. Popov. Web-assisted annotation, semantic indexing and search of television and radio news. Proceedings of the 14th International World Wide Web Conference (WWW2005), 2005

208. Rinaldi, F., G. Schneider, K. Kaljurand, J. Dowdall, A. Persidis, and O. Konstanti. Mining relations in the GENIA corpus. Second European Workshop on Data Mining and Text Mining for Bioinformatics, 2004

209. Maynard, D., M. Yankova, N. Aswani, and H. Cunningham. Automatic creation and monitoring of semantic metadata in a dynamic knowledge portal. Proceedings on the Artificial Intelligence: Methodology, Systems, Applications (AIMSA 2004), 2004

210. Svab, O., M. Labsky, and V. Svatek. RDF-based retrieval of information extracted from web product catalogues. Proceedings of the SIGIR'04 Semantic Web Workshop, 2004

211. Carr, L., T. Miles-Board, A. Woukeu, G. Wills, and W. Hall. The case for explicit knowledge in documents. Proceedings of the ACM Symposium on Document Engineering (DocEng '04), 2004

212. Lanfranchi, V., F. Ciravegna, and D. Petrelli. Semantic Web-based document: editing and browsing in AktiveDoc. Proceedings of the 2nd European Semantic Web Conference, 2005

213. Tallis, M. SemanticWord processing for content authors. Proceedings of the Knowledge Markup and Semantic Annotation Workshop (SEMANNOT 2003), 2003

214. Dzbor, M., E. Motta, and J. Domingue. Opening up magpie via semantic services. Proceedings of the 3rd International Semantic Web Conference, 2004

215. Hogue, A., D. Karger. Thresher: automating the unwrapping of semantic content from the world wide web. Proceedings of the 14th International World Wide Web Conference (WWW2005), 2005

216. Huynh, D., D. Kerger, and D. Quan. Haystack: a platform for creating, organizing and visualizing information using RDF. Proceedings of the 11th International World Wide Web Conference (WWW2002), 2002

217. Rahm, E., and P. Bernstein. A survey of approaches to automatic schema matching. The VLDB Journal. Vol. 10, No. 4, 2001

218. Shvaiko, P., and J. Euzenat. A Survey of Schema-based Matching approaches. Journal of Data Semantics. Vol. 4, 2005

219. Cohen, W., P. Ravikumar, and S. Fienberg. A comparison of string metrics for matching names and records. Proceedings of the workshop on Data Cleaning and Object Consolidation, 2003

220. Aumuller, D., H. H. Do, S. Massmann, and E. Rahm. Schema and ontology matching with COMA++. In Proceedings of the International Conference on Management of Data (SIGMOD), 2005

221. Bouquet, P., L. Serafini, and S. Zanobini. Semantic coordination: A new approach and an application. Proceedings of the International Semantic Web Conference (ISWC), 2003

222. Castano, S., V. De Antonellis, and S. De Capitani di Vimercati. Global viewing of heterogeneous data sources. IEEE Transactions on Knowledge and Data Engineering, Vol. 13, No. 2, 2001

223. Di Noia, T., E. Di Sciascio, F. M. Donini, and M. Mongiello. A system for principled matchmaking in an electronic marketplace. Proceedings of the World Wide Web Conference (WWW), 2003

224. Dieng, R., and S. Hug. Comparison of "personal ontologies" represented through conceptual graphs. Proceedings of the European Conference on Artificial Intelligence (ECAI), 1998

225. Do, H. H., and E. Rahm. COMA - a system for flexible combination of schema matching approaches. Proceedings of the Very Large Data Bases Conference (VLDB), 2001

226. Ehrig, M., and S. Staab. QOM: Quick ontology mapping. Proceedings of the International Semantic Web Conference (ISWC), 2004

227. Ehrig, M., and Y. Sure. Ontology mapping - an integrated approach. In Proceedings of the European Semantic Web Symposium (ESWS), 2004

228. Euzenat, J., and P.Valtchev. Similarity-based ontology alignment in OWL-lite. Proceedings of the European Conference on Artificial Intelligence (ECAI), 2004.

229. Giunchiglia, F., and P. Shvaiko. Semantic matching. The Knowledge Engineering Review Journal (KER), Vol. 18, No. 3, 2003

230. Giunchiglia, F., P. Shvaiko, and M. Yatskevich. S-Match: an algorithm and an implementation of semantic matching. Proceedings of the European Semantic Web Symposium (ESWS), 2004

231. Giunchiglia, F., P. Shvaiko, and M. Yatskevich. Semantic schema matching. Technical Report DIT-05-014, University of Trento, 2005

232. Madhavan, J., P. Bernstein, and E. Rahm. Generic schema matching with Cupid. Proceedings of the Very Large Data Bases Conference (VLDB), 2001

233. Melnik, S., H. Garcia-Molina, and E. Rahm. Similarity flooding: A versatile graph matching algorithm. Proceedings of the International Conference on Data Engineering (ICDE), 2002

234. Noy, N., and M. Musen. Anchor-PROMPT: using non-local context for semantic matching. Proceedings of the workshop on Ontologies and Information Sharing, 2001

235. Rahm, E., H. H. Do, and S. Maßmann. Matching large XML schemas. SIGMOD Record, Vol. 33, No. 4, 2004

236. Sotnykova, A., C. Vangenot, N. Cullot, N. Bennacer, and M.-A. Aufaure. Semantic mappings in description logics for spatio-temporal database schema integration. Journal on Data Semantics (JoDS). Special Issue on Semantic-based Geographical Information Systems, III, 2005

237. Miller, A. G. WordNet: A lexical database for English. Communications of the ACM, Vol. 38, No. 11, 1995

238. Bergamaschi, S., S. Castano, and M. Vincini. Semantic integration of semistructured and structured data sources. SIGMOD Record, Vol. 28, No. 1, 1999

239. Uschold, M., and M. Gruniger. Ontologies: Principles, methods and applications. Knowledge Engineering Review, Vol. 11, No. 2, 1996
240. Wache, H., T. Vogele, U. Visser, H. Stuckenschmidt, G. Schuster, H. Neumann and S. Hubner, IJCAI Workshop on Ontologies and Information Sharing, 2001
241. Arens, Y., C. Y. Chee, C. Hsu, and C. A. Knoblock. Retrieving and integrating data from multiple information sources. International Journal of Intelligent and Cooperative Information Systems. Vol. 2, No. 2, 1993
242. Garcia-Molina, H., Y. Papakonstantinou, D. Quass, A. Rajaraman, Y. Sagiv, J. Ullman, and J. Widom. The TSIMMIS Approach to Mediation: Data Models and Languages. Proceeding of NGITS (Next Generation Information Technologies and System), 1995
243. Bayardo, R., W. Bohrer, R. Brice, A. Cichocki, G. Fowler, A. Helal, V. Kashyap, T. Ksiezyk, G. Martin, M. Nodine, M. Rashid, M. Rusinkiewicz, R. Shea, C. Unnikrishnan, A. Unruh, and D. Woelk. Infosleuth: Semantic Integration of Information in Open and Dynamic Environments. Proceedings of the 1997 ACM International Conference on the Management of Data (SIGMOD), 1997
244. Preece, A. D., K.-J. Hui, W.A. Gray, P. Marti, T.J.M. Bench-Capon, D.M. Jones, and Z. Cui. The KRAFT architecture for knowledge fusion and transformation. Proceedings of the 19th SGES International Conference on Knowledge-Based Systems and Applied Artificial Intelligence (ES'99), 1999.
245. Goasdoue, F., V. Lattes, and M. Rousset. The use of Carin language and algorithms for Information Integration: The PICSEL project. International Journal of Cooperative Information Systems (IJCIS). Vol. 9, No. 4, 1999
246. Calvanese, D., G. DeGiacomo, and M. Lenzerini. Description logics for information integration. Computational Logic: From Logic Programming into the Future (In honour of Bob Kowalski). Lecture Notes in Computer Science #2408, 2001
247. Decker, S., M. Erdmann, D. Fensel, and R. Studer. Ontobroker: Ontology based access to distributed and semi-structured information. R. Meersman, Z. Tari, S. Stevens (editors): Semantic Issues in Multimedia Systems. Proceedings of DS-8, 1999
248. Heflin, J., and J Hendler. Dynamic ontologies on the web. Proceedings of American Association for Artificial Intelligence Conference (AAAI-2000), 2000
249. Goh, C. H. Representing and Reasoning about Semantic Conflicts in Heterogeneous Information Sources. Phd Thesis, MIT, 1997
250. Wache, H., Th. Scholz, H. Stieghahn and B. Konig-Ries. An integration method for the specification of rule–oriented mediators. Y. Kambayashi and H. Takakura (editors): Proceedings of the International Symposium on Database Applications in Non-Traditional Environments (DANTE'99), 1999

251. Stuckenschmidt, H., H. Wache, T. Vogele, and U. Visser. Enabling technologies for interoperability. U. Visser and H. Pundt (editors): Workshop on the 14th International Symposium of Computer Science for Environmental Protection, 2000
252. Kashyap, V., and A. Sheth, Schematic and semantic semilarities between database objects: A context-based approach. The International Journal on Very Large Data Bases. Vol. 5, No. 4, 1996
253. MacGregor, R. M. Using a description logics classifier to enhance deductive inference. Proceedings of the Seventh IEEE Conference on AI Applications, 1991
254. Rector, A. L., S. Bechofer, C. A. Goble, I. Horrocks, W. A. Nowlan, and W. D. Solomon. The GRAIL concept modelling language for medical terminology. Artificial Intelligence in Medicine, Volume 9, 1997
255. Stuckenschmidt, H., and H. Wache. Context modeling and transformation for semantic interoperability. Knowledge Representation Meets Databases (KRDB), 2000
256. Fensel, D., I. Horrocks, F. Van Harmelen, S. Decker, M. Erdmann, and M. Klein. OIL in a nutshell. 12th International Conference on Knowledge Engineering and Knowledge Management EKAW2000, 2000
257. Donini, F., M. Lenzerini, D. Nardi, and A. Schaerf. AL-log: Integrating datalog and description logics. Journal of Intelligent Information Systems (JIIS), Vol. 27, No. 1, 1998
258. Arens, Y., C. Hsu, and C. A. Knoblock. Query processing in the SIMS information mediator. Advanced Planning Technology, 1996
259. Hwang, C. H. Incompletely and imprecisely speaking: Using dynamic ontologies for representing and retrieving information. Technical Report, Microelectronics and Computer Technology Corporation (MCC), 1999
260. Pazzaglia, J.-C. R., and S.M. Embury. Bottom-up integration of ontologies in a database context. In KRDB'98 Workshop on Innovative Application Programming and Query Interfaces, 1998
261. Ashish, N., and C. A. Knoblock. Semi-automatic wrapper generation for internet information sources. Second IFCIS International Conference on Cooperative Information Systems, 1997
262. Heflin, J., and J. Hendler. Semantic interoperability on the web. Extreme Markup Languages 2000, 2000
263. Bussler, C. B2B Integration. Concepts and Architecture. Springer Verlag, 2003
264. Fensel, D., and C. Bussler. The Web Service Modelling Framework WSMF. Electronic Commerce Research and Applications, Vol. 1, Issue 2, Elsevier Science B.V., Summer 2002
265. Fielding, R. Architectural Styles and the Design of Network-based Software Architectures. Dissertation. Information and Computer Science, University of California, Irvine, 2000

266. Alves, A., A. Arkin, S. Askary, C. Barreto, B. Bloch, F. Curbera, M. Ford, Y. Goland, A. Guízar, M. Kartha, C. K. Liu, R. Khalaf, D. König, M. Marin, V. Mehta, S. Thatte, D. van der Rijn, P. Yendluri, and A. Yiu (editors): Web Services Business Process Execution Language, Version 2.0. OASIS Committee Draft, wsbpel-specification-cd_Jan_25_2007, 2007. http://docs.oasis-open.org/wsbpel/2.0/ (Accessed September 10, 2007)

267. Carman, M., L. Serafini, and P. Traverso. Web Service Composition as Planning. ICAPS'03 Workshop on Planning for Web Services, 2003

268. Du, W., J. Davis, M.-C. Shan, U. Dayal. Flexible Compensation of Workflow Processes. HPL-96-72 (R.1), Software Technology Laboratory, Hewlett Packard, 1997. http://www.hpl.hp.com/techreports/96/HPL-96-72r1.pdf (Accessed September 10, 2007)

269. Jablonski, S., and C. Bussler. Workflow Management - Modeling, Concepts, Architecture and Implementation, International Thomson Computer Press, 1996

270. Kavantzas, N., D. Burdett, G. Ritzinger, T. Fletcher, Y. Lafon, C. Barreto. Web Services Choreography Description Language Version 1.0. W3C Candidate Recommendation. World Wide Web Consortium, 2005

271. Leymann, F. Web Services Flow Language (WSFL 1.0). IBM Software Group, IBM, 2001

272. Martin, D., M. Paolucci, S. McIlraith, M. Burstein, D. McDermott, D. McGuinness, B. Parsia, T. Payne, M. Sabou, M. Solanki, N. Srinivasan, and K. Sycara. Bringing Semantics to Web Services: The OWL-S Approach. Proceedings of the First International Workshop on Semantic Web Services and Web Process Composition (SWSWPC 2004), 2004

273. Martinez, E., and Y. Lesperance. Web Servcie Composition as a Planning Task: Experiments using Knowledge-Based Planning. Proceedings of the ICAPS-2004 Workshop on Planning and Scheduling for Web and Grid Services, 2004

274. Business Process Modeling Notation (BPMN) Specification. Final Adopted Specification. dtc/06-02-01. Object Managemenet Group (OMG), 2006

275. Oracle BPEL Process Manager Developer's Guide 10g Release 2 (10.1.2) B14448-03. Oracle Corporation, 2006. http://download-west.oracle.com/docs/cd/B14099_19/integrate.1012/b14448/toc.htm (Accessed September 10, 2007)

276. Roman, D., U. Keller, H. Lausen, J. de Bruijn, R. Lara, M. Stollberg,A. Polleres, C. Feier, C. Bussler, and D. Fensel. Web Service Modeling Ontology. Applied Ontology. Vol. 1, No. 1, 2005

277. Ten-Hove, R., P. Walker (editors): Java Business Integration (JBI) 1.0. Final Release. Sun Microsystems, USA, 2005

278. Thatte, S. XLANG - Web Services for Business Process Design. Microsoft Corporation, 2001

279. Tilkov, S. Choreography vs. Orchestration. Blog discussion. http://www.innoq.com/blog/st/2005/02/16/choreography_vs_orchestration.html (Accessed September 10, 2007)
280. Traverso, P., M. Pistore. Automated Composition of Semantic Web Services into Executable Processes. International Semantic Web Conference (ISWC 2004), 2004
281. Wikipedia 2007. http://www.wikipedia.org (Accessed September 10, 2007)
282. Wohed, P., W. van der Aalst, M. Dumas, and A. ter Hofstede. Pattern Based Analysis of BPEL4WS. Technical Report FIT-TR-2002-04, Queensland University of Technology, 2002
283. Wu, D., B. Parsia, E. Sirin, J. Hendler, D. Nau. Automating DAML-S Web Services Composition Using SHOP2. Proceedings of 2nd International Semantic Web Conference (ISWC2003), 2003
284. WSIF 2006. http://ws.apache.org/wsif/ (Accessed September 10, 2007)
285. Genesereth, M., and R. Filkes. Knowledge Interchange Format (KIF). Stanford University Logic Group. Logic-92-1, 1992
286. Fensel, D., I. Horrocks, F. V. Harmelen, S. Decker, M. Erdmann, and M. Klein. OIL in a Nutshell. Proceedings of the European Knowledge Acquisition Conference (EKAW-2000), 2000
287. Brickley, D., and R. V. Guha. Resource Description Framework (RDF). Schema Specification 1.0. W3C Candidate Recommendation, 2000
288. MacKenzie, C. M., K. Laskey, F. McCabe, P. F. Brown, and R. Metz (editors): Reference Model for Service Oriented Architecture 1.0. OASIS Standard, 2006
289. McGuinness, D. L., R. Fikes, L. A. Stein, and J. A. Hendler. DAML-ONT: An Ontology Language for the Semantic Web. D. Fensel, J. A. Hendler, H. Lieberman, and W. Wahlster (editors): Spinning the Semantic Web: Bringing the World Wide Web to Its Full Potential. The MIT Press, 2003
290. Burstein, M., C. Bussler, M. Zaremba, T. Finin, M. N. Huhns, M. Paolucci, A. P. Sheth, and S. Williams. A Semantic Web Services Architecture. IEEE Internet Computing. Vol. 9, 2005. http://dx.doi.org/10.1109/MIC.2005.96 (Accessed September 10, 2007)
291. Hendler, J., and D. L. McGuinness. DARPA Agent Markup Language. IEEE Intelligent Systems. Vol. 15, 2001
292. Fensel, D., I. Horrocks, F. V. Harmelen, S. Decker, M. Erdmann, and M. Klein. OIL in a nutsell. Proceedings of the European Knowledge Acquisition Conference (EKAW-2000), 2000
293. Gill, A. Introduction to the Theory of Finite-state Machines. McGraw-Hill, 1962
294. Moore, E. F. (editor): Gedanken-experiments on Sequential Circuits. Automata Studies. Annals of Mathematical Studies. No. 34, Princeton University Press, 1956
295. Hendler, J. Agents and the Semantic Web. IEEE Intelligent Systems. Vol. 2, 2001

296. Berardi, D., D. Calvanese, G. D. Giacomo, M. Lenzerini, and M. Mecella. Automatic Composition of E-services That Export Their Behavior. Proceedings of the First International Conference of Service-Oriented Computing (ICSOC 2003), 2003

297. Bultan, T., X. Fu, R. Hull, and J. Su. Conversation specification: a new approach to design and analysis of e-service composition. Proceedings of the World Wide Web Conference, 2003

298. Harel, D. Statecharts: A Visual Formalism for Complex Systems. Science of Computer Programming. Vol. 8, 1987

299. Fielding, R. T., and R. N. Taylor. Principled design of the modern Web architecture. Proceedings of the 22nd International Conference on Software Engineering, 2000

300. Czajkowski, K., D. F. Ferguson, I. Foster, J. Frey, S. Graham, I. Sedukhin, D. Snelling, S. Tuecke, and W. Vambenepe. The WS-Resource Framework. Version 1.0, 2004

301. Box, D., E. Christensen, F. Curbera, D. Ferguson, J. Frey, M. Hadley, C. Kaler, D. Langworthy, F. Leymann, B. Lovering, S. Lucco, S. Millet, N. Mukhi, M. Nottingham, D. Orchard, J. Shewchuk, E. Sindambiwe, T. Storey, S. Weerawarana, and S. Winkler. Web Services Addressing. W3C Member Submission, 2004

302. The Object Management Group. The Common Object Request Broker Architecture (CORBA): Architecture and Specifications, 2002

303. Microsoft Corporation. Microsoft Distributed Common Object Model (DCOM) v1.3, 1998

304. Petri, C. A. Kommunikation mit Automaten. Institut fuer Instrumentelle Mathematik, 1962

305. Best, E. Weighted basic Petri Nets. Proceedings of the International Conference on Concurrency, 1988

306. Jensen, K. An Introduction to the Theoretical Aspects of Coloured Petri Nets. A Decade of Concurrency, Lecture Notes in Computer Science #803, 1994

307. Merlin, P. M. A study of the recoverability of computing systems. Department of Information and Computer Science, 1974

308. Stork, D. G., and R. J. v. Glabbeek. Token-Controlled Place Refinement in Hierarchical Petri Nets with Application to Active Document Workflow. Proceedings of the 23rd International Conference on Applications and Theory of Petri Nets (ICATPN '02), 2002

309. v. d. Aalst, W. M. P. The Application of Petri Nets to Workflow Management. The Journal of Circuits, Systems and Computers. Vol. 8, 1998

310. The OWL Services Coalition. OWL-S: Semantic Markup for Web Services. Version 1.1, 2004. http://www.daml.org/services/owl-s/1.1/ (Accessed September 10, 2007)

311. Balzer, S., T. Liebig, and M. Wagner. Pitfalls of OWL-S: A practical semantic web use case. Proceedings of the 2nd international conference on Service oriented computing, 2004
312. The Semantic Web Rule Language (SWRL). http://www.w3.org/Submission/SWRL/ (Accessed May 27, 2008)
313. The Knowledge Interchange Format (KIF). http://www-ksl.stanford.edu/knowledge-sharing/kif/ (Accessed September 10, 2007)
314. McDermott, D. The 1998 AI planning systems competition. The AI Magazine. Vol. 21, 2000
315. Battle, S., A. Bernstein, H. Boley, B. Grosof, M. Gruninger, R. Hull, M. Kifer, D. Martin, S. McIlraith, D. McGuinness, J. Su, and S. Tabet (editors): Semantic Web Services Framework (SWSF). Overview. W3C Submission, 2005. http://www.w3.org/Submission/SWSF/ (Accessed September 10, 2007)
316. Michel, J. J., and A. F. Cutting-Decelle. The Process Specification Language. International Standards Organization ISO TC184/SC5 Meeting, 2004
317. Brodie, M., C. Bussler, J. deBruijn, T. Fahringer, D. Fensel, M. Hepp, H. Lausen, D. Roman, T. Strang, H. Werthner, and M. Zaremba. Semantically Enabled Service-Oriented Architectures: A Manifesto and a Paradigm Shift in Computer Science. DERI Technical Report, 2005
318. Grosof, B. N. IBM Common Rules Report: A Courteous Compiler From Generalized Courteous Logic Programs To Ordinary Logic Programs, 1999. http://www.research.ibm.com/rules/paps/gclp_report1.pdf (Accessed September 10, 2007)
319. Chen, W., M. Kifer, and D. S. Warren. HiLog: A Foundation for Higher Order Logic Programming. Journal of Logic Programming. Vol. 15, 1993
320. Kifer, M., G. Lausen, and a. J. Wu. Logical foundations of object oriented and frame based languages. JACM. Vol. 42, 1995
321. Lloyd, J. W. Foundations of logic programming. Springer Verlag, 1987
322. Akkiraju, R., J. Farrell, J.Miller, M. Nagarajan, M. Schmidt, A. Sheth, and K. Verma. Web Service Semantics - WSDL-S, 2005. http://lsdis.cs.uga.edu/projects/meteor-s/wsdl-s/ (Accessed September 10, 2007)
323. The Web Service Business Process Execution Language. OASIS Specification. http://www.oasis-open.org/committees/wsbpel (Accessed September 10, 2007)
324. Balzer, S., T. Liebig, and M. Wagner. Pitfalls of OWL-S: A practical semantic web use case. Proceedings of the 2nd international conference on Service oriented computing, 2004
325. Roman, D., U. Keller, H. Lausen, J. deBruijn, R. Lara, M. Stollberg, A. Polleres, D. Fensel, and C. Bussler. Web Service Modeling Ontology. Applied Ontology Journal. Vol. 1, No. 1, 2005
326. The Object Management Group (OMG). Meta Object Facility (MOF). http://www.omg.org/mof/ (Accessed September 10, 2007)

327. Gurevich, Y. Evolving algebras 1993: Lipari guide in Specification and validation methods. Oxford University Press, 1993
328. deBruijn, J., H. Lausen, A. Polleres, and D. Fensel. The Web Service Modeling Language WSML: An Overview. Proceedings of the European Semantic Web Services Conference (ESWC), 2006
329. Grosof, B. N., I. Horrocks, R. Volz, and S. Decker. Description logic programs: Combining logic programs with description logic. Proceedings of the Twelfth International World Wide Web Conference (WWW 2003), 2003
330. Haller, A., E. Cimpian, A. Mocan, E. Oren, and C. Bussler. WSMX - A Semantic Service-Oriented Architecture. Proceedings of the International Conference on Web Service (ICWS 2005), 2005
331. Mocan, A., M. Moran, E. Cimpian, and M. Zaremba. Filling the Gap - Extending Service Oriented Architectures with Semantics. Proceedings of the ICEBE, 2006
332. Haselwanter, T., P. Kotinurmi, M. Moran, T. Vitvar, and M. Zaremba. WSMX: A Semantic Service Oriented Middleware for B2B Integration. Proceedings of the International Conference on Service Oriented Computing, 2006
333. Moran, M., Michal. Zaremba, A. Mocan, E. Cimpian, T. Haselwanter, and Maciej Zaremba. DIP Deliverable 6.11. Semantic Web Services Architecture and Information Model, 2006
334. Preist, C. A Conceptual Architecture for Semantic Web Services. Proceedings of the Third International Semantic Web Services Conference (ISWC), 2004
335. Baida, Z., J. Gordijn, and B. Omelayenko. A shared service terminology for online service provisioning. Proceedings of the 6th international conference on Electronic commerce, 2004
336. Fensel, D., U. Keller, H. Lausen, A. Polleres, and I. Toma. What is wrong with Web Services Discovery. Proceedings of the W3C Workshop on Frameworks for Semantics in Web Services, 2005
337. Keller, U., R. Lara, H. Lausen, A. Polleres, and D. Fensel. Automatic Location of Services. Proceedings of the 2nd European Semantic Web Symposium (ESWS2005), 2005
338. Voskob, M. UDDI Spec TC V4 Requirement - Taxonomy support for semantics, 2004
339. Benatallah, B., M. S. Hacid, C. Rey, and F. Toumani. Request rewriting-based web service discovery. Proceedings of the International Conference on The Semantic Web, 2003
340. Bernstein, A., and M. Klein. Discovering services: Towards high-precision service retrieval. Web Services, E-Business, and the Semantic Web. Lecture Notes in Computer Science #2512, 2002
341. Albert, P., L. Henocque, and M. Kleiner. A Constrained Object Model for Configuration Based Workflow Composition. Proceedings of the Business Process Management Workshops (BPM 2005), 2005

342. McIlraith, S., and T. Son. Adapting Golog for composition of semantic Web Services. Proceedings of the 8th International Conference on Principles of Knowledge Representation and Reasoning, 2002
343. Sirin, E., B. Parsia, D. Wu, J. Hendler, and D. Nau. HTN planning for web service composition using SHOP2. Web Semantics Journal, 2004
344. Mayer, H., H. Overdick, and M. Weske. Plængine: A System for Automated Service Composition and Process Enactment. Proceedings of the Business Process Management Workshop, 2005
345. v. d. Aalst, W. M. P., M. Dumas, and A. H. M. t. Hofstede. Web Service Composition Languages: Old Wine in New Bottles? Proceedings of the 29th EUROMICRO Conference 2003, 2003
346. Osman, T., D. Thakker, and D. Al-Dabass. Bridging the Gap between Workflow and Semantic-based Web Services Composition. Proceedings of the Business Process Management Workshop 2005, 2005
347. Casati, F., S. Ilnicki, L. Jin, V. Krishnamoorthy, and M. Shan. Adaptive and Dynamic Service Composition in eFlow. HP Technical Report HPL-200039, 2000
348. Norton, B., and C. Pedrinaci. 3-Level Service Composition and Cashew: A Model for Orchestration and Choreography in Semantic Web Services. Proceedings of the On the Move to Meaningful Internet Systems 2006 OTM 2006 Workshops, 2006
349. Mocan, A., E. Cimpian, and M. Kerrigan. Formal Model for Ontology Mapping Creation. Proceedings of the International Semantic Web Conference, 2006
350. Cimpian, E., and A. Mocan. WSMX Process Mediation Based on Choreographies. Proceedings of the Business Process Management Workshops 2005, 2005
351. Vu, L. H., M. Hauswirth, and K. Aberer. Towards P2P-Based Semantic Web Service Discovery with QoS Support. Proceedings of the Business Process Management Workshops, BPM 2005, 2005
352. XML RPC. http://www.xmlrpc.com/ (Accessed September 10, 2007)
353. SOAP. http://en.wikipedia.org/wiki/SOAP (Accessed September 10, 2007)
354. Gudgin, M., M. Hadley, N. Mendelsohn, J.-J. Moreau, and H. F. Nielsen (editors): SOAP Version 1.2 Part 1: Messaging Framework. W3C Recommendation. World Wide Web Consortium, 2003
355. Christensen, E., F. Curbera, G. Meredith, and S. Weerawarana (editors): Web Services Description Language (WSDL) 1.1. W3C Note. World Wide Web Consortium, 2001
356. Chinnici, R., J.-J. Moreau, A. Ryman, and S. Weerawarana (editors): Web Services Description Language (WSDL) Version 2.0 Part 1: Core Language. W3C Recommendation. World Wide Web Consortium, 2007
357. Boubez, T., M. Hondo, C. Kurt, J. Rodriguez, and D. Rogers (editors): UDDI Programmer's API 1.0. UDDI Published Specification. UDDI.org, 2002

358. Clement, L., A. Hately, C. von Riegen, and T. Rogers (editors): UDDI Version 3.0.2. UDDI Spec Technical Committee Draft. OASIS, 2004

359. Alonso, G., F. Casati, H. Kuno, and V. Machiraju. Web Services. Springer Verlag, 2003

360. Chinnici, R., H. Haas, A. Lewis, J.-J. Moreau, D. Orchard, and S. Weerawarana (editors): Web Services Description Language (WSDL) Version 2.0 Part 2: Adjuncts. W3C Recommendation. World Wide Web Consortium, 2007

361. Chinnici, R., H. Haas, A. Lewis, J-J. Moreau, D. Orchard, and S. Weerawarana (editors): Web Services Description Language (WSDL) Version 2.0 Part 2: Adjuncts. W3C Recommendation. World Wide Web Consortium. March 2007

362. Zapthink. Key XML Specifications and Standards. Poster. Document ID: ZTS-GI101. ZapThink LLC. http://www.zapthink.com/report.html?id=ZTS-GI101. 2002 (Accessed September 7, 2007)

363. MacKenzie, C. M., K. Kaskey, F. McCabe, P. Brown, and R. Metz. Reference Model for Service Oriented Architecture 1.0. Committee Specification 1. OASIS, 2006

364. Wilkes, R. The Web Services Protocol Stack. CBDI Forum, 2005. http://roadmap.cbdiforum.com/reports/protocols/ (Accessed September 10, 2007)

365. Mocan, A., and E. Cimpian. An ontology-based data mediation framework for semantic environments. International Journal on Semantic Web and Information Systems (IJSWIS). Vol. 3, No. 2, 2007

366. Kilic, O., and A. Dogac. Achieving Clinical Statement Interoperability using RMIM and Archetypebased Semantic Transformations. IEEE Transactions on Information Technology in Biomedicine, 2007. http://www.srdc.metu.edu.tr/webpage/projects/ride/publications/KilicDogac.pdf (Accessed September 10, 2007)

367. Paolucci, M., T. Kawamura, T. Payne, and K. Sycara. Semantic Matching of Web Services Capabilities. Proceedings of the 1st International Semantic Web Conference (ISWC), 2002

368. Goldberg, H., M. Vashevko, A. Postilnik, K. Smith, N. Plaks, and B. Blumenfeld. Evaluation of a Commercial Rules Engine as a basis for a Clinical Decision Support Service. Proceedings of the Annual Symposium on Biomedical and Health Informatics, 2006

369. Kashyap, V., A. Morales and T. Hongsermeier. Implementing Clinical Decision Support: Achieving Scalability and Maintainability by combining Business Rules with Ontologies. Proceedings of the Annual Symposium on Biomedical and Health Informatics, 2006

370. W3C Semantic Web Best Practices and Deployment Working Group. http://www.w3.org/2001/sw/BestPractices/ (Accessed September 10, 2007)

371. Schulz, S., and U. Hahn. A Knowledge Representation view on Biomedical Structure and Function. Proceedings of AMIA, 2002

372. Bussler, C. B2B Protocol Standards and their Role in Semantic B2B Integration Engines. Bulletin of the IEEE Computer Society Technical Committee on Data Engineering, 2001
373. Williams, S. K., S. A. Battle, and J. E. Cuadrado. Protocol Mediation for Adaptation in Semantic Web Services. Technical Report HPL-2005-78. Digital Systems Media Laboratory 2005
374. Bochmann, G. V. Higher-level Protocols are not Necessary End-to-End. ACM SIGCOMM. Computer Communications Review. Vol. 13, 1983
375. Calvert, L., and S. S. Lam. Deriving a Protocol Converter: A Top-Down Method. ACM SIGCOM Computer Communications Review. Vol. 19, 1989
376. Kopecký, J., D. Roman, T. Vitvar, M. Moran, and A. Mocan. WSMO Grounding. WSMO Working Draft v0.1, 2007. http://wsmo.org/TR/d24/d24.2/v0.1/ (Accessed September 10, 2007)
377. Kerrigan, M., A. Mocan, M. Tanler, and D. Fensel. The Web Service Modeling Toolkit - an Integrated Development Environment for Semantic Web Services. Proceedings of the 4th European Semantic Web Conference (ESWC 2007), 2007
378. Dimitrov, M., A. Simov, V. Momtchev, and M. Konstantinov. WSMO Studio — A Semantic Web Services Modelling Environment for WSMO. Proceedings of the 4th European Semantic Web Conference (ESWC 2007), 2007
379. Web Services: Been There Done That. Intelligent Systems. Vol. 18, No. 1, 2003. http://hcs.science.uva.nl/semanticweb/literatuur/x1072.pdf (Accessed September 10, 2007)
380. Sheth, A. Blog entry at http://lsdis.cs.uga.edu/~amit/blog/index.php?title=why_are_we_still_pushing_semantic_web&more=1&c=1&tb=1&pb=1 [3] A. Sheth; Blog; http://lsdis.cs.uga.edu/~amit/blog/ (Accessed September 10, 2007)
381. Van Damme, C., M. Hepp, and K. Siorpaes. FolksOntology: An Integrated Approach for Turning Folksonomies into Ontologies. Proceedings of Bridging the Gap between Semantic Web and Web 2.0 (SemNet 2007), 2007
382. Han, S.-K., and D. Roman. Towards Semantic Service-Oriented Systems on the Web. Slide 90 of tutorial at Web Intelligence 2006, http://www.wsmo.org/TR/d17/resources/200612-WI06/wi2006-tutorial.pdf (Accessed September 10, 2007)
383. European Union IST Integrated Project "SUPER". http://www.ip-super.org/ (Accessed September 10, 2007)
384. Pressman, R. Software Engineering, A Practitioner's Approach. 4th Edition. ISBN 0077094115, 1997
385. Semantic Web Service Challenge. http://sws-challenge.org (Accessed September 10, 2007)
386. Braga, D., A. Campi and S. Ceri. XQBE (XQuery By Example): A Visual Interface to the Standard XML Query Language. ACM Transactions on Database Systems. Vol. 30, No. 2, June 2005

387. Baeten, J. C. M. A Brief History of Process Algebra. Journal of Theoretical Computer Science (Elsevier Publishing Ltd.), Vol. 335, 2005

388. v. d. Aalst, W. M. P. Pi calculus versus petri nets: Let us eat humble pie rather than further inflate the pi hype. http://tmitwww.tm.tue.nl/research/patterns/download/pi-hype.pdf (unpublished discussion paper, accessed March 27, 2008), 2003

389. Milan, C., U. Milan. Modelling and Simulation of Parallel Systems Using CCS and Petri Nets: Major Concepts. In Proceedings of XXIst International Colloquium ASIS 1999, Krnov, Czech Republic, p. 371-377, ISBN 80-85988-41-0, 1999

390. Milner, R. A Calculus of Communicating Systems. Lecture Notes in Computer Science. Vol. 92, Springer-Verlag, Berlin, 1980

391. Milner, R. Communicating and Mobile Systems: The Pi-Calculus. Cambridge University Press, Cambridge, UK, 1999

392. Milner, R. The Polyadic pi-Calculus: A Tutorial. In F. L. Hamer, W. Brauer and H. Schwichtenberg, editors, Logic and Algebra of Specification. Springer-Verlag, 1993

393. Salaun, G., L. Bordeaux, M. Schaerf. Describing and Reasoning on Web Services using Process Algebra. Proceedings of the IEEE International Conference on Web Services (ICWS), Washington DC, USA, 2004

394. Magee, J., and J. Kramer. Concurrency: State Models and Java Programs. Wiley, New York, NY, USA, Second Edition, 2006

395. The Rule Markup Initiative, http://www.ruleml.org (Accessed March 27, 2008)

396. The Ontology Definition Metamodel (ODM). http://www.omg.org/docs/ptc/07-09-09.pdf (Accessed March 27, 2008)

397. Open Knowledge Base Connectivity 2.0.3 — Proposed —. http://www.ai.sri.com/~okbc/spec/okbc2/okbc2.html (Accessed March 27, 2008)

398. DIG Interface Standard (DIG 2.0). http://dl.kr.org/dig/interface.html (Accessed March 27, 2008)

399. OWL API. http://owlapi.sourceforge.net/ (Accessed March 27, 2008)

400. Bechhofer, S., R. Moller and P. Crowther. The DIG description logic interface. Proceedings of the International Description Logics Workshop (DL), 2003

401. Haarslev, V., R. Moller and M. Wessel. Querying the Semantic Web with Racer+ nRQL. Procedings of the KI-04 Workshop on Applications of Description Logics, 2004

402. Sirin, E., and B. Parsia. 3rd OWL Experiences and Directions Workshop (OWLED), 2007

403. Simple Knowledge Organization System (SKOS). http://www.w3.org/2004/02/skos/ (Accessed March 27, 2008)

404. Fensel, D., H. Lausen, A. Polleres, J. de Bruijn, M. Stollberg, D. Roman, J. Domingue. Enabling Semantic Web Services: The Web Service Modeling Ontology, Springer Verlag, 2006

405. ILOG Business Rules Management Systems. http://www.ilog.com/products/businessrules/ (Accessed May 27th , 2008)
406. NeOn Toolkit Portal. http://www.neon-toolkit.org/ (Accessed May 30, 2008)
407. OWL 1.1 Web Ontology Language Overview. http://www.w3.org/Submission/owl11-overview/ (Accessed May 30, 2008)
408. Jena 2 Inference Support. http://jena.sourceforge.net/inference/ (Accessed May 30, 2008)
409. Jess, the Rule Engine for the Java Platform. http://www.jessrules.com/jess/index.shtml (Accessed May 30, 2008)
410. Forgy, C. L. Rete: A Fast Algorithm for the Many Pattern/Many Object Pattern Match Problem. Artificial Intelligence 19, p 17-37, 1982
411. Staab, S., and R. Studer (editors). Handbook on Ontologies. Birkhauser Publishers, 2004
412. Kifer, M., G. Lausen and J. Wu. Logical Foundations of Object-Oriented and Frame-Based Languages. Journal of the ACM, May 1995
413. Concepts of Crossvision Information Integrator. http://documentation.softwareag.com/crossvision/xei/concepts/conceptsover.htm (Accessed June 1, 2008)

Index

A

B

C

D

E

F

T

Printing: Krips bv, Meppel, The Netherlands
Binding: Stürtz, Würzburg, Germany